T0074552

Mosquitoes of the Southeastern United States

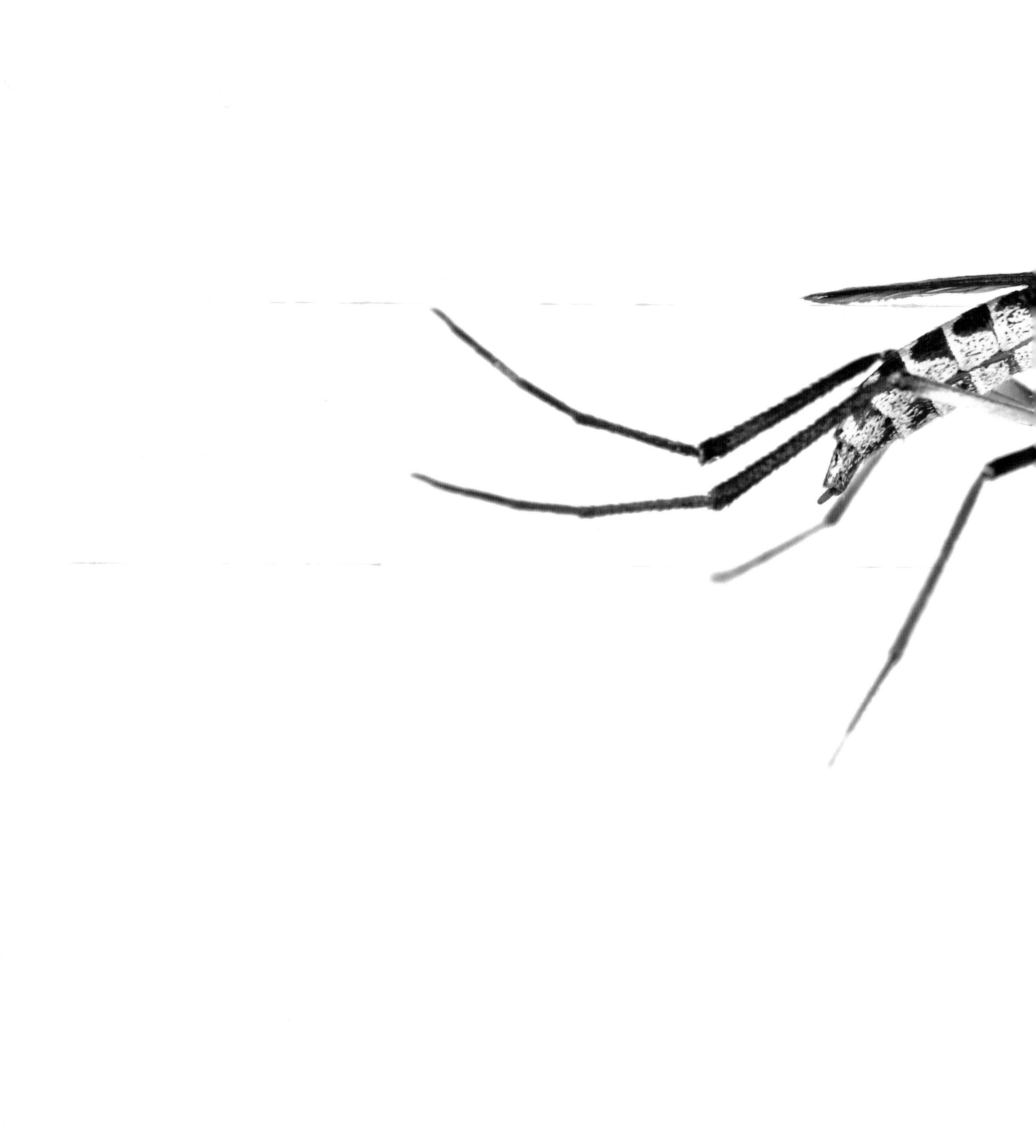

Nathan D. Burkett-Cadena

Mosquitoes

of the Southeastern United States

• • •

The University of Alabama Press / Tuscaloosa

Printed in Korea

Cover photographs (clockwise from top): *Psorophora ferox* female; *Aedes fulvus pallens* female; *Psorophora horrida* female; *Aedes canadensis* female; *Culex restuans* female. Courtesy of the author.

Frontispiece: *Psorophora cyanescens* female. Courtesy of Gayle and Jeanell Strickland.

Cover design: Todd Lape / Lape Designs

All images are courtesy of the author unless otherwise indicated.

The paper on which this book is printed meets the minimum requirements of American National Standard for Information Sciences—Permanence of Paper for Printed Library Materials, ANSI Z39.48-1984.

Library of Congress Cataloging-in-Publication Data

Burkett-Cadena, Nathan D.
Mosquitoes of the southeastern United States / Nathan D. Burkett-Cadena.
p. cm.
Includes bibliographical references and index.
ISBN 978-0-8173-1781-2 (hardcover : alk. paper) —
ISBN 978-0-8173-8648-1 (ebook)
1. Mosquitoes—Southern States—Identification. I. Title.
QL536.B916 2013
595.77'2—dc23 2012024177

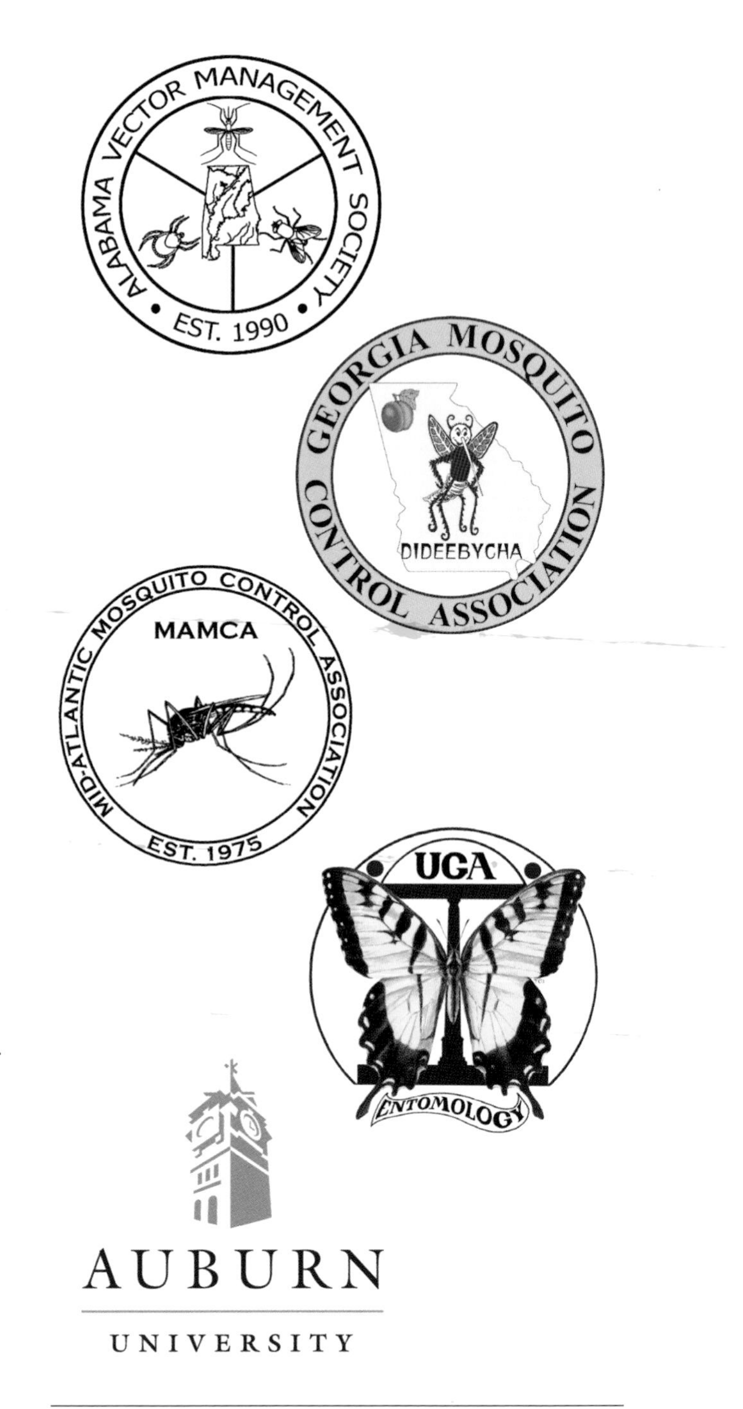

Publication of this book was made possible by support from the Alabama Vector Management Society, the Georgia Mosquito Control Association, the Mid-Atlantic Mosquito Control Association, Auburn University Department of Entomology and Plant Pathology, Auburn University College of Agriculture, and the University of Georgia Department of Entomology.

For my family

Contents

Preface

Mosquitoes of the Southeastern United States is intended to be a user-friendly guide to the identification and biology of the mosquito species that are known to occur in the southeastern United States. The purpose of this book is to provide a comprehensive resource that bridges the gap between technical scientific literature and popular information regarding the mosquitoes found in our region. The material is treated in a way that will make it comprehendible to virtually anyone. Overtechnical terms are avoided when possible, but those that are unavoidable are defined in the glossary. While most of the existing "mosquito books" focus on *either* identification *or* biology, *Mosquitoes of the Southeastern United States* attempts to synthesize biological information, identification keys, and images for all of the mosquito species that occur in our region.

Mosquitoes of the Southeastern United States is *not* intended to be a systematic treatise of mosquito classification, nor is it an annotated bibliography of mosquito literature. The illustrations and terms used throughout are simplified to facilitate the identification process. Rather than portray every taxonomic character, this guide deals only with those characters directly relevant to the identification of the mosquitoes of the southeastern United States. The distribution maps are for reference purposes and indicate probable presence or absence of a species. No map can accurately portray the distribution of a mosquito species, since mosquito populations in a given location fluctuate between years and even between seasons. In addition, mosquito species may be accidentally introduced and become established in a new area, as in the case of *Aedes albopictus.* Other species may suddenly expand their distribution over a short period of time, as in the case of *Culex coronator.* Therefore, caution should be exercised when attempting to use any map to determine whether or not a species occurs in your area. For references on the recorded collection locales of a species, users can refer to fully annotated technical works, such as *Identification and Geographical Distribution of the Mosquitoes of North America, North of Mexico* (2005) by Richard Darsie Jr. and Ronald Ward, or *The Mosquitoes of North America* (1955) by Stanley Carpenter and Walter LaCasse.

Although *Mosquitoes of the Southeastern United States* may appeal to nature lovers and homeowners, this book is likely be of greatest use to mosquito control professionals and vector biologists, as they attempt to identify the mosquito species that they encounter during their surveillance and research activities. Users of the guide may take advantage of the fully illustrated key to identify mosquito specimens and then refer to the photographs and distribution maps accompanying the species accounts to validate (or invalidate) their identifications. Habitat associations of the larvae and adults may also be used to confirm identification. *Mosquitoes of the Southeastern United States* covers all of the mosquitoes that are known to inhabit the states of Alabama, Georgia, Mississippi, North Carolina, South Carolina, and Tennessee. The mosquito fauna of northern

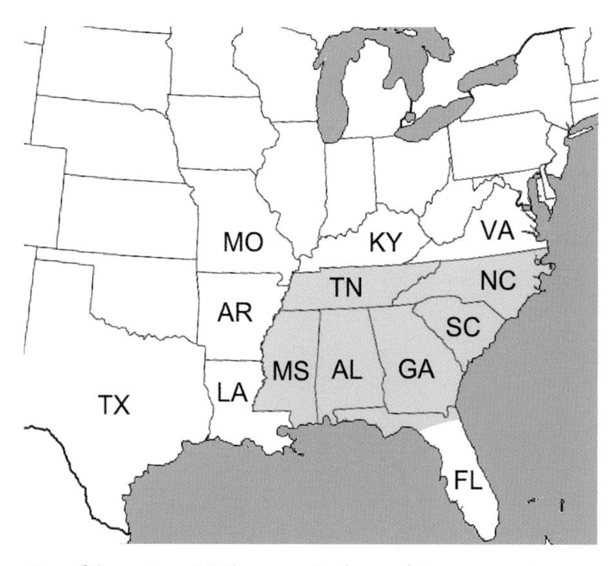

Map of the eastern U.S. The mosquito fauna of the gray area is covered by this book.

Florida is also covered by this book. Those who wish to identify mosquitoes in neighboring states, particularly Louisiana, Arkansas, Kentucky, and (southern) Florida, will also find this guide useful but should interpret the material with caution, since several species known from those states are not covered here (see p. 73–74 for a synopsis of those species). Southern Florida was excluded from this guide for two reasons: First and foremost, a number of mosquito species that occur in southern Florida do not occur in any of the other southern states. Second, an excellent guide to the mosquitoes of Florida already exists (*Keys to the Adult Females and Fourth Instar Larvae of the Mosquitoes of Florida* by Richard Darsie Jr. and Charlie Morris). The terminology used in this book follows that of the *Taxonomists' Glossary of Mosquito Anatomy* by Ralph Harbach and Kenneth Knight (1980), with a few exceptions (anal segment instead of segment X; maxillary palp instead of maxillary palpus).

• • •

The idea for this guide came about several years ago (2003), shortly after I started studying mosquitoes for my master's degree at Auburn University in Alabama. My thesis research revolved around mosquito trapping and required the identification of large numbers of field-collected female mosquitoes. I rapidly became intrigued by the diversity of mosquitoes in my surroundings, and I developed a fondness for photographing the different species that I encountered. After a couple of years I found that I had amassed an extensive collection of photographs of the mosquitoes that occurred in Alabama. At that point I felt that a guide to the mosquitoes of Alabama might be a worthwhile project. I talked to my adviser at the time, Dr. Gary Mullen, to see what he thought about the idea. He was encouraging and felt that such a guide would be a substantial contribution to the literature dealing with the mosquitoes of the United States. As I became more familiar with the distributions of the various species that occurred in Alabama, I noticed that almost all of these same species were also found in neighboring states and that the southeastern states seemed to share a common mosquito fauna. By adding a few more species to my list of "Alabama mosquitoes," I found that I could broaden the scope of the book to cover the mosquitoes of the southeastern United States.

Mosquitoes of the Southeastern United States has drawn upon and benefited greatly from other works. Since my first days working with mosquitoes, I have been impressed by the identification keys in *Identification and Geographical Distribution of the Mosquitoes of North America, North of Mexico* (2005) by Darsie and Ward, which itself built upon other great works, particularly *The Mosquitoes of North America* (1955) by Carpenter and LaCasse. I have previously used (and still use) Darsie and Ward's identification keys extensively in my own research and have always felt fortunate to have such an excellent resource for identifying mosquitoes. However, many of the mosquito species covered by Darsie and Ward's guide do not occur in the southeastern states. By not including these western and northern species in *Mosquitoes of the Southeastern United States,* I have been able to keep the identification keys much shorter and, as a consequence, easier to use. That being said, anyone using this book to identify mosquitoes of the Southeast should be aware (especially near the edges of the region) that some species, such as *Aedes stimulans,* which are not included here, may occasionally be collected in the Southeast (see p. 74). If you are in doubt, I suggest using Darsie and Ward's keys as a more inclusive resource.

Much of the information included in this guide came from my own field experience. For species with which I have little or no field experience, I have relied largely on published accounts. Some mosquito species, however, are rarely collected anywhere and little biological information is available in the scientific literature. In such cases I have used anecdotal information. If you have spent much time working with mosquitoes of the United States, you may have more experience than I have with some of the species that are covered in this guide and thus your knowledge may be slightly different from the information published here. I welcome input from experienced mosquito workers for future editions of the book.

It has been a labor of love to produce this guide. I hope that users will find it useful, informative, and full of the diversity that inspired me to study this intriguing group of insects.

Acknowledgments

I am deeply indebted to all of the people who have directly or indirectly helped to produce this book. Special thanks go to Dr. Gary R. Mullen, who offered continual encouragement, as well as helpful comments on the text and layout of the book. Elizabeth Motherwell, natural history editor at The University of Alabama Press, continually offered feedback and helpful suggestions. Dr. Geoff Hill of Auburn University suggested collaboration with The University of Alabama Press, which turned out to be a very important event in the publishing of the book. Several excellent photographers allowed the use of images of species that I had not encountered myself. Sean McCann provided photographs of *Aedes cinereus* (p. 88) and *Mansonia dyari* (p. 154). Bo Zaremba contributed photographs of *Aedes cantator* (p. 87). Tom Murray provided photos of *Anopheles walkeri* (p. 125). Treasure Tolliver contributed a photo of a *Psorophora ferox* female gathering nectar (p. 4) and collaborated with me to photograph a female of *Culiseta inornata* (p. 150). Gayle and Jeanell Strickland provided photos of a female *Psorophora cyanescens* (p. 165). Several generous people provided live or preserved specimens that enabled me to photograph those species. Linda McCuiston of Rutgers University provided eggs of *Aedes atropalpus.* Larry Hribar, Florida Keys Mosquito Control, provided live and preserved specimens of *Aedes aegypti* and *Anopheles atropos,* respectively. Sean Graham captured a toad with a female *Uranotaenia lowii* feeding upon it. Thanks to Alan Dupuis for providing specimens of *Culex tarsalis.* Cuauhtemoc Villarreal supplied specimens of *Anopheles pseudopunctipennis.* Steve Sullivan provided a location of *Aedes taeniorhynchus* larvae. Laurel Anderton meticulously edited the first full draft. My direct supervisors during my PhD and postdoctoral assignments, Micky Eubanks and Tom Unnasch, indulged my passion for mosquitoes and allowed me enough latitude in my regular work activities that I could incorporate time for working on this volume. My wife, Marleny, gave constant support and listened with patient interest to my prattling each time I encountered and photographed a species for the first time. Kelly Stevens, Steven McDaniel, Rosmarie Kelly, Brian Byrd, Joe Andrews, Bruce Harrison, Arthur Appel, Ray Noblet, and others helped obtain funding from various sources. Richard Darsie Jr. and Bruce Harrison reviewed and edited the original version of the manuscript and offered invaluable and indispensable comments and corrections that greatly improved the final version of the book. Sincere thanks go to all who have helped to produce this book. It is so much better because of your contributions.

Mosquitoes of the Southeastern United States

Introduction

Mosquito Diversity

Mosquitoes are found in just about every region of the planet—including the icy tundra of northern Canada, the deserts of the American Southwest, and the humid tropics of Africa, Asia, and South America. One of the reasons that mosquitoes are found in such a wide variety of habitats is their incredible diversity. More than 3,500 species of mosquitoes are known from around the world, and each species has very particular biological requirements, such as the habitat in which the females lay eggs and the larvae develop. All mosquito larvae are aquatic and the larvae of each mosquito species are adapted to a specific aquatic habitat. The larvae of the floodwater mosquito, *Aedes vexans,* for example, are found in rainwater pools but not in tree holes or man-made containers (birdbaths, discarded automobile tires, and so forth). Larvae of some other mosquitoes, such as the eastern treehole mosquito, *Aedes triseriatus,* are found in water-filled natural and man-made containers but not in rainwater pools, marshes, or ponds. Adult mosquitoes are also somewhat habitat specific. *Aedes thibaulti* is a wetland mosquito and is rarely encountered away from the forested wetlands where it breeds. The Asian tiger mosquito, *Aedes albopictus,* on the other hand, is very common in urban and suburban areas but is rather rare in forests and other "wild" areas.

Body of *Uranotaenia sapphirina*

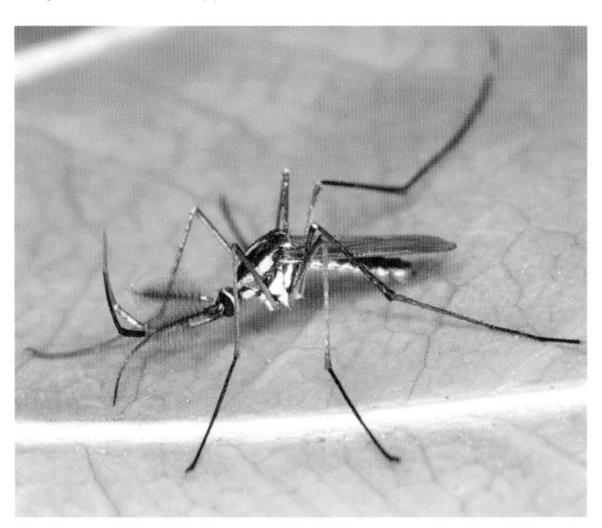

Adult male of the cannibal mosquito, *Toxorhynchites rutilus*

Mosquitoes are as diverse in size and coloration as they are in habitat and biology. *Uranotaenia sapphirina* is one of the smallest mosquitoes in our area, with a total body length (tip of proboscis to tip of abdomen) of about 6 mm (less than a quarter of an inch). *Psorophora ciliata* (also known as the gallinipper), on the other hand, has a total body length of about 18 mm (nearly three-quarters of an inch). Most mosquitoes are far more colorful than people realize. While some mosquitoes, especially members of the genus *Culex,* are rather drab, many other mosquitoes have bold markings in a variety of colors. Several members of the genus *Psorophora* have brilliant purple scales on the body and legs, including *Psorophora ferox,* featured in the cover photo. Many species of *Uranotaenia* have stripes of iridescent blue scales. The body of the adult cannibal mosquito, *Toxorhynchites rutilus,* is covered in iridescent scales that can appear purple, blue, green, and silvery white, depending on the light. These variations in size and coloration are useful in mosquito identification.

Mosquito Life Cycle

For the vast majority of mosquito species, the life cycle is very similar. The larva lives in water and is the stage of growth and development. The pupa is also aquatic, but it does not feed; it is the stage of transformation from the larva to the adult. The adult is the winged stage, and the stage of blood feeding and reproduction. The female lays eggs in a suitable habitat, thus completing the life cycle. This general life cycle describes the development of most mosquitoes, but there are many variations and exceptions.

The adult mosquito is the form that we are most familiar with, because this is the form that seeks out our blood. In fact, it is really only the adult female mosquito that most people are familiar with, since only female mosquitoes feed on blood. Females need the resources that they receive from blood in order to produce eggs. Some mosquitoes, such as the cannibal mosquito, *Toxorhynchites rutilus,* do not need to feed on blood since they receive enough resources from their larval diet to produce eggs. Larvae of the cannibal mosquito are predators that kill and consume other mosquito larvae (and other small aquatic organisms). Adult cannibal mosquitoes feed on nectar from flowers. A few other species found in our area, such as the pitcher plant mosquito, *Wyeomyia smithii,* and the rock pool mosquito, *Aedes atropalpus,* may not need blood in order to produce their first batch of eggs but may need it to produce later batches. While not all mosquito species feed on blood, all mosquitoes (males and females) feed on nectar. From nectar, mosquitoes receive the energy needed for flying and other activities.

Female *Psorophora ferox* feeding on flower nectar. Courtesy of Treasure Tolliver.

While it is no secret that mosquitoes feed on human blood, people are often surprised to learn that mosquitoes also feed on birds, reptiles, and amphibians as well as mammals other than humans. Of the mosquito species found in the southeastern United

Larva of the cannibal mosquito, *Toxorhynchites rutilus,* preying upon the pupa of another mosquito

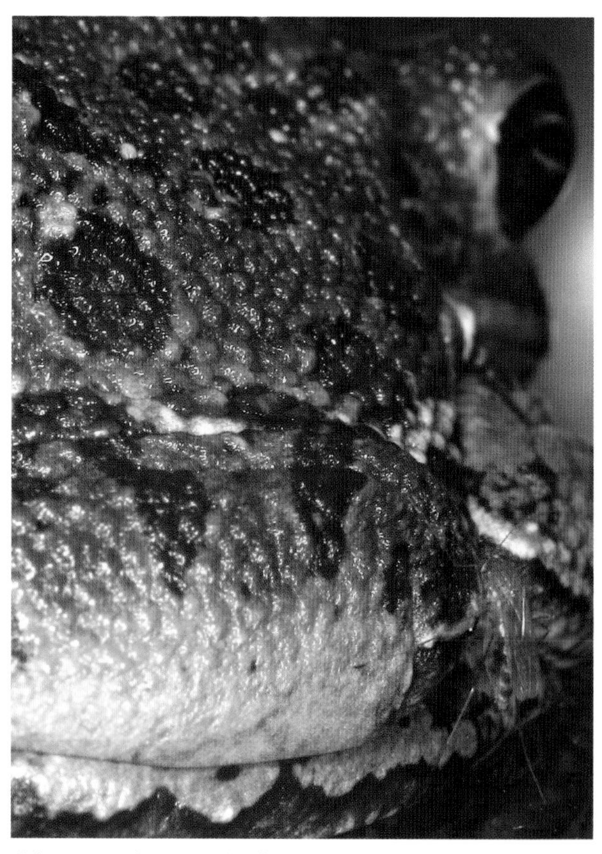

Culex territans (bottom right) feeding on a barking tree frog (*Hyla gratiosa*)

Aedes canadensis female feeding on a box turtle (*Terrapene carolina*)

States, about half feed mostly on mammals; about a third feed mostly on birds; and a few feed mostly on reptiles and amphibians. Some mosquito species feed on virtually any animal they encounter. Through observations and genetic analyses of the blood in the stomachs of wild-caught mosquitoes, scientists have found that mosquitoes of the Southeast feed on an incredible (and sometimes surprising) array of animals. Examples include white-tailed deer, ruby-throated hummingbirds, chipmunks, green tree frogs, armadillos, box turtles, and even venomous cottonmouth snakes. In fact, any vertebrate animal that is not totally aquatic is likely to be fed upon by mosquitoes. Mosquitoes generally do not bite other invertebrates, such as other insects, arachnids, worms, and mollusks (slugs and snails), although a few bizarre behaviors have been documented in which mosquitoes and other blood-sucking flies puncture the flesh of another insect and drink either its bodily fluids or the contents of its stomach.

Mosquitoes usually mate in the air. Males will form a swarm in some conspicuous location (in a sunlit patch in a shady forest, over a bush in a marsh) and wait for females to fly into the swarm. When a female enters the swarm, she will be approached

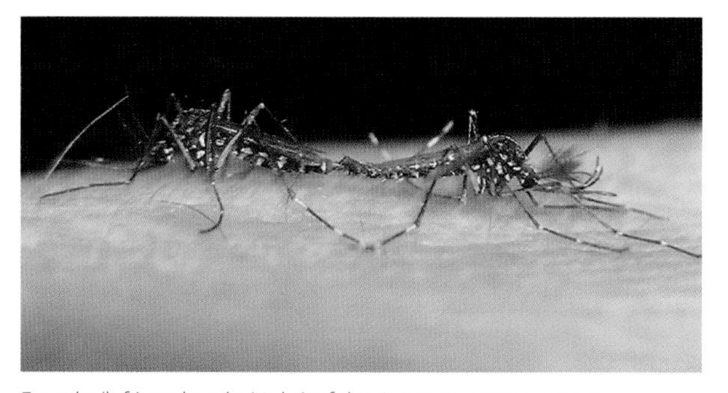

Female (left) and male (right) of the Asian tiger mosquito in copulation

by males, which will then attempt to copulate with her. Mating is usually very brief and lasts just a few seconds. In general, a female mosquito mates before feeding on blood. In some species males will form a swarm over a host animal. Females will mate with a male from the swarm and then fly down to bite the host animal. A swarm of male mosquitoes may hover above a stationary or slow-moving host (a cow, for example) for an hour or more.

Egg laying in mosquitoes is extremely variable, but some generalizations can be made, depending on the biology of the mosquito and the group to which it belongs. Mosquitoes of the genus *Culex,* for example, lay their eggs in floating clusters (called egg rafts or egg boats) on the surface of the water. Egg rafts can contain hundreds of eggs. Members of the genus *Aedes* do not lay their eggs in clusters but instead lay many single eggs in a place that will one day be submerged in water. Some *Aedes* species lay their eggs on the walls of container-like habitats, such as cavities in trees, while others lay their eggs in depressions on the ground. Later, when the tree cavities and ground depressions fill with rainwater, the eggs will hatch

Eggs of *Aedes sollicitans*

An egg raft of *Culex restuans*

and the larvae will begin to develop. In some cases, eggs of *Aedes* mosquitoes may lie dormant for years, until they are finally submerged. *Anopheles* mosquitoes lay individual eggs with lateral floats on the surface of the water. A number of other egg-laying behaviors may be found in other mosquito groups. As mentioned earlier, egg laying follows blood feeding (in the species that feed on blood).

The mosquito larva is the legless, wormlike, aquatic stage also known as a "wiggler." The larval stage is the time of growth and development for mosquitoes. Mosquito larvae grow rapidly and must eat nearly continuously. Most larvae are filter feeders and use brushlike appendages on their heads to sweep small food particles, which consist mostly of decomposing plant and animal matter, into their mouths. Larvae of some mosquito species are predators of other small aquatic insects (including other mosquito larvae) and crustaceans. They catch their prey with grasping mouthparts. Although mosquito larvae are aquatic, they must breathe air. Larvae of most mosquito species (except for *Anopheles*) have a snorkel-like tube near the end of the abdomen, called a siphon, through which they breathe. To breathe, larvae touch their siphon to the surface of the water and maintain their connection to the surface using the expanded lobes of the spiracular apparatus. The length of the larval stage (from egg to pupa) depends on the temperature, availability of food, and the mosquito species. Under optimal conditions, some mosquitoes can complete their development (egg to pupa) in as little as four days. Others may take a month or more to complete their development.

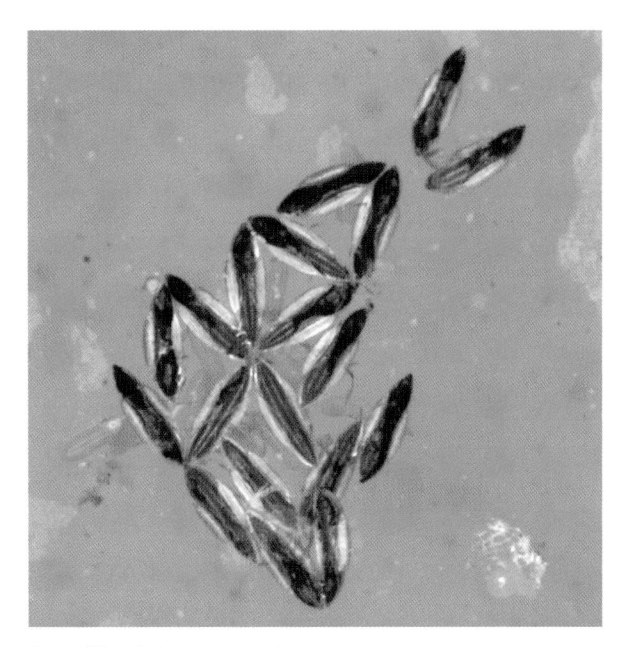

Eggs of *Anopheles punctipennis*

Mosquito larvae live in a wide variety of aquatic habitats, but they cannot survive in turbulent or fast-flowing water. Each species is adapted to a specific type of larval habitat. Ditches and puddles are common places to encounter mosquito larvae of some species, while other species are adapted to water-filled cavities in trees, pools in rock outcroppings, or man-made containers that hold water. Larvae of some mosquitoes are adapted to very specific habitats. Pitcher plant mosquito larvae, for example, thrive in the water-filled leaves of the purple pitcher plant, and nowhere else. Considering that the pitcher plant uses these same water-filled leaves to catch, kill, and consume small insects, this is no small feat. Larvae of the pitcher plant mosquito are specially adapted not only to resist being killed in this environment but also to benefit by sharing the food of the pitcher plant (trapped insects).

The mosquito pupa, also known as a "tumbler," follows the larval stage and is the stage during which the insect transforms from larva to adult. The comma-shaped pupa is very mobile but does not feed. Like the larva, the pupa is aquatic but must breathe air. Instead of a single long siphon at the rear end, the pupa has two short tubes on the thorax through which it breathes. These tubes are called "trumpets"

Larvae of *Anopheles quadrimaculatus* (left), *Uranotaenia sapphirina* (center), and *Culex erraticus* (right)

A vegetated wetland

Water-filled cavity in an oak tree (tree hole)

A discarded cup fragment harboring larvae of *Aedes albopictus*

A water-filled cavity in a streamside boulder (rock pool)

The purple pitcher plant (*Sarracenia purpurea*)

because of their resemblance to the bell of a trumpet. The pupal stage is generally rather short, lasting between one and three days. It ends when the adult mosquito emerges from the pupal skin (exoskeleton) at the surface of the water. At this point the aquatic stages are over and the aerial life of the mosquito begins.

Adult of *Aedes canadensis* emerging from its pupal skin

Mosquitoes and the Environment

What is the purpose of a mosquito? This is a question that is often asked of entomologists and it is difficult to answer in just a sentence or two. Most people know of no beneficial properties of mosquitoes and are only aware of the spectrum of problems that mosquitoes can cause, from being backyard nuisances to transmitting deadly pathogens, like malaria and West Nile virus. Given the negative attributes of mosquitoes, many people would likely find it hard to imagine any beneficial property of a mosquito that could justify its existence. Whether or not you think that mosquitoes deserve to exist, they serve many purposes in our ecosystems, probably even in ways that benefit you.

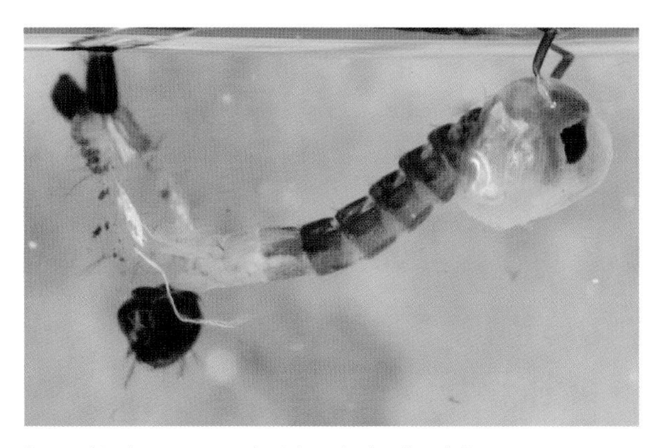

Pupa of *Aedes tormentor* shedding the last larval skin

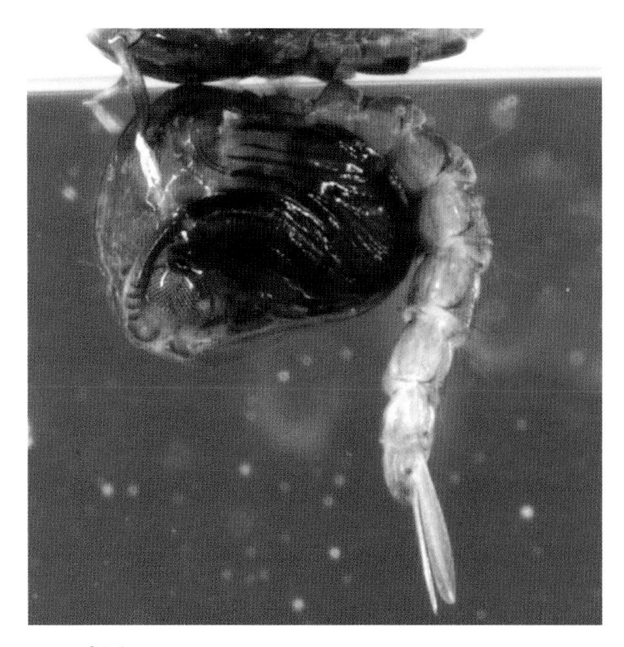

Pupa of *Culex restuans*

Mosquitoes are very abundant in most ecosystems and are very important food sources for other animals, especially as larvae. Each female mosquito can lay hundreds of eggs in her lifetime. If each and every mosquito egg that was laid resulted in an adult mosquito, the world would quickly be completely overrun with mosquitoes. This is clearly not the case. Of the hundreds of offspring produced by a female mosquito, only a few will survive to adulthood. Most of the others are eaten in the larval or pupal stage. Since mosquito larvae and pupae are aquatic, they are eaten by small fish, tadpoles, and aquatic insects and crustaceans. These animals are, in turn, food for larger fish and birds that humans appreciate very much—as wildlife, food, game animals, and dispersers of seeds. Largemouth bass, for example, probably eat very few mosquito larvae, since their diet consists mostly of smaller fish. However, the smaller fish that largemouth bass prey upon are heavily dependent upon mosquito larvae as a food source. So, without mosquito larvae we would likely have fewer largemouth bass. This sort of example likely applies to many species of birds, as well as reptiles and amphibians. Adult mosquitoes are food for birds, bats, frogs, spiders, dragonflies, and a host of other animals.

Mosquitoes are pollinators of flowering plants. As a mosquito feeds on nectar, it transfers pollen from other flowers. Although there are no known plants that are pollinated only by mosquitoes, it is certain that mosquitoes contribute to plant pollination. Eliminating mosquitoes would reduce pollination

Pygmy sunfish (*Elassoma zonatum*), a predator of mosquito larvae

Male nursery web spider (*Pisaurina*) preying upon a female treehole mosquito

rates of many types of plants, with negative effects on plant diversity and abundance.

Mosquitoes probably affect our environment in ways that we cannot yet imagine. Whether or not humans would be better off without mosquitoes, these fascinating insects are important components of the biological diversity of the southeastern United States, and our region would be diminished without them.

Form and Function of the Mosquito Body

Adult mosquitoes, like other insects, have three body regions: the head, the thorax, and the abdomen. Each of these regions is further subdivided into segments, which may or may not be discernible as distinct units. In the head and thorax, the segments are mostly fused and not easily distinguished. Segments of the abdomen are generally evident.

The mosquito head is the body's sensory center. The head is nearly spherical and is dominated by two large compound eyes, which are excellent visual organs, even in low-light situations. The surface of the eye is divided into many small units, called facets. The paired antennae arise between the eyes and serve as both chemosensory and mechanosensory (sound-detecting) organs. The antennae are divided into three regions. The flagellum is the long, segmented, whiplike portion. Each segment of the flagellum (flagellomere) bears a whorl of sensory setae. The pedicel is basal to the flagellum and appears as a swollen or bulbous segment. Neurosensory cells within the pedicel receive vibratory signals from the sensory setae on the flagellum. The scape is the ringlike or cuplike basal segment of the antennae. Below the antennae is the clypeus, which covers the forward-projecting portion of the head that gives rise to the paired maxillary palps and the proboscis. The maxillary palps (often referred to simply as the palps) are jointed chemosensory and mechanosensory appendages that flank the proboscis. In most mosquitoes, the palps are shorter in the females than in the males. The proboscis is the conspicuous elongate, projecting mouthparts of the adult mosquito. It is composed of a ventral sheath that holds the styliform (needlelike) elements that pierce host flesh, deliver mosquito saliva, and transport blood. At the tip of the proboscis are the labella, two sensory lobes (usually appearing fused) that mosquitoes use to locate host blood vessels.

The thorax, located between the head and the abdomen, bears the legs and wings and is therefore the locomotory center of the adult mosquito. Adult mosquitoes have six legs, of which the hind legs are the longest. The legs are divided into five segments. The coxa is the basal segment, followed by the trochanter, the femur, the tibia, and finally the tarsus (plural tarsi). The tarsus is further divided into five subunits, called tarsomeres. The apical tarsomere terminates in a claw. Mosquitoes technically have four wings, but only the front wings are used for flying. The hind wings, called "halteres," are small and do not resemble true wings at all. The halteres are short

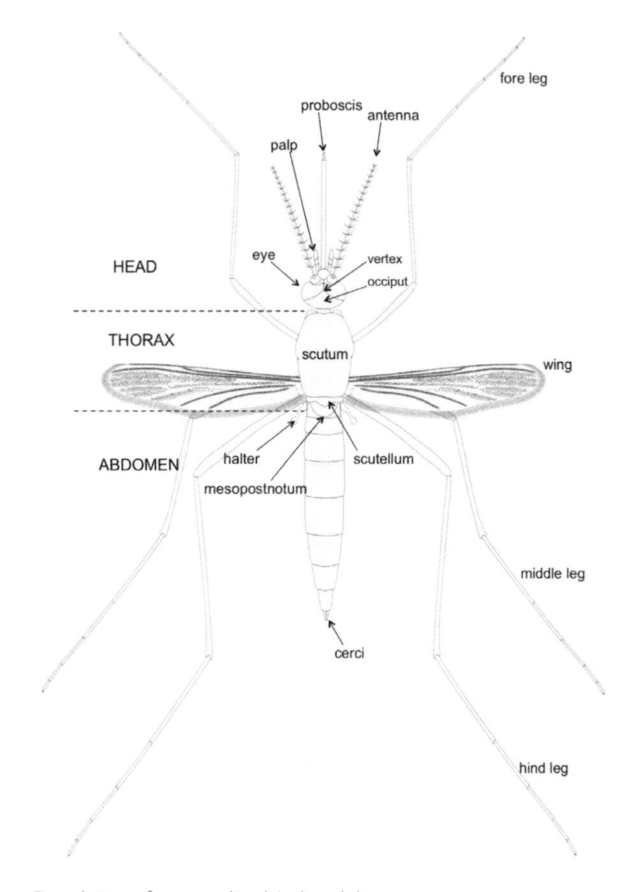

Dorsal view of a generalized *Aedes* adult

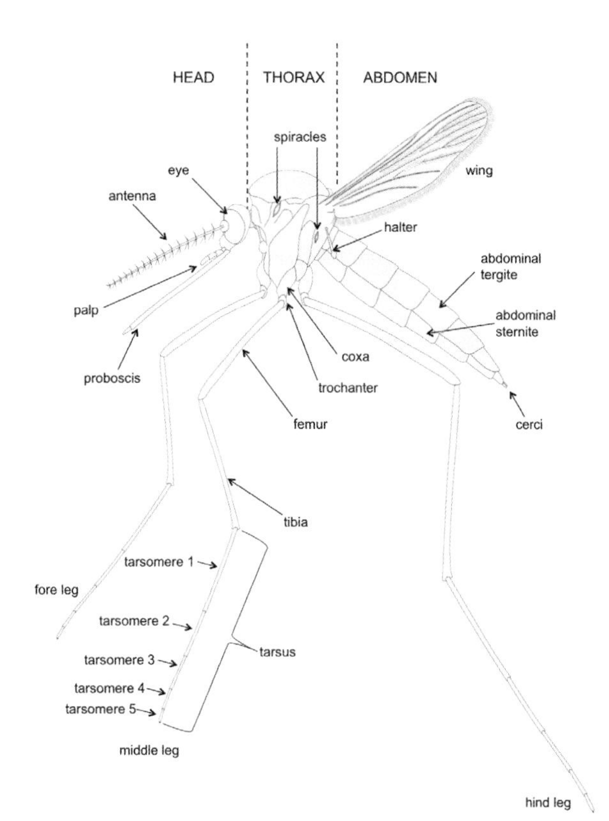

Lateral view of a generalized *Aedes* adult

and knob-like and are used to help maintain balance during flight. The front wings have long thickenings, called veins, which give the wing rigidity. The veins are covered with scales, which can be dark or light colored. There are six major veins, with several subdivisions and crossveins. The major veins are the costal, subcostal, radial, medial, cubital, and anal veins. The membranous portions of the wing between the veins are called cells and are named after the vein that they follow—for example, radial cell, costal cell. The apical tip and posterior margins of the wings have borders of long, narrow setae, called (collectively) the wing fringe. The major dorsal portion of the thorax is the scutum. The scutum of some mosquitoes is covered in dark and light scales that can form striking patterns. Posterior to the scutum is the scutellum, and posterior to the scutellum is the mesopostnotum. The lateral portion of the thorax is the pleuron. The pleuron has several exoskeletal plates, called sclerites. Two of the larger sclerites are the mesokatepisternum and mesepimeron. The arrangements of setae and scales on these sclerites are often used in mosquito identification. The pleuron also bears two large spiracles, openings in the exoskeleton through which insects breathe.

The abdomen, the most posterior region of the body, is the primary site for digestion, excretion, and reproduction. It is divided into ten segments, each composed of a dorsal and ventral plate. The dorsal plates are called tergites, and the ventral plates are called sternites. Tergites and sternites are connected by membranous exoskeleton that can expand and stretch during feeding. The abdomen terminates in two fingerlike appendages, the cerci, which function in egg laying and copulation. In *Aedes* and *Psorophora* females, the cerci are visible, protruding from the tip of the abdomen. In many other genera, the cerci are retracted within the body and are not visible.

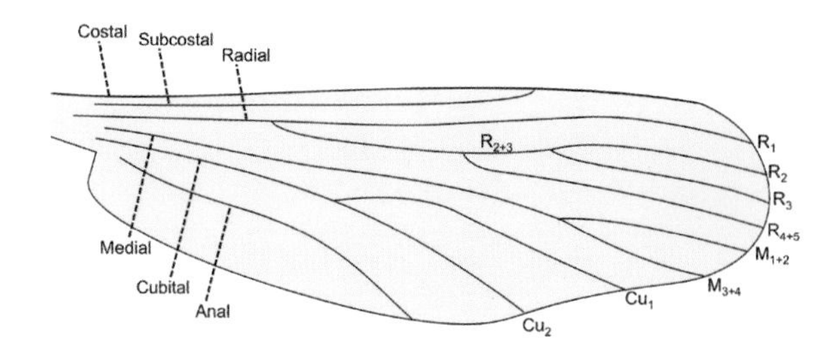

Generalized mosquito wing (scales not shown)

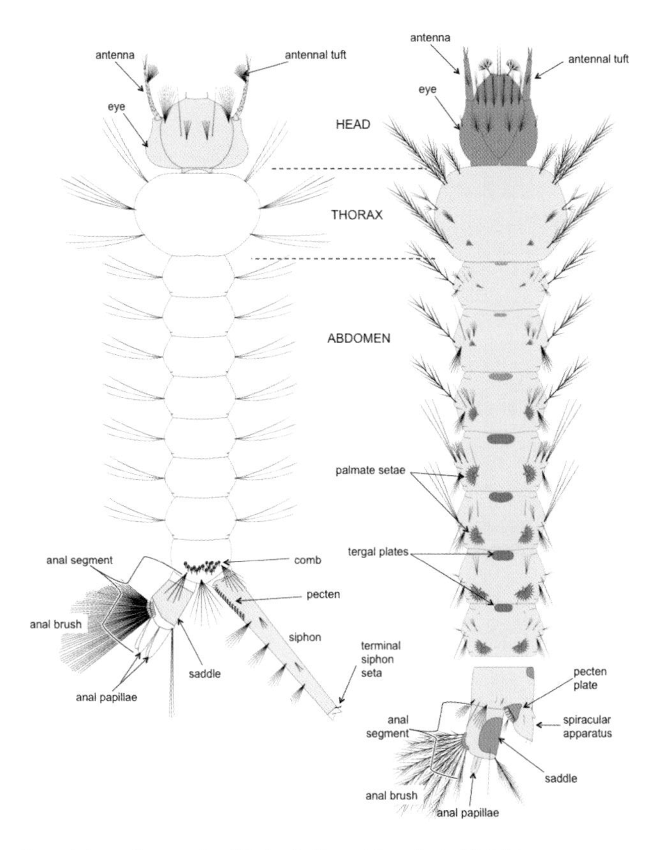

Dorsal view of gereralized *Culex* (left) and *Anopheles* (right) larvae

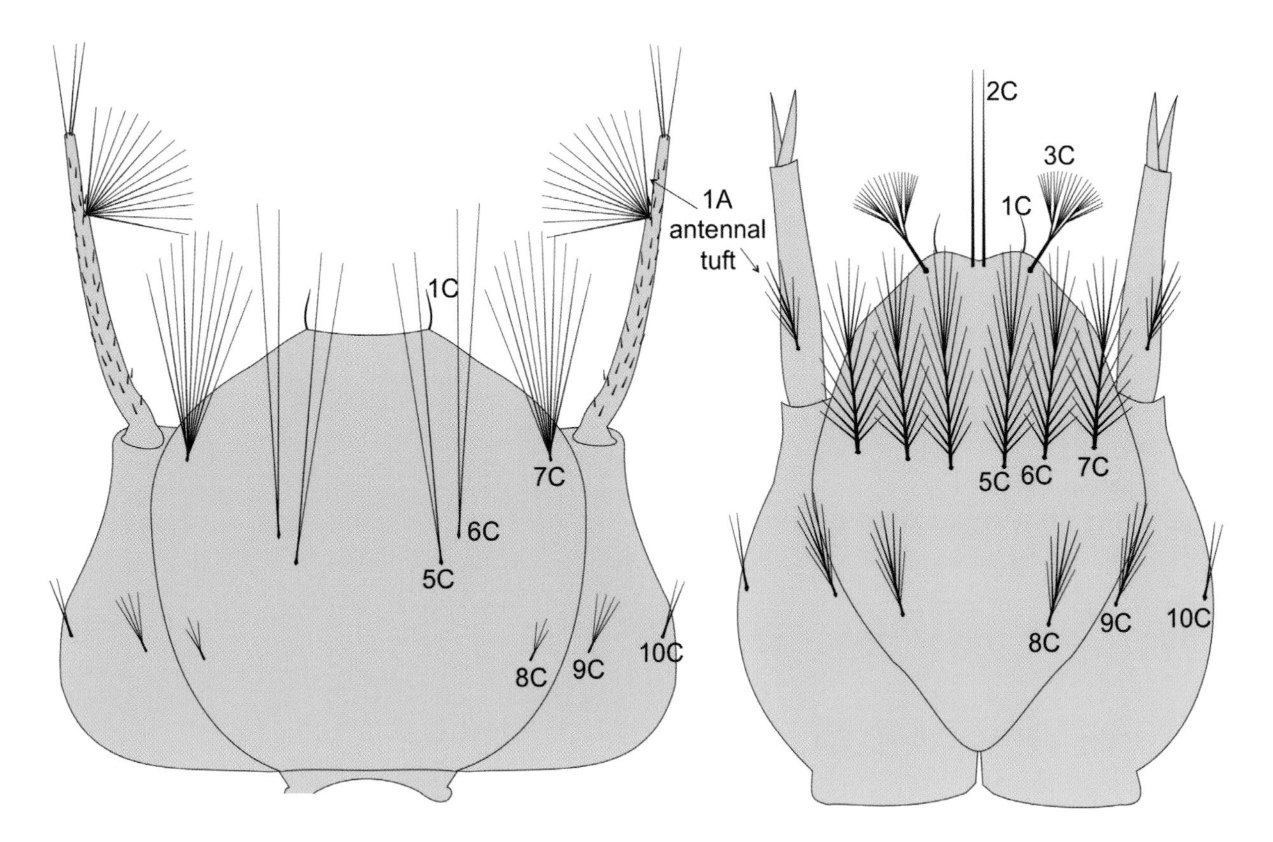

Heads of *Culex* (left) and *Anopheles* (right) larvae

Like adults, mosquito larvae also have three body regions: the head, thorax, and abdomen. However, larval mosquitoes are aquatic and wormlike. They lack the legs, wings, and proboscis that are characteristic of adults.

The head of a mosquito larva is large and sclerotized (made of hardened exoskeleton). The shape of the head may be elongate (as in *Anopheles* and *Uranotaenia*) or broad (*Aedes* and *Culex*). The head bears two eyes, two antennae, and brushlike or comblike mouthparts. The eyes are generally small and simple (not compound) and are found on either side of the head. The antennae are quite variable and may be very short to quite long. One or more setae are borne along the length of the antenna and may be branched or unbranched. The mouthparts are composed of articulating appendages of the mandible and maxilla. Setae of the head are numerous and variable in length and form. The arrangement, length, branching, and shape of head setae are used in the identification of larvae.

The thorax is elliptical, usually wider than the head, and lacks appendages. The numerous setae of the thorax are arranged in three rows that correspond to the three subdivisions of the thorax. Thoracic setae are often useful in identifying larvae.

The abdomen is elongate and cylindrical and made up of ten segments. Segments 1–7 are fairly uniform in size and shape. Segment 8 is usually smaller than the seven preceding segments and bears the comb scales (when present) and the respiratory siphon (when present). The comb scales are spine-like projections that occur in a row or patch and are sometimes borne on a sclerotized plate, called the comb plate. The number, shape, and arrangement of comb scales are useful in identifying larvae. The respiratory siphon (or simply siphon) is a sclerotized dorsal breathing tube that bears the respiratory spiracles. In most mosquito species of our region, the siphon bears a pecten, a row of spines (spicules) extending from the ventral base of the siphon to some point along its length. The size, shape, and length of the siphon and the pecten vary from one species to the next and are extremely useful in identification. Members of the genus *Anopheles* have no siphon but breathe through a flattened spiracular apparatus on segment 8. Segment 9 is reduced in mosquito larvae and is not discernible as a distinct segment. The anal segment (segment 10) bears the anal papillae, saddle, and ventral brush. The anal papillae are bulbous, membranous protrusions of the exoskeleton that function primarily in osmoregulation. The saddle, a sclerotized plate, may cover only the dorsal portion of the anal segment or may encircle it completely. The ventral brush is a row of paired setae extending along the ventral midline of the anal segment.

Part One

An Illustrated Key for Identifying the Mosquitoes of the Southeastern United States

The following section is a "key" for identifying the adult females and mature (fourth-instar) larvae of the mosquitoes of the southeastern states (Alabama, Georgia, Mississippi, North Carolina, South Carolina, and Tennessee). "Key" is a term for a tool designed to aid in the identification of members of a group of organisms. A key offers statements or questions that describe two or more alternatives of characteristics of organisms, such as "legs with bands" versus "legs without bands." Users of the key must decide which statement correctly matches the organism that they are attempting to identify. Deciding that the correct "answer" is "legs with bands" will take you to the next set of questions or statements (called a couplet) until you arrive at a species name. The type of key used in this book is called a dichotomous key, meaning that only two options (as opposed to three or more) for each characteristic are given at a time.

The key uses nontechnical terminology when possible. As with any key, however, users will need to acquaint themselves with some terms that have no simple equivalent. A glossary (p. 183) and illustrations of mosquito morphology (p. 10-11) are provided to aid users of this key. The following keys use terminology established by Harbach and Knight (1980) in the *Taxonomists' Glossary of Mosquito Anatomy*, except where otherwise noted.

Users of this key will need a magnifying tool (microscope) to effectively identify specimens. The features that one must see in order to correctly identify mosquitoes are often far too small for the human eye to see clearly without magnification. This is particularly true for mosquito larvae. The relative length and arrangement of setae on the head and body of larvae are often the only characteristics that allow one to accurately identify mosquitoes in the larval stage. Coloration of larvae is often highly variable and is therefore usually an unreliable characteristic to use in identification.

Field identification with the naked eye can be achieved for the adults of some species, such as the Asian tiger mosquito (the only mosquito in our area that is black and white with a bold white stripe down the center of its scutum and white bands on the legs). However, in most cases field identification is not reliable unless you are already an expert mosquito biologist and have spent numerous hours in the field and laboratory identifying the mosquitoes of a region.

Mosquitoes are bilaterally symmetrical; meaning that one half of the body is more or less a mirror image of the other. This is an important consideration when interpreting the following key.

Key to Genera of Adult Female Mosquitoes of the Southeastern United States

1a Proboscis strongly curved downward and tapering to a sharp point. ***Toxorhynchites rutilus*** (p. 175).
1b Proboscis straight (or slightly curved) and not tapering to a sharp point. Go to **2.**

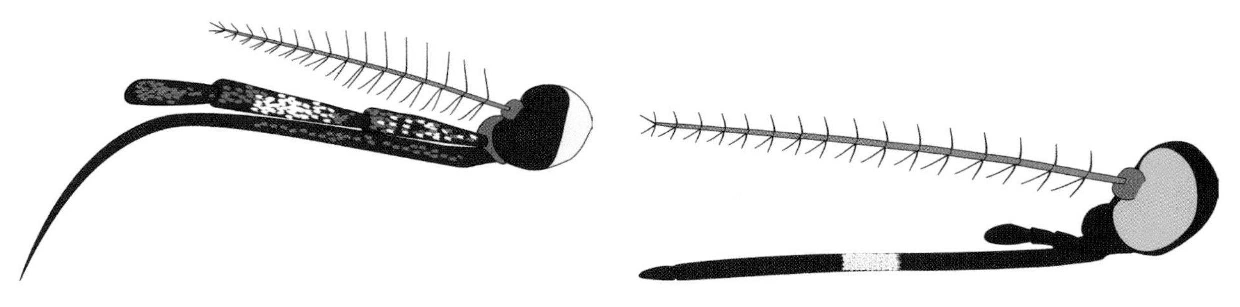

Head of *Tx. rutilus* female

Head of *Ae. sollicitans* female

2a Maxillary palps about as long as proboscis. ***Anopheles*** (p. 31).
2b Maxillary palps much shorter than proboscis. Go to **3.**

Head of *An. walkeri* female

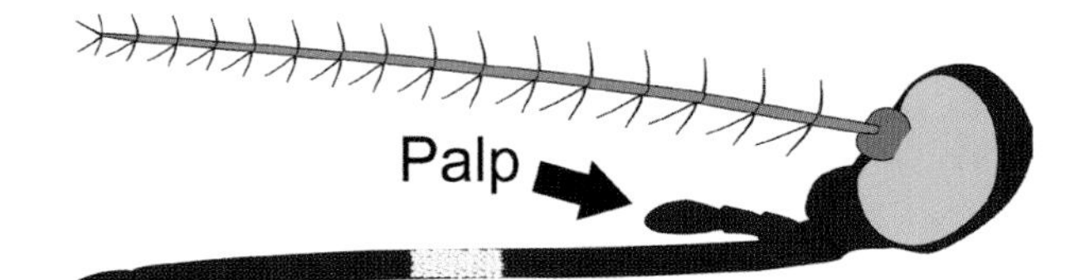

Head of *Ae. sollicitans* female

3a Mesopostnotum with tuft of setae. ***Wyeomyia smithii*** (p. 181).
3b Mesopostnotum without tuft of setae. Go to **4.**

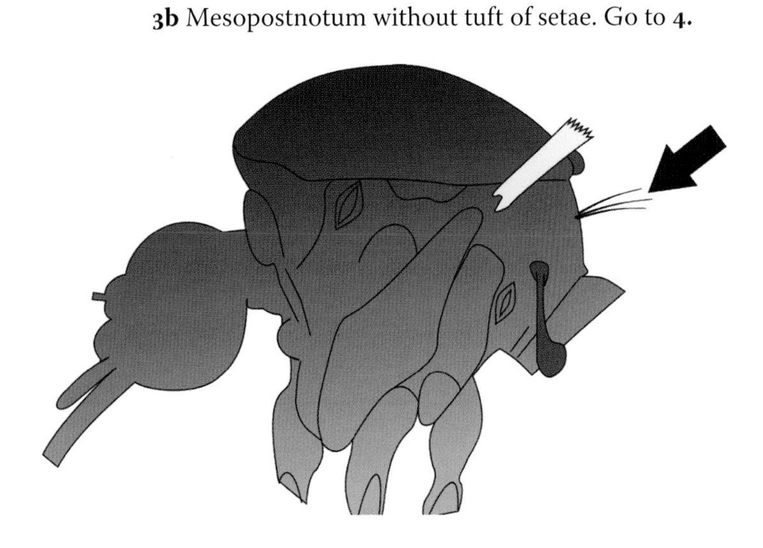

Thorax of *Wy. smithii* female

Thorax of *Cx. salinarius* female

4a Thorax with lines of iridescent blue scales. ***Uranotaenia*** (p. 43).
4b Thorax without lines of iridescent blue scales. Go to **5.**

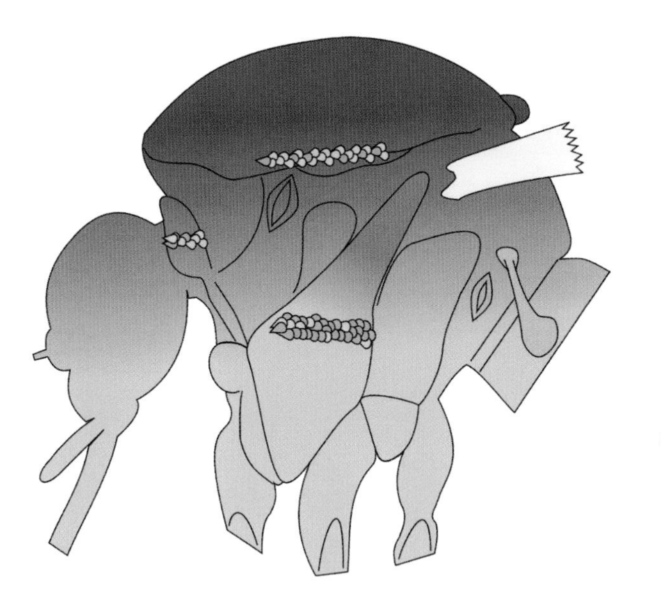

Thorax of *Ur. sapphirina* female

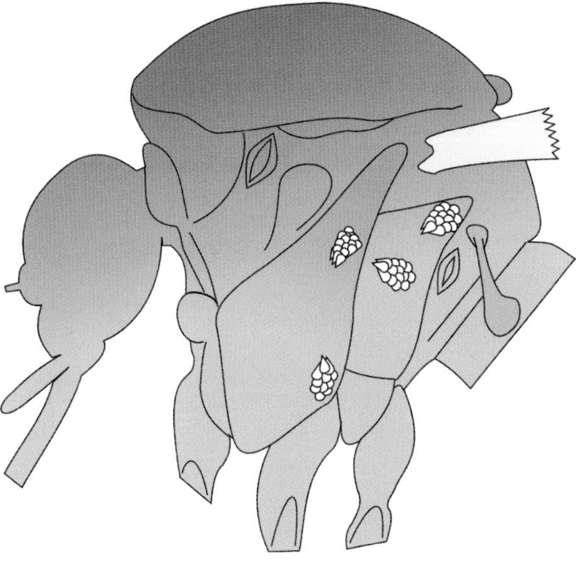

Thorax of *Cx. salinarius* female

5a Postspiracular setae present. Go to **6.**
5b Postspiracular setae absent. Go to **8.**

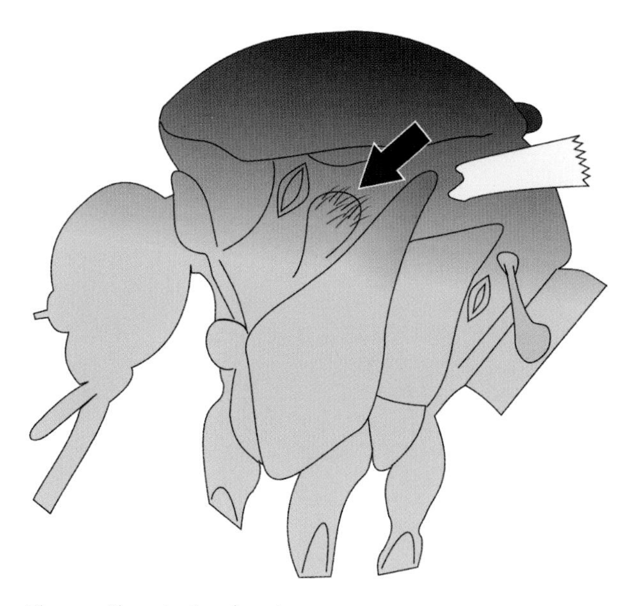

Thorax with postspiracular setae

Thorax without postspiracular setae

6a Apex of abdomen not tapering to point (in dorsal view). ***Mansonia*** (p. 39).
6b Apex of abdomen tapering to point (in dorsal view). Go to **7.**

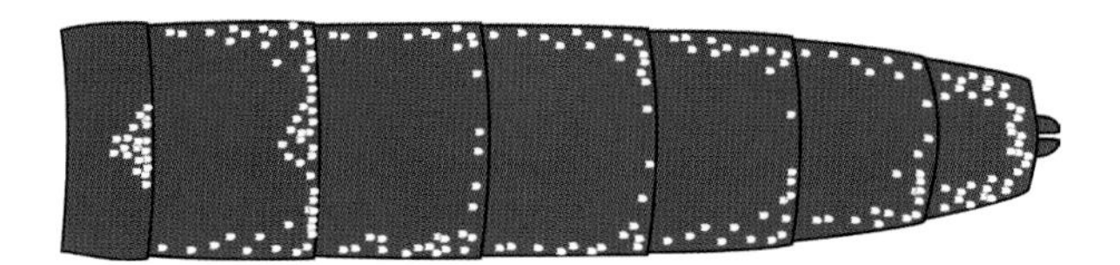

Abdomen of *Mn. dyari* female

Abdomen of *Ps. cyanescens* female

7a Prespiracular setae present; pale bands and lateral spots of abdomen on apical edge of segments. ***Psorophora*** (p. 40).
7b Prespiracular setae absent; pale bands and lateral spots of abdomen on basal edge of segments. ***Aedes*** (p. 19).

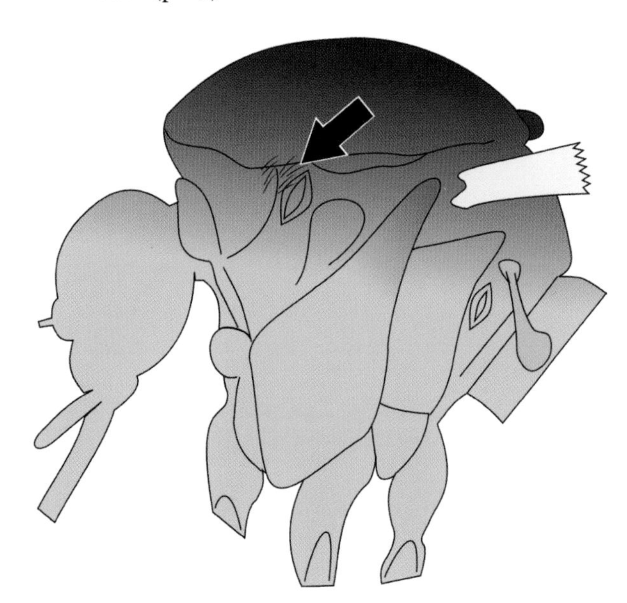
Thorax with prespiracular setae

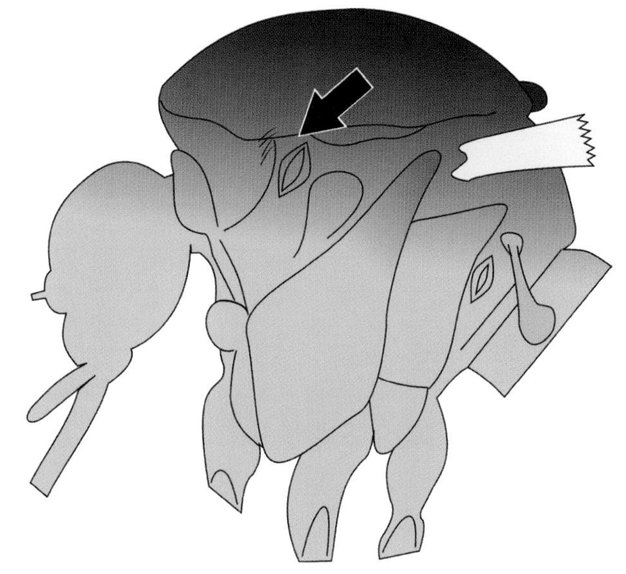
Thorax without prespiracular setae

Abdomen of *Ps. cyanescens* female

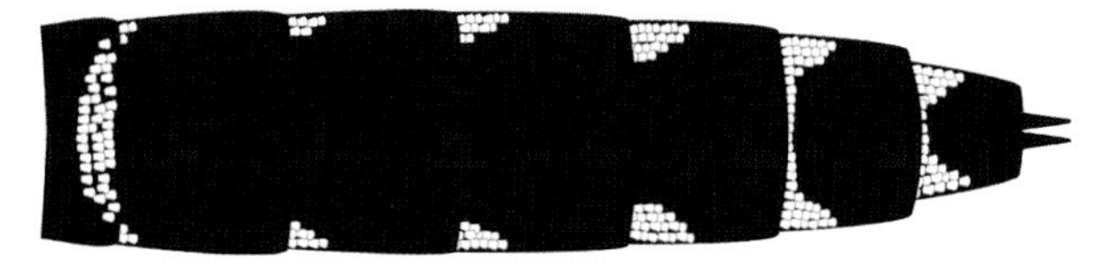
Abdomen of *Ae. thibaulti* female

8a Prespiracular setae present; underside of wing with tuft of setae at base of subcostal vein.
Culiseta (p. 38).

8b Prespiracular setae absent; tuft of setae at base of subcostal vein absent. Go to **9.**

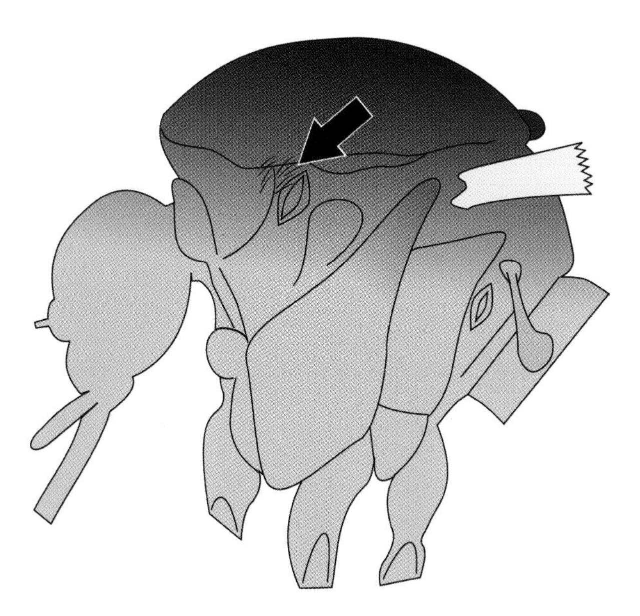

Thorax with prespiracular setae

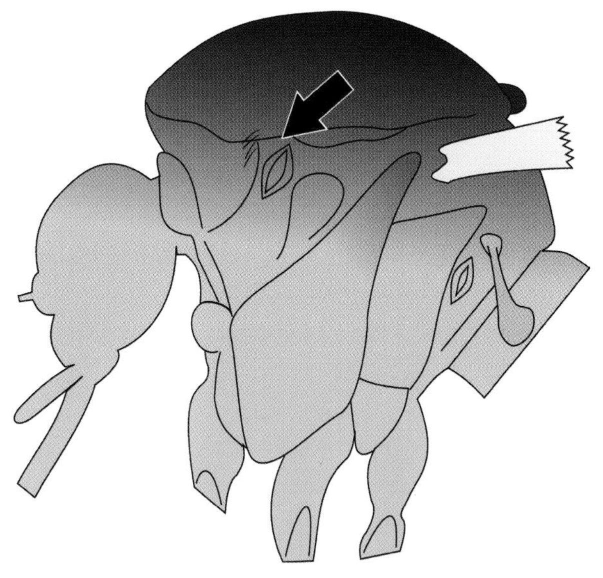

Thorax without prespiracular setae

Wing base of *Cs. melanura*

Wing base of *Cx. pipiens*

9a Segment 1 of tarsus of forelegs and middle legs longer than other tarsal segments combined; scutum black, with narrow lines of pale scales. ***Orthopodomyia*** (p. 39).

9b Segment 1 of tarsus of forelegs and middle legs not longer than other tarsal segments combined; scutum brown, without narrow lines of pale scales. Go to **10.**

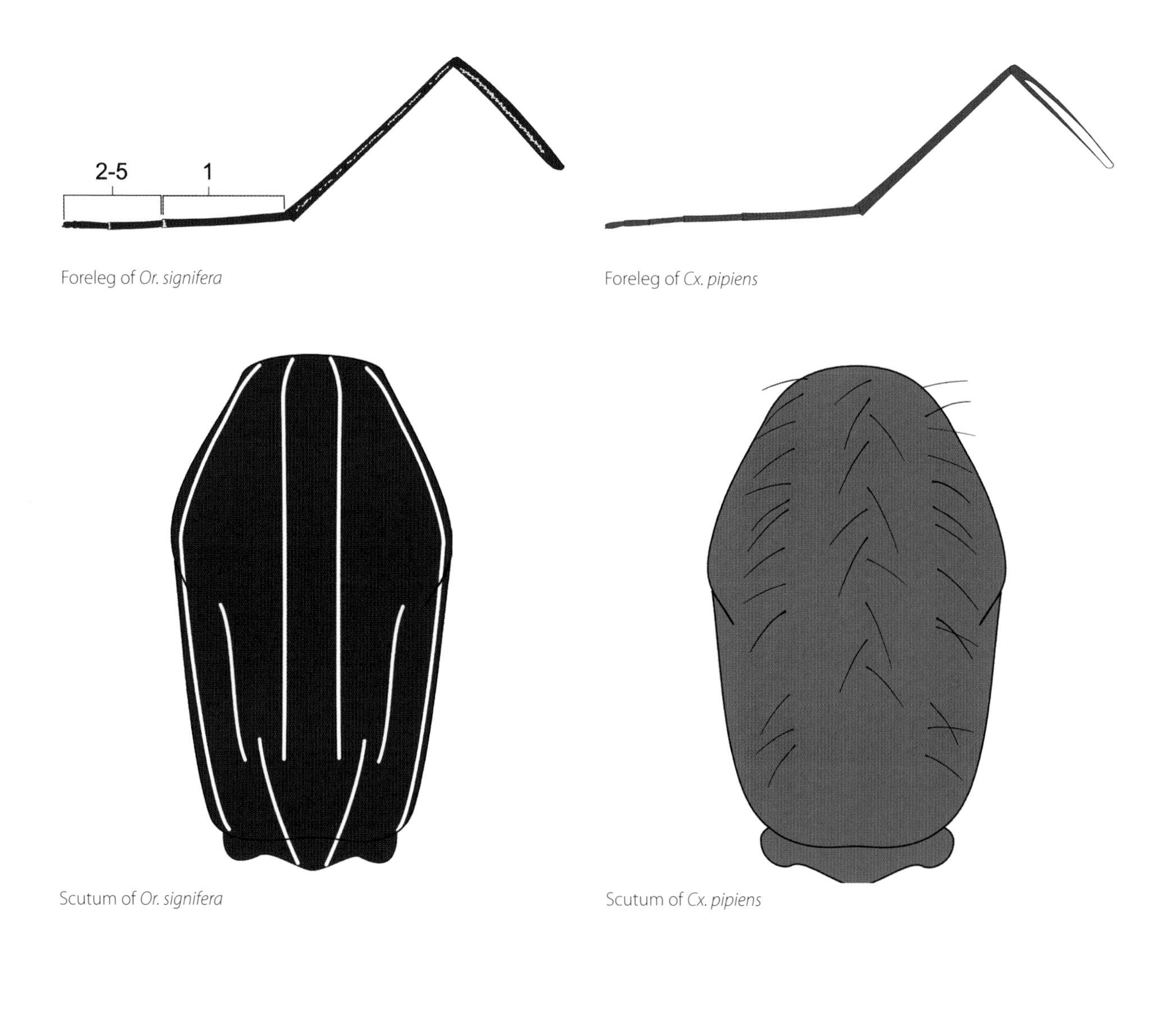

Foreleg of *Or. signifera*

Foreleg of *Cx. pipiens*

Scutum of *Or. signifera*

Scutum of *Cx. pipiens*

10a Scales on dorsal surface of wing very broad, with dark and pale scales intermixed. ***Coquillettidia perturbans*** (p. 127).

10b Scales on dorsal surface of wing long and narrow on most veins, all dark. ***Culex*** (p. 33).

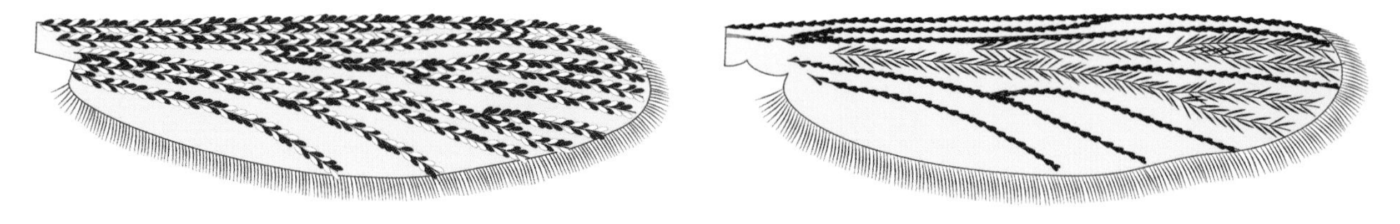

Wing of *Cq. perturbans*

Wing of *Cx. pipiens*

Aedes (Ae.)

1a Hind tarsomeres with distinct pale bands. Go to **2.**
1b Hind tarsomeres without pale bands. Go to **13.**

Hind leg of *Ae. albopictus*

Hind leg of *Ae. triseriatus*

2a Hind tarsomeres with pale bands on basal portion of segment only. Go to **3.**
2b Hind tarsomeres with pale bands on basal and apical portion, on at least 2 segments. Go to **11.**

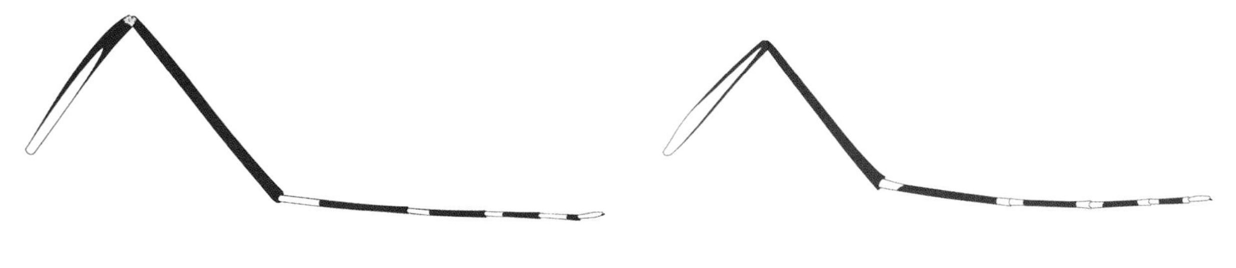

Hind leg of *Ae. aegypti*

Hind leg of *Ae. canadensis*

3a Proboscis with definite pale-scaled band near middle. Go to **4.**
3b Proboscis lacking definite pale-scaled band near middle. Go to **6.**

Head of *Ae. sollicitans* female

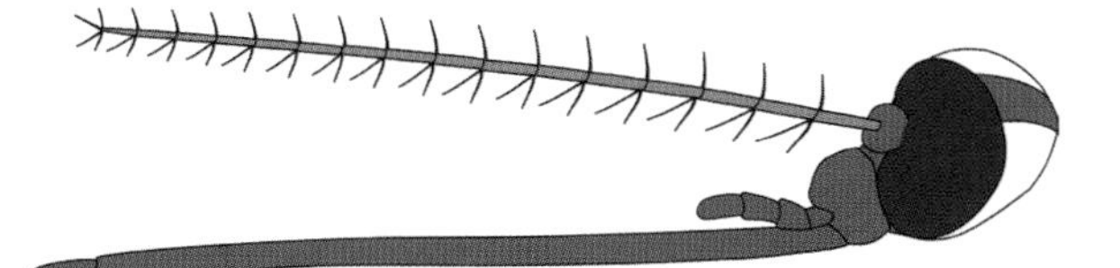

Head of *Ae. vexans* female

4a Wing with dark and pale scales intermixed. ***Ae. sollicitans*** (p. 101).
4b Wing with only dark scales. Go to **5.**

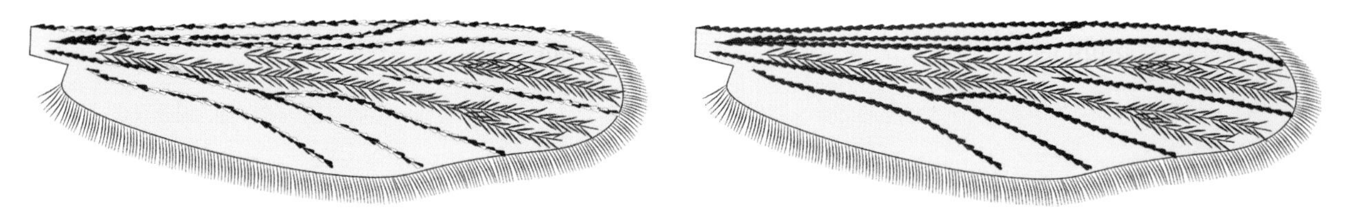

Wing of *Ae. sollicitans*

Wing of *Ae. mitchellae*

5a Abdominal terga with a pale-scaled median longitudinal stripe or row of disconnected spots. ***Ae. mitchellae*** (p. 99).
5b Abdominal terga with pale transverse basal bands but lacking a pale-scaled median longitudinal stripe. ***Ae. taeniorhynchus*** (p. 104).

Abdomen of *Ae. mitchellae* female

Abdomen of *Ae. taeniorhynchus* female

6a Scutum with a median stripe of dark scales, pale scales laterally; wing with large triangular dark and pale scales intermixed. ***Ae. grossbecki*** (p. 93).
6b Scutum with other markings; wing with only dark and mostly narrow scales. Go to **7.**

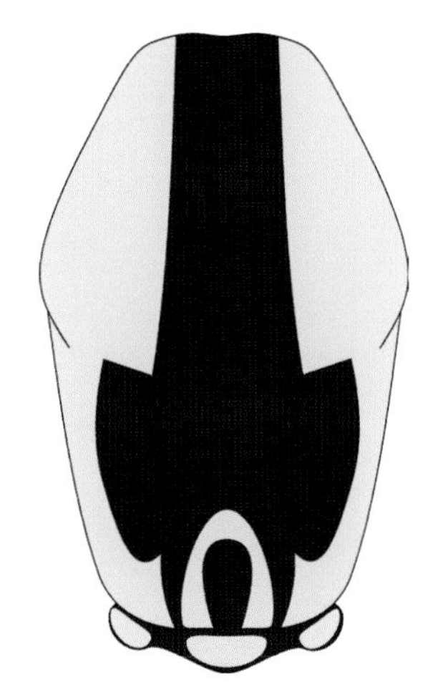

Scutum of *Ae. grossbecki*

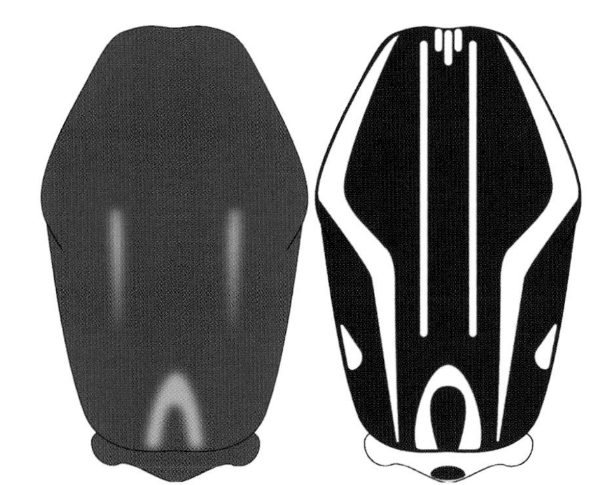

Scutum of *Ae. vexans* and *Ae. aegypti*

Wing of *Ae. grossbecki*

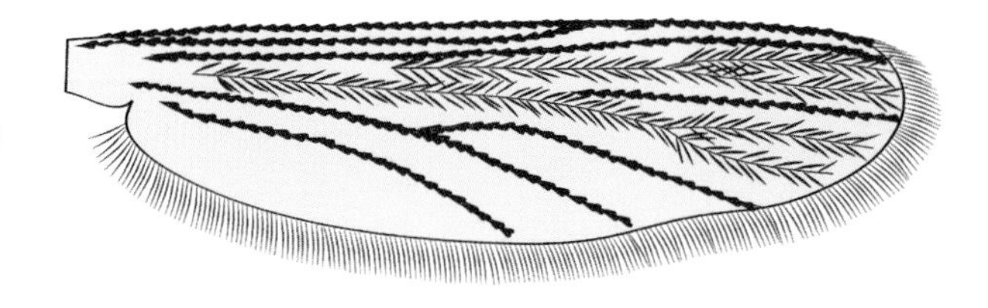

Wing of *Ae. vexans*

7a Pale bands of hind tarsomeres narrow, the pale band of tarsomere 2 less than one-quarter the length of the segment; scutum without a distinct pattern of dark and light scales. Go to **8.**

7b Pale bands of hind tarsomeres broad, the pale band of tarsomere 2 one-third or more the length of the segment; scutum with a distinct pattern of dark and light scales. Go to **9.**

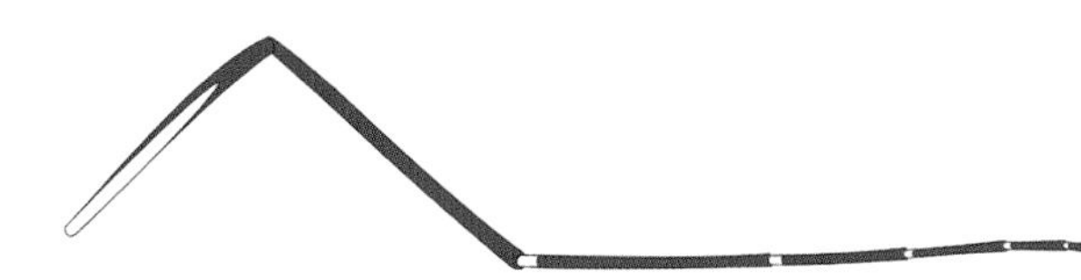

Hind leg of *Ae. vexans*

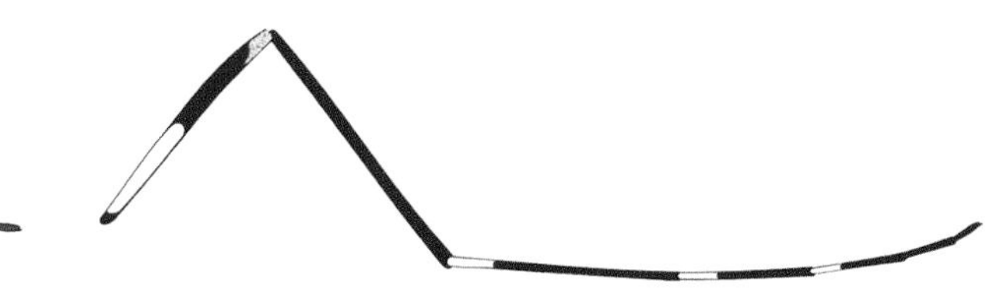

Hind leg of *Ae. japonicus*

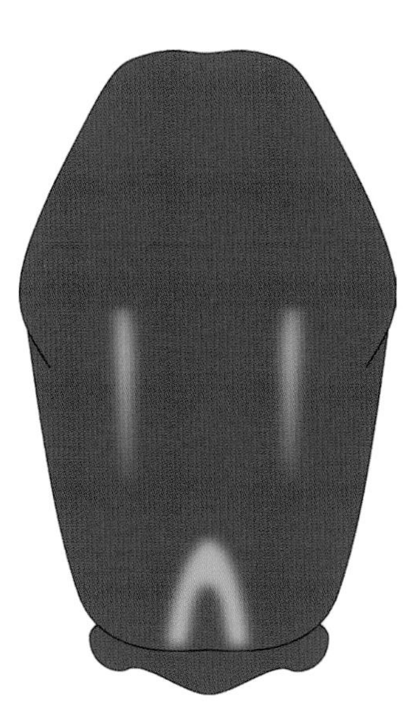

Scutum of *Ae. vexans*

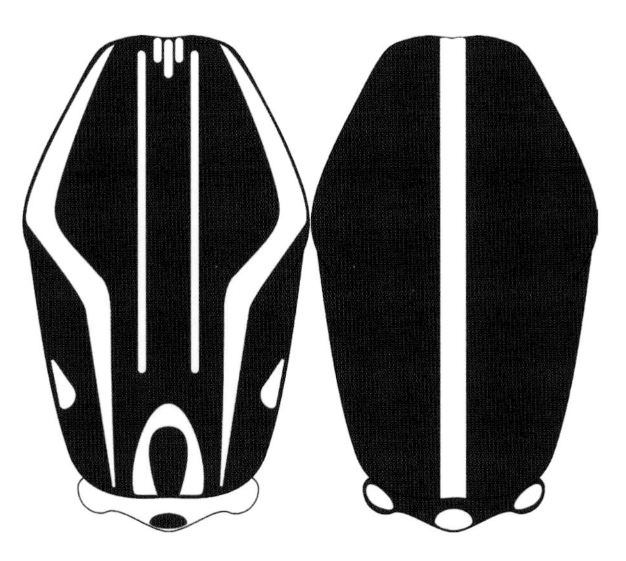

Scutum of *Ae. aegypti* (left) and *Ae. albopictus* (right)

8a Pale bands of abdominal segments 2–6 subdivided into lobes, appearing as 2 chevrons (V shapes), segment 7 mostly dark scaled; scutum brown. ***Ae. vexans*** (p. 111).

8b Pale bands of abdominal segments 2–6 not subdivided, but usually narrowest near middle, segment 7 mostly pale scaled; scutum reddish brown. ***Ae. cantator*** (p. 87).

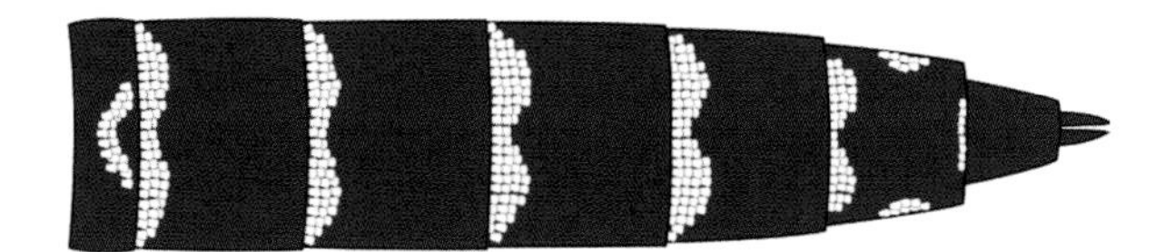

Abdomen of *Ae. vexans* female

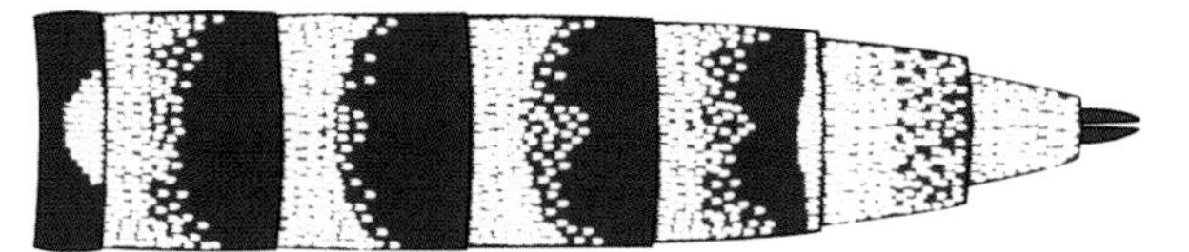

Abdomen of *Ae. cantator* female

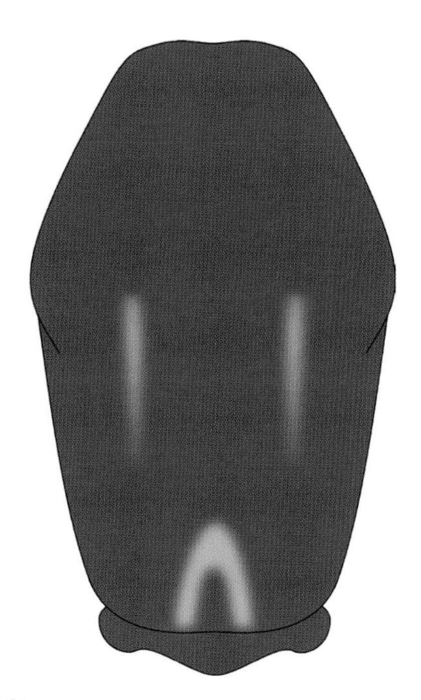

Scutum of *Ae. vexans*

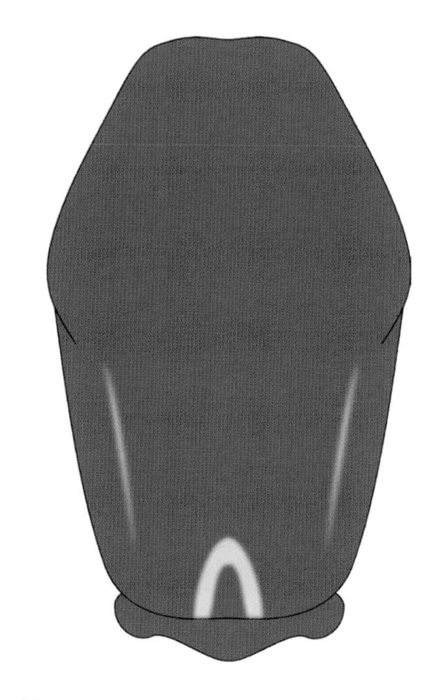

Scutum of *Ae. cantator*

9a Palps dark, without pale scales; tarsomeres 4 and 5 of hind leg completely dark scaled. ***Ae. japonicus*** (p. 97).

9b Palps with terminal segment white; tarsomere 4 of hind leg with pale basal band, tarsomere 5 solid white. Go to **10.**

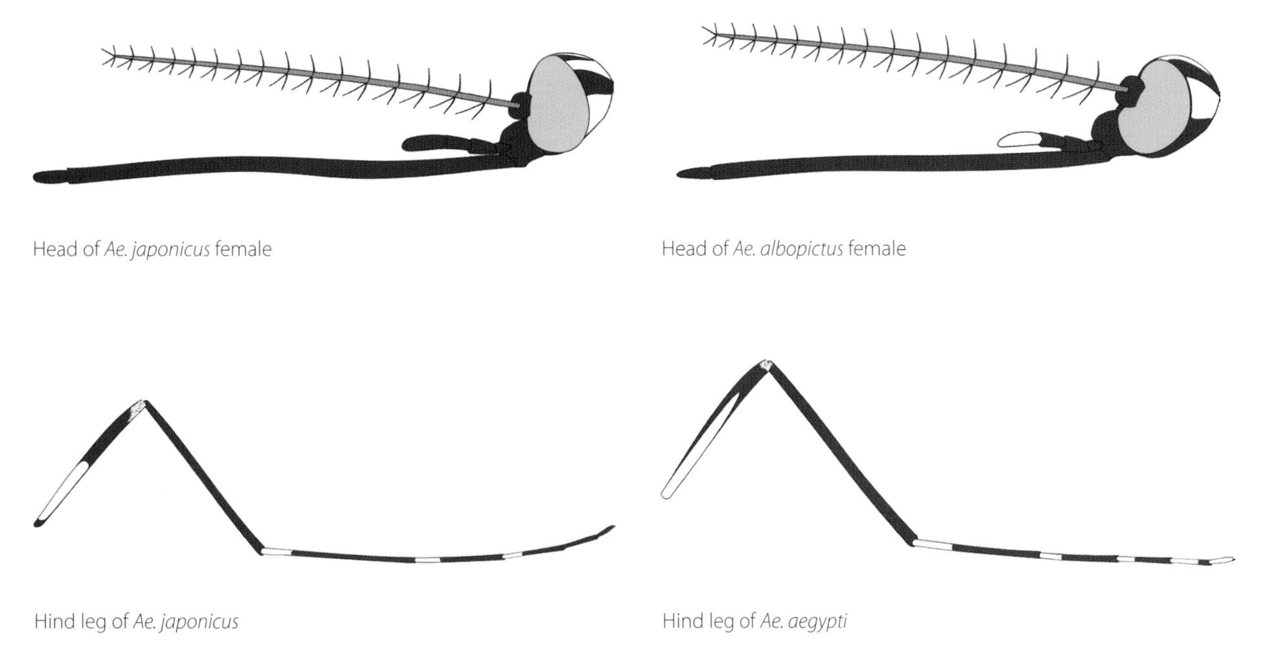

Head of *Ae. japonicus* female

Head of *Ae. albopictus* female

Hind leg of *Ae. japonicus*

Hind leg of *Ae. aegypti*

10a Scutum with a single bold median stripe of white scales, otherwise black. ***Ae. albopictus*** (p. 77).

10b Scutum with silvery-white lyre-shaped markings on black background. ***Ae. aegypti*** (p. 75).

Scutum of *Ae. albopictus*

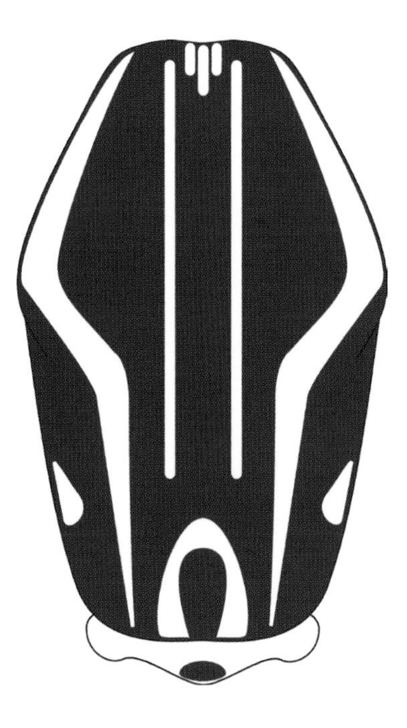

Scutum of *Ae. aegypti*

11a Wing with prominent patch of pale scales on base of costal vein; scutum with broad dark median stripe bordered by a patch of golden scales on each side. ***Ae. atropalpus*** (p. 82).

11b Wing entirely dark scaled; scutum without dark median stripe, usually evenly reddish or golden brown. Go to **12.**

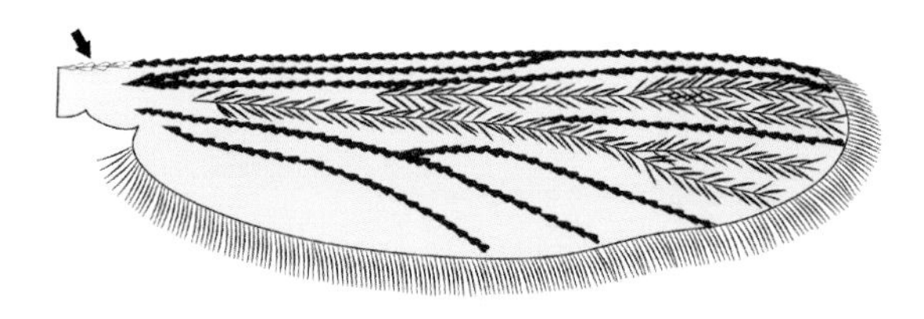

Wing of *Ae. atropalpus*

Wing of *Ae. canadensis*

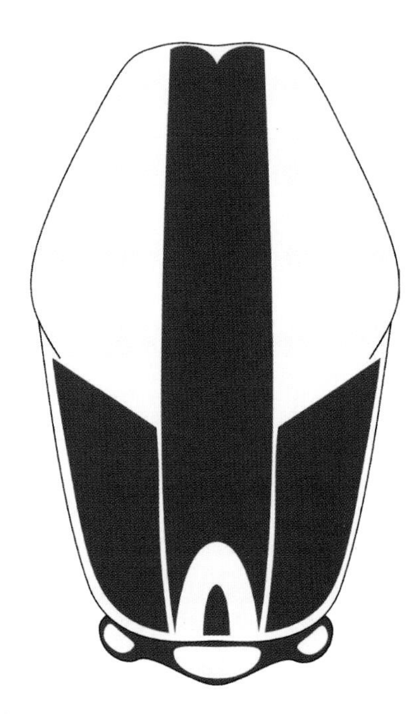

Scutum of *Ae. atropalpus*

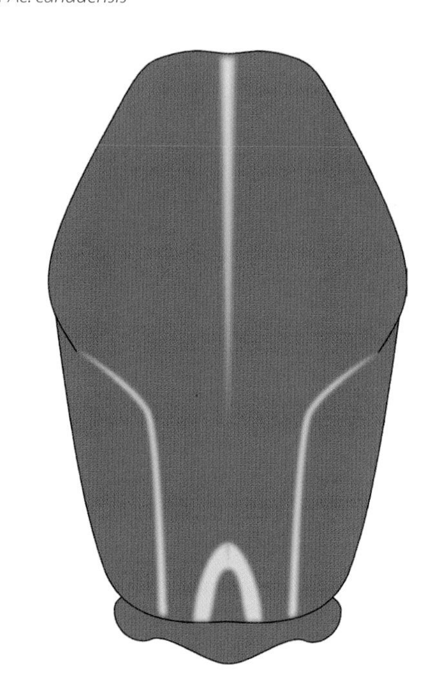

Scutum of *Ae. canadensis mathesoni*

12a Segments 1–4 of hind tarsus with broad basal and apical rings, segment 5 entirely pale scaled; scutum golden brown. ***Ae. canadensis canadensis*** (p. 85).

12b Segments of hind tarsus with narrow pale rings, segment 5 dark scaled; scutum dark brown, with indistinct median stripe of pale scales. ***Ae. canadensis mathesoni*** (p. 85).

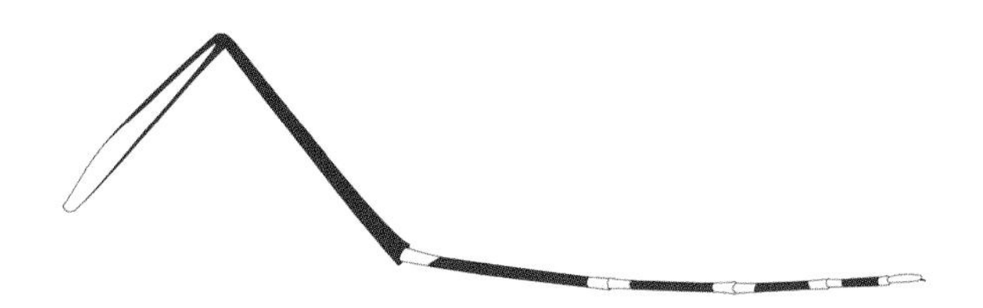

Hind leg of *Ae. canadensis canadensis*

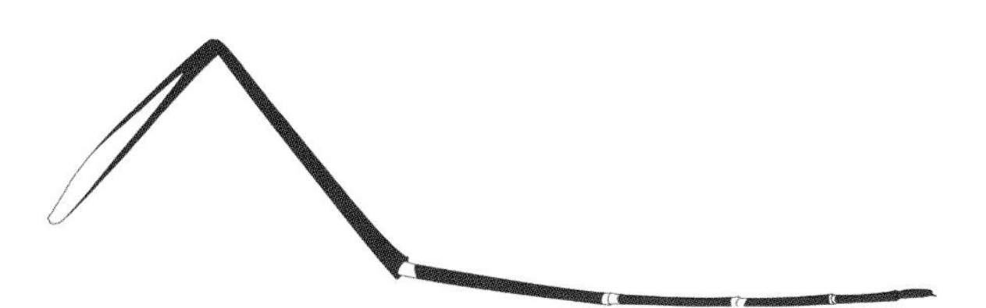

Hind leg of *Ae. canadensis mathesoni*

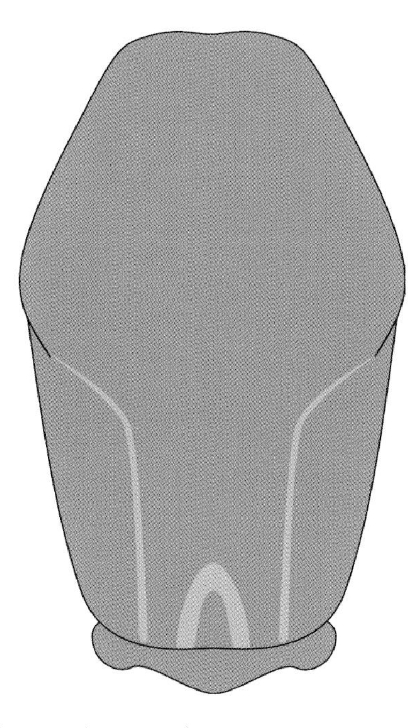

Scutum of *Ae. canadensis canadensis*

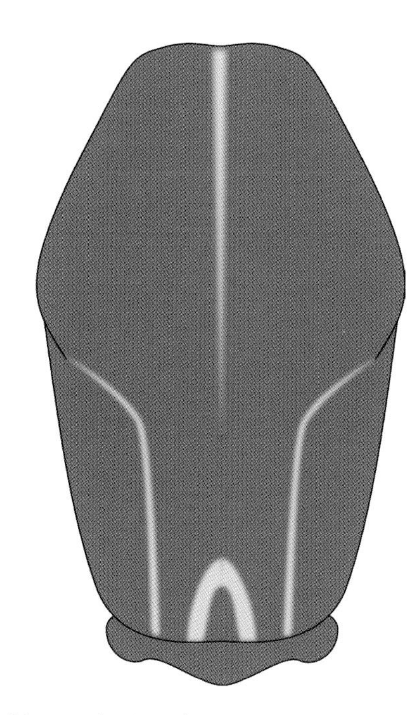

Scutum of *Ae. canadensis mathesoni*

13a Integument of scutum golden yellow, with a pair of large dark spots. ***Ae. fulvus pallens*** (p. 92).
13b Integument of scutum not golden yellow and lacking large dark spots. Go to **14.**

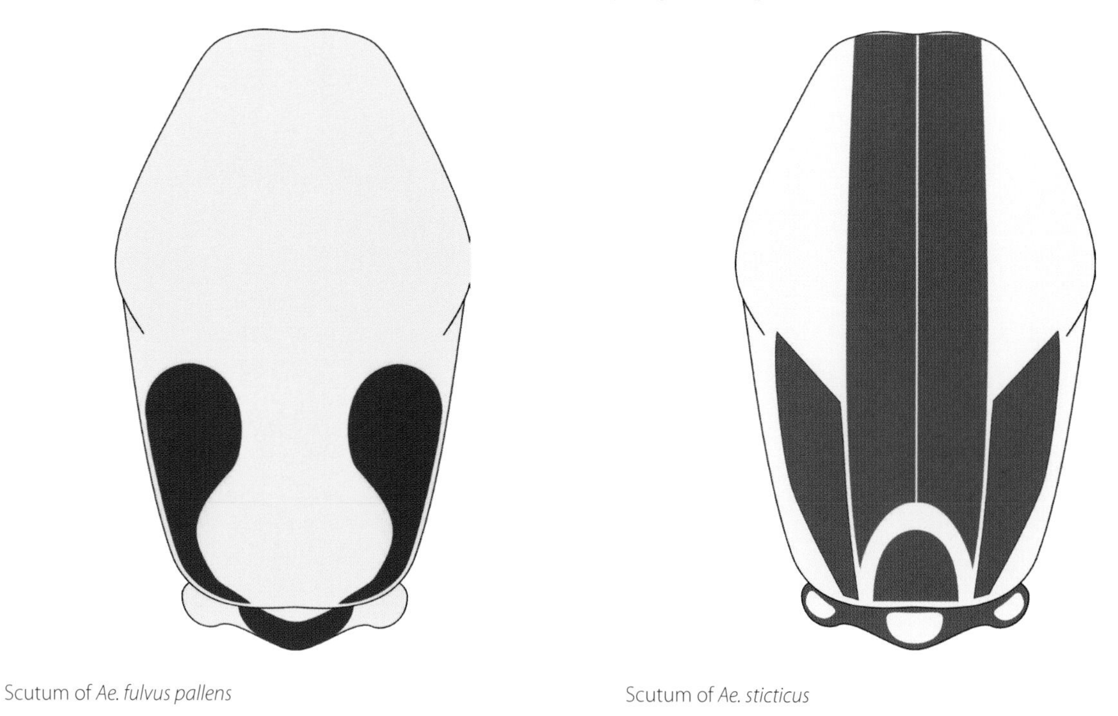

Scutum of *Ae. fulvus pallens*

Scutum of *Ae. sticticus*

14a Scutum with pale-scaled median stripe, dark scales laterally. Go to **15.**
14b Scutum with ornamentation other than above. Go to **17.**

Scutum of *Ae. atlanticus*

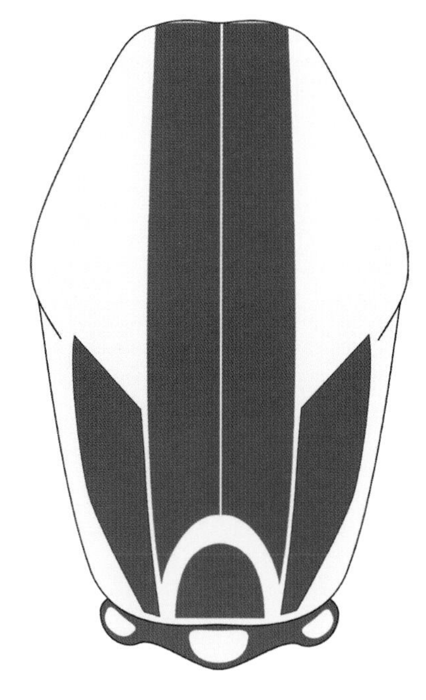

Scutum of *Ae. sticticus*

15a Pale median stripe extending from anterior border to middle of scutum or a little beyond, much broader than lateral dark-scaled areas. ***Ae. infirmatus*** (p. 95).

15b Pale median stripe extending full length of scutum, usually same width as lateral dark-scaled areas. Go to **16.**

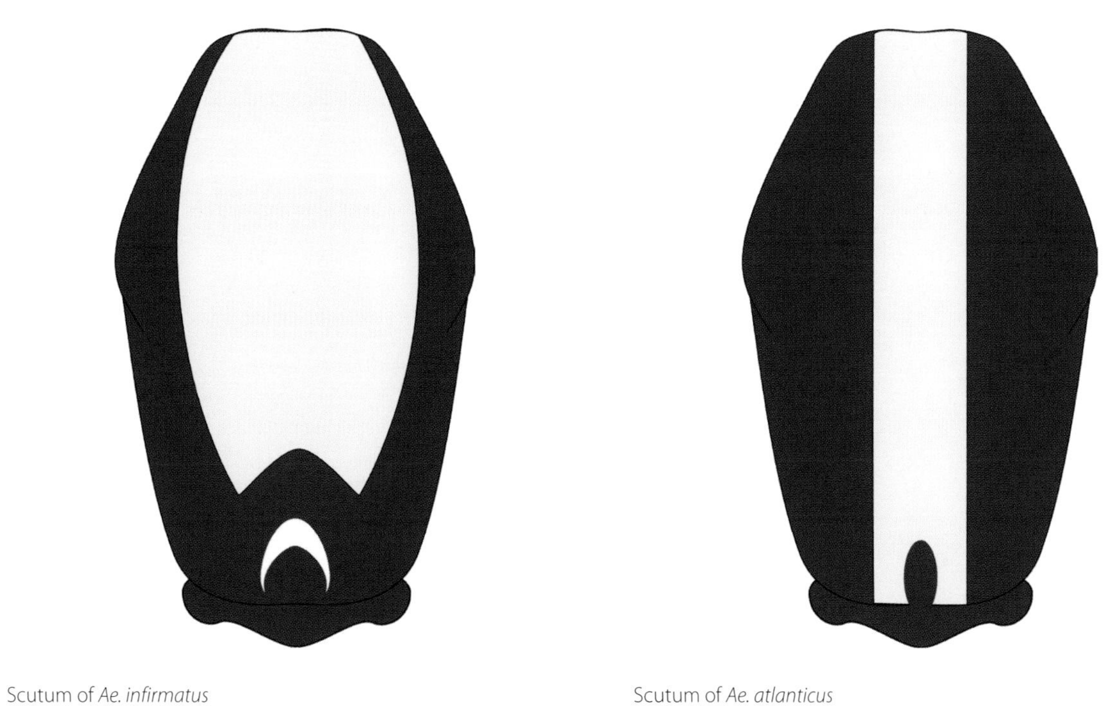

Scutum of *Ae. infirmatus* Scutum of *Ae. atlanticus*

16a Occiput with patches of dark scales behind eyes; medium-sized species, wing length 3.0–4.0 mm. ***Ae. atlanticus*** and ***Ae. tormentor*** (p. 79).

16b Occiput without dark scales; small species, wing length about 2.5 mm. ***Ae. dupreei*** (p. 90).

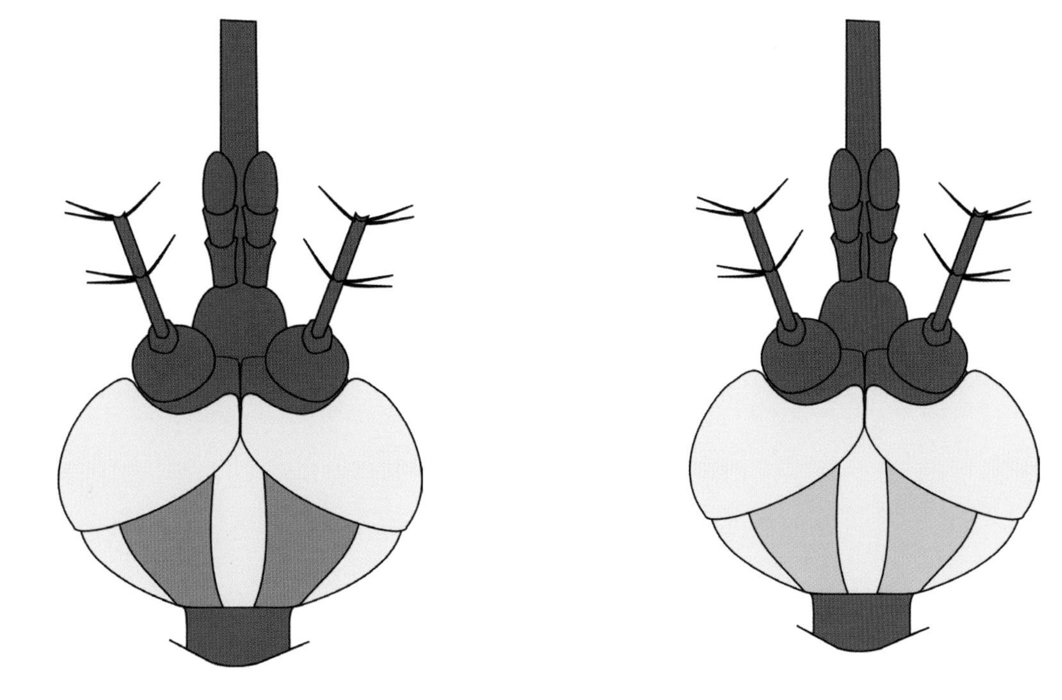

Head of *Ae. atlanticus* female Head of *Ae. dupreei* female

17a Scutum red brown, without distinct pattern. ***Ae. cinereus*** (p. 88).
17b Scutum with distinct median stripe of dark scales, flanked by areas of pale scales. Go to **18.**

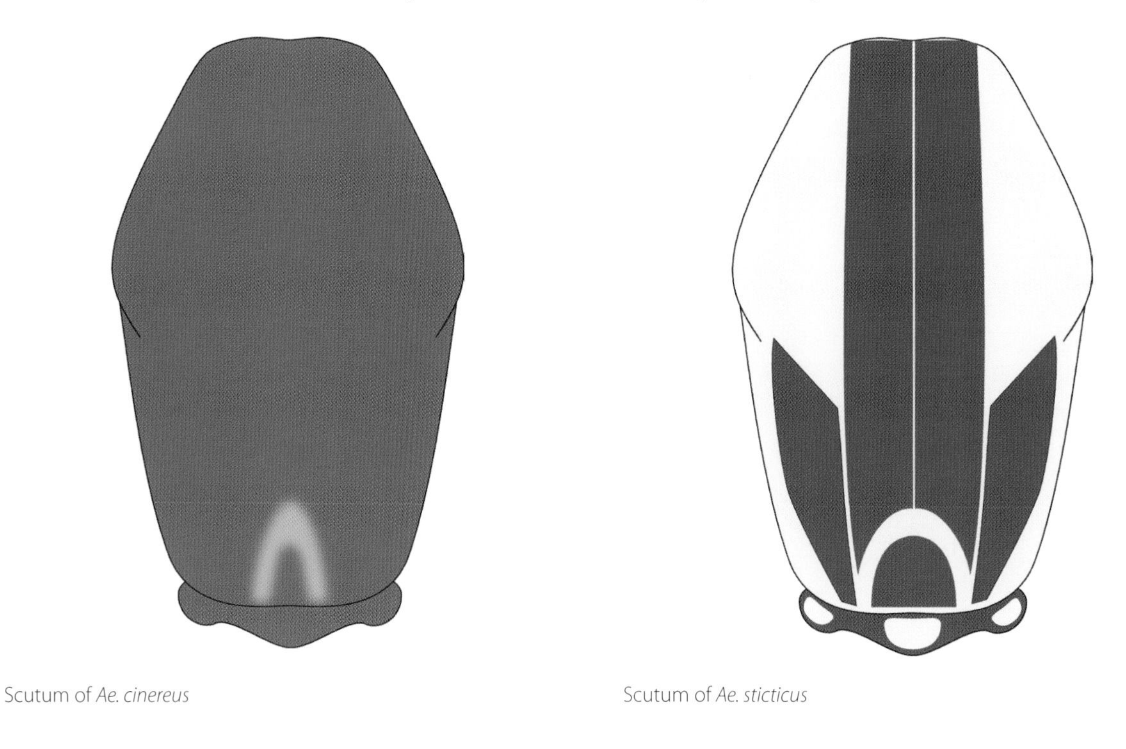

Scutum of *Ae. cinereus*

Scutum of *Ae. sticticus*

18a Abdominal tergites with broad pale basal bands. ***Ae. sticticus*** (p. 103).
18b Abdominal tergites without complete pale basal bands, but with lateral basal patches of pale scales. Go to **19.**

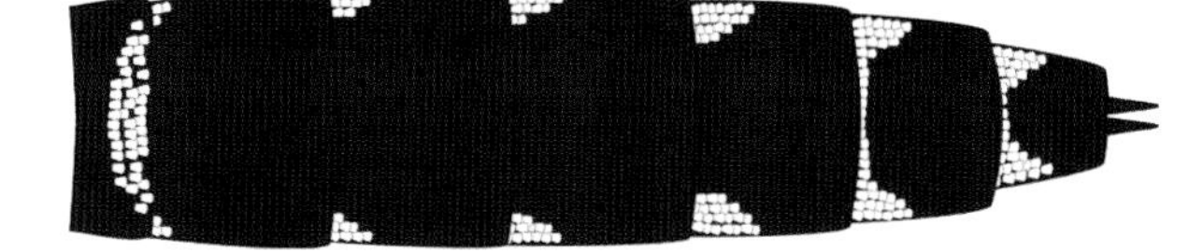

Abdomen of *Ae. sticticus* female

Abdomen of *Ae. thibaulti* female

19a Scutum with 2 silvery white stripes and 3 dark stripes, all of about the same width. ***Ae. trivittatus*** (p. 110).
19b Scutum with single broad dark median stripe, lateral areas pale. Go to **20.**

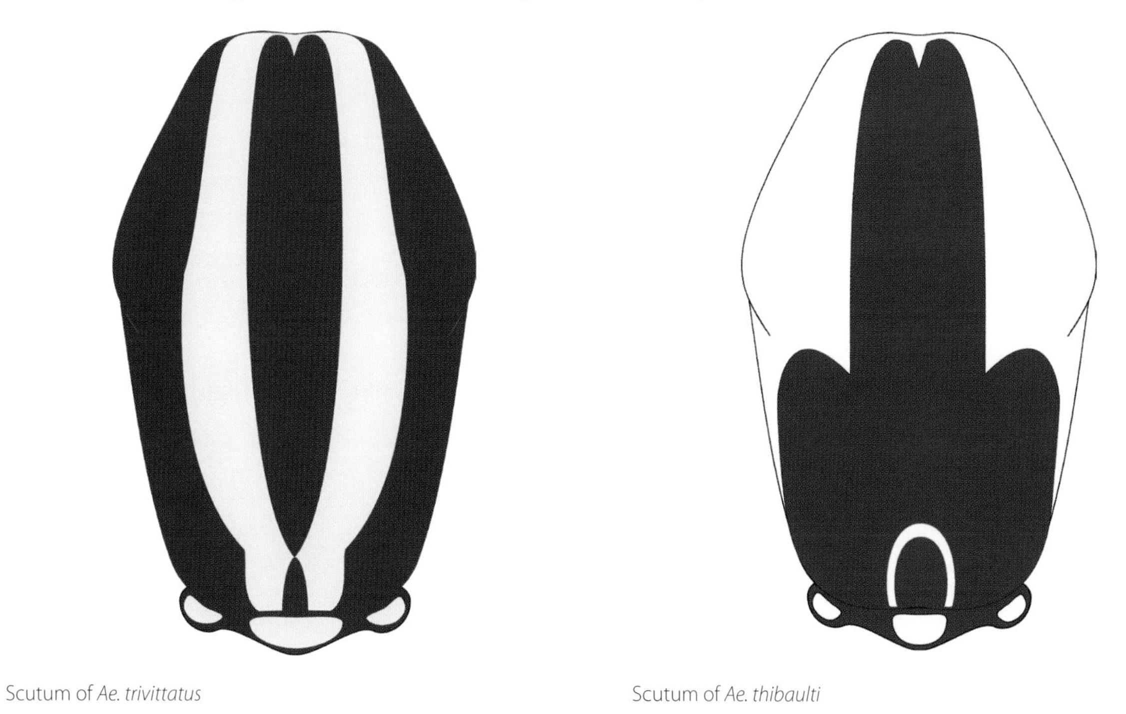

Scutum of *Ae. trivittatus*

Scutum of *Ae. thibaulti*

20a Scutum with pale golden-scaled lateral areas; apex of abdomen tapering to point in lateral view. Go to **21.**
20b Scutum with silvery-white lateral areas; apex of abdomen not tapering to point in lateral view, appearing blunt. Go to **22.**

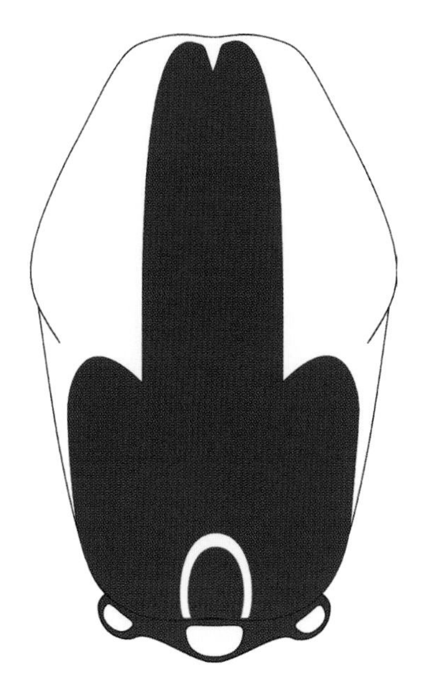

Scutum of *Ae. thibaulti*

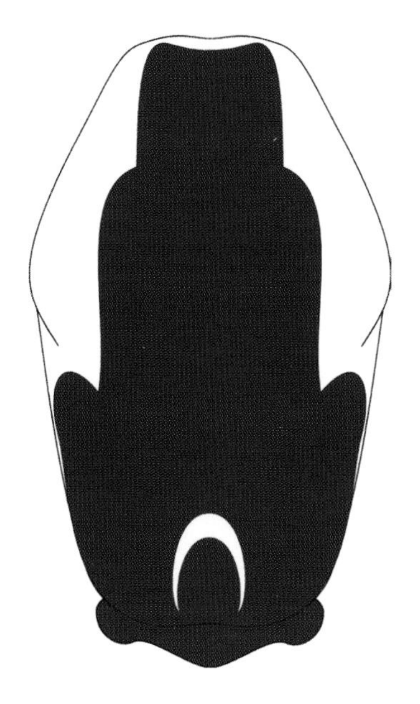

Scutum of *Ae. triseriatus*

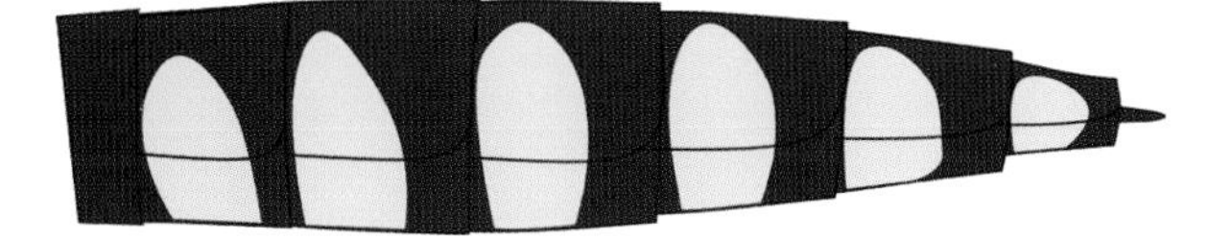

Abdomen of *Ae. thibaulti* female (lateral view)

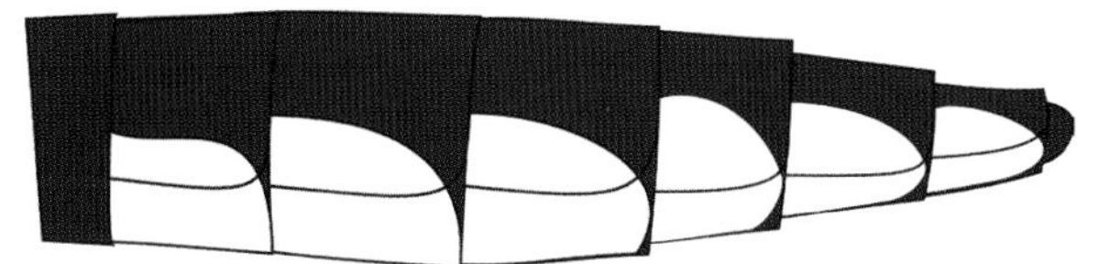

Abdomen of *Ae. triseriatus* (lateral view)

21a Abdominal sternites pale on anterior half and dark on posterior half. ***Ae. thibaulti*** (p. 106).
21b Abdominal sternites entirely pale. ***Ae. aurifer*** (p. 84).

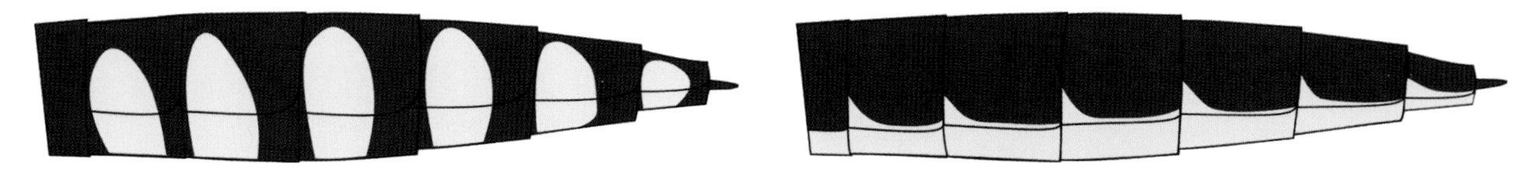

Abdomen of *Ae. thibaulti* female (lateral view)

Abdomen of *Ae. aurifer* female (lateral view)

22a Silvery-scaled lateral areas of scutum narrow, separated by broad black stripe that is much broader than pale lateral areas. ***Ae. triseriatus*** (p. 108).
22b Silvery-scaled areas of scutum usually broad, separated by black stripe of same width as pale lateral areas. ***Ae. hendersoni*** (p. 108).

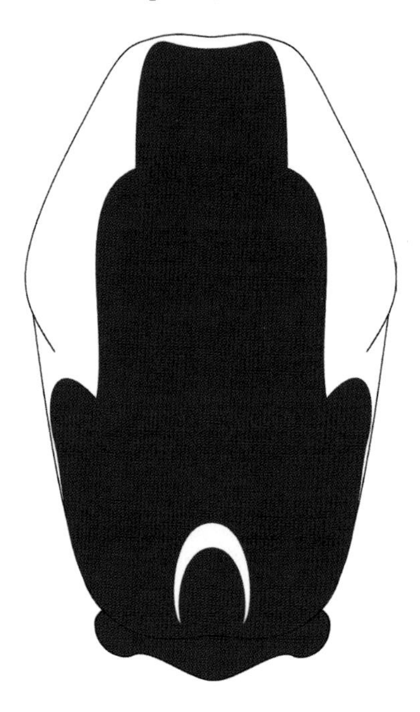

Scutum of *Ae. triseriatus*

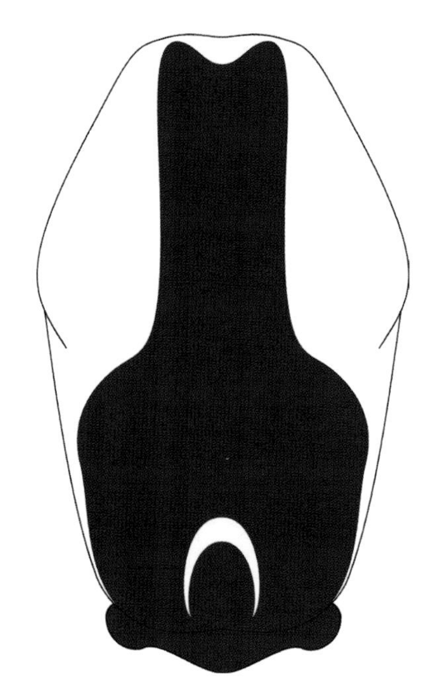

Scutum of *Ae. hendersoni*

Anopheles (An.)

1a Wing with pale spots. Go to **2.**
1b Wing without pale spots. Go to **5.**

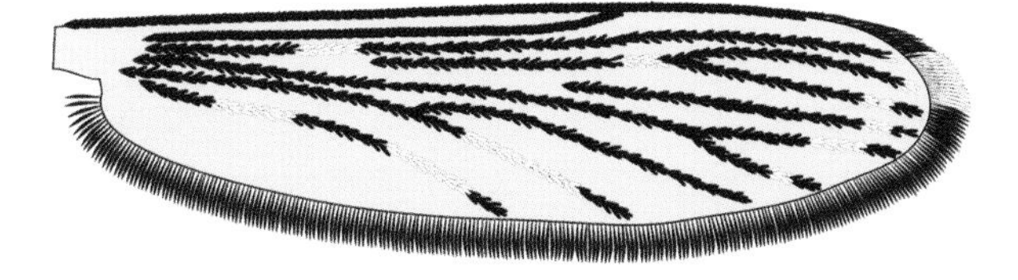

Wing of *An. crucians*

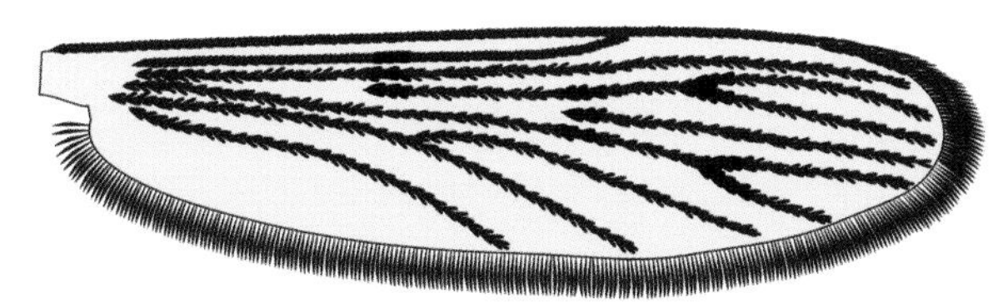

Wing of *An. quadrimaculatus*

2a Palps with pale rings. Go to **3.**
2b Palps dark, without pale rings. Go to **4.**

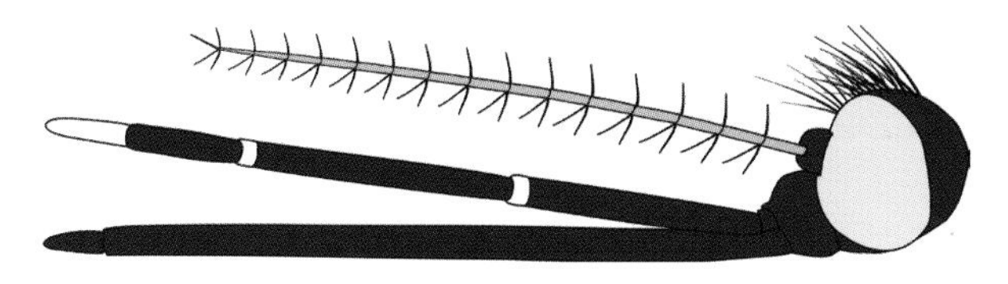

Head of *An. pseudopunctipennis* female

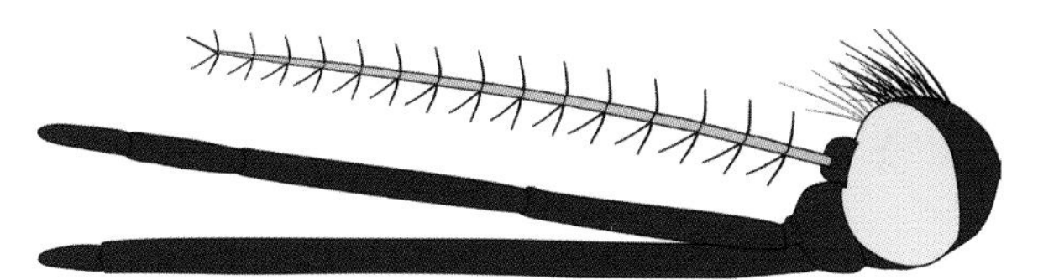

Head of *An. quadrimaculatus* female

3a Costal vein with apical pale spot, otherwise dark scaled. ***An. crucians* complex** (p. 116).
3b Costal vein with 2 pale spots (apical and subcostal pale spots). ***An. pseudopunctipennis*** (p. 120).

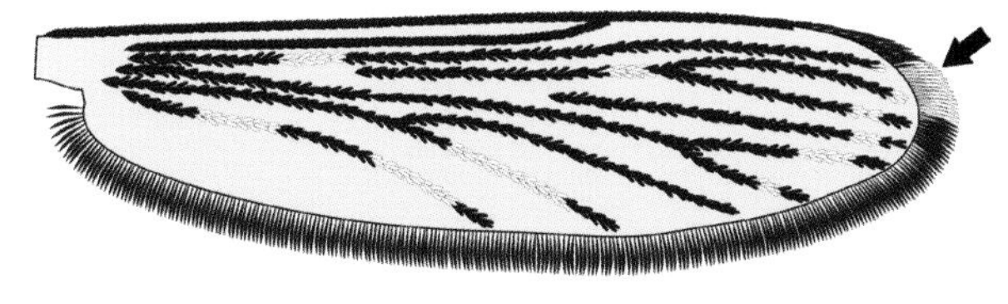

Wing of *An. crucians*

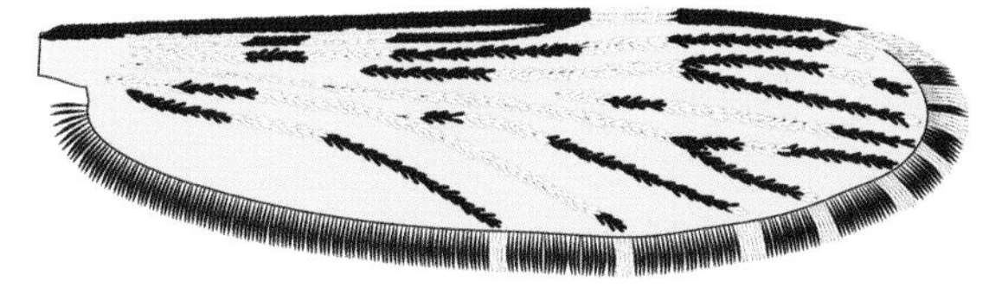

Wing of *An. pseudopunctipennis*

4a Costal pale spot large, one-half or more the length of preapical dark spot. ***An. punctipennis*** (p. 121).
4b Costal pale spot small, less than one-third the length of preapical dark spot. ***An. perplexens*** (p. 118).

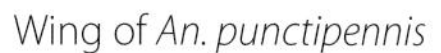

Wing of *An. punctipennis*

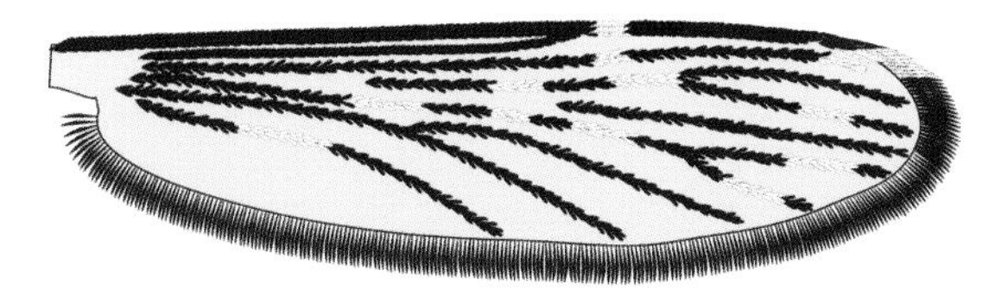

Wing of *An. pseudopunctipennis*

5a Wing without pattern of dark spots; small species, wing length about 3 mm. ***An. barberi*** (p. 115).

5b Wing with scales forming dark spots; medium to large species, wing length 5 mm or more. Go to **6.**

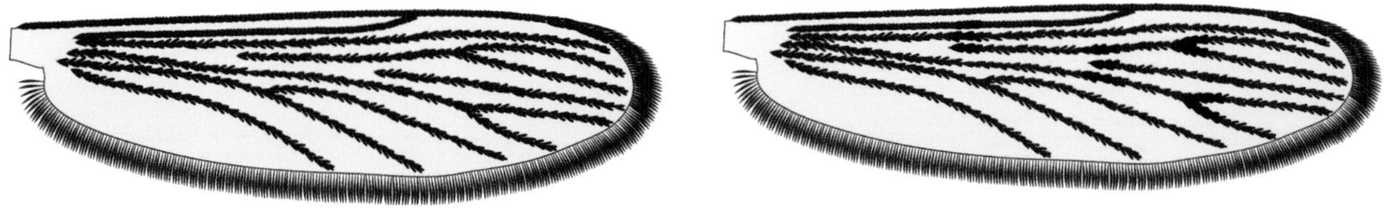

Wing of *An. barberi*

Wing of *An. quadrimaculatus*

6a Tuft of setae between eyes with pale dorsal setae; wing with 4 distinct dark spots. ***An. quadrimaculatus* group** (p. 123).

6b Tuft of setae between eyes with dark dorsal setae; wing with 4 indistinct dark spots. Go to **7.**

Head of *An. quadrimaculatus* female

Head of *An. atropos* female

Wing of *An. quadrimaculatus*

Wing of *An. walkeri*

7a Capitellum (knob) of halter pale; femur of hind leg with pale scales apically. ***An. walkeri*** (p. 125).

7b Capitellum (knob) of halter dark; femur of hind leg with few or no pale scales apically. ***An. atropos*** (p. 113).

Halter and hind leg of *An. walkeri*

Halter and hind leg of *An. atropos*

Culex (Cx.)

1a Scutum with mid-dorsal row of setae (acrostichal setae); occiput with narrow scales bordering eye; medium-sized mosquitoes with light brown thorax. Go to **2.**

1b Scutum without mid-dorsal row of setae; occiput with broad scales bordering eye; small mosquitoes with dark brown thorax. Go to **8.**

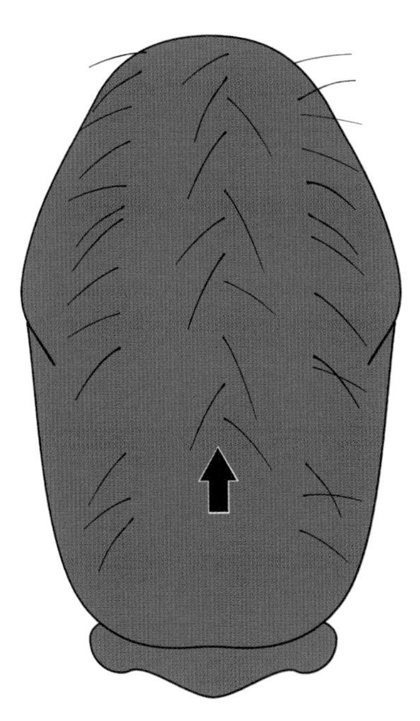

Scutum of *Cx. pipiens*

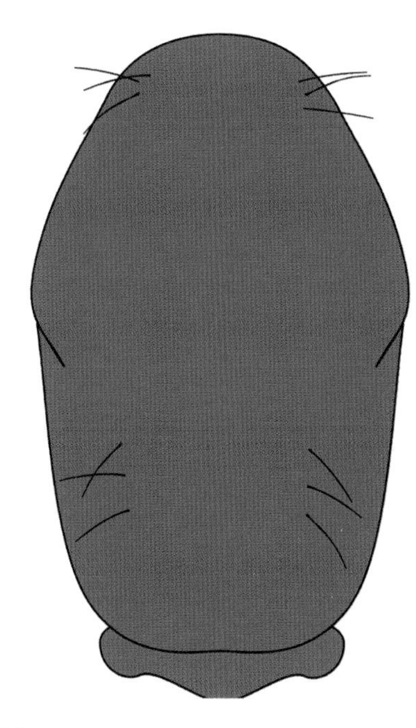

Scutum of *Cx. erraticus*

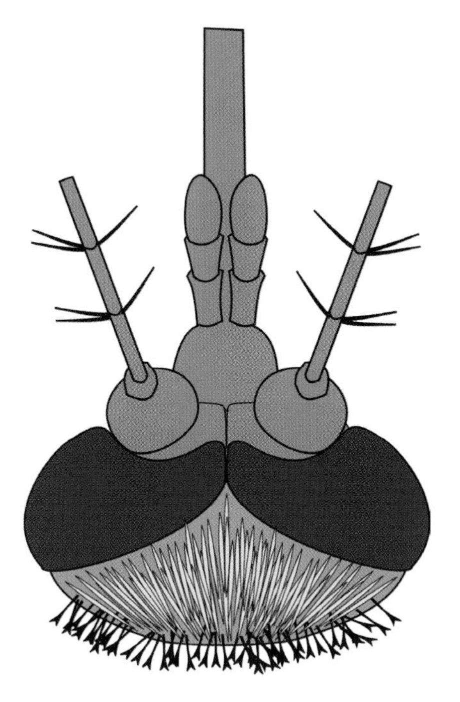

Head of *Cx. restuans* female

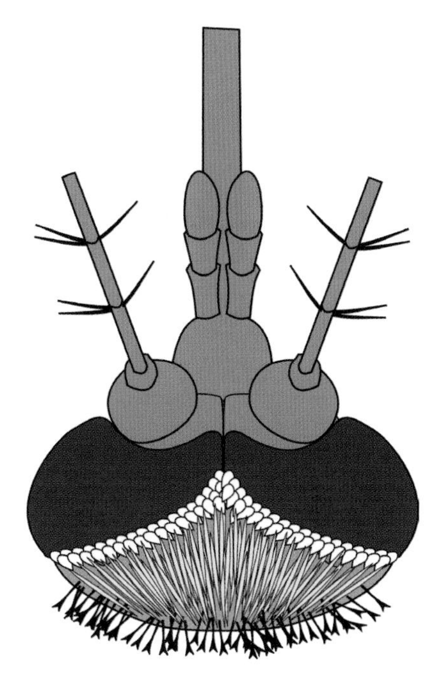

Head of *Cx. erraticus* female

2a Abdominal tergites with pale lateral patches or bands along apical border of segments. ***Cx. territans*** (p. 148).
2b Abdominal tergites with pale lateral patches or bands along basal border of segments. Go to **3.**

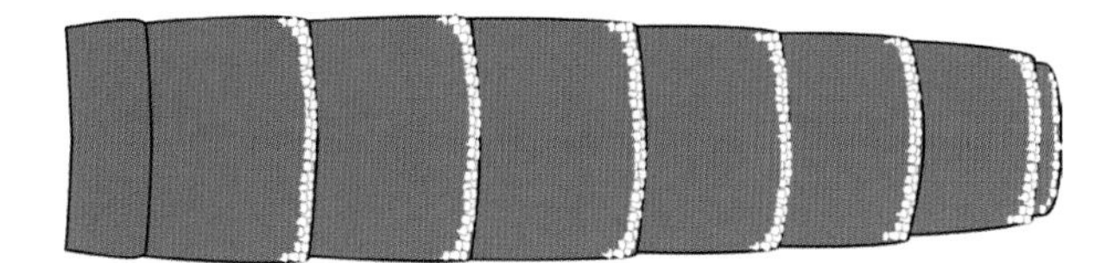

Abdomen of *Cx. territans* female

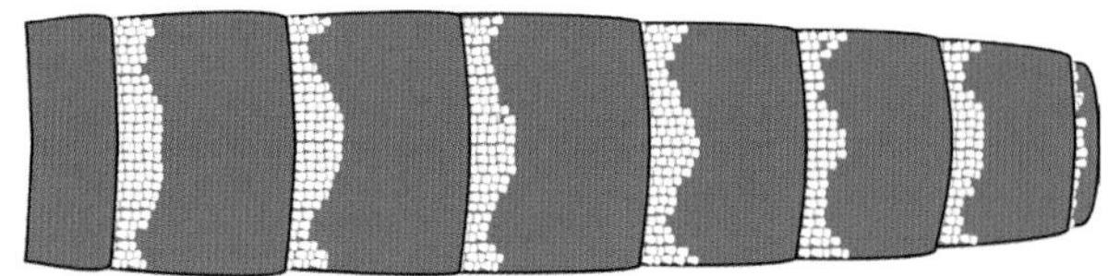

Abdomen of *Cx. pipiens* female

3a Hind tarsomeres with distinct basal and apical bands of pale scales. Go to **4.**
3b Hind tarsomeres dark scaled. Go to **5.**

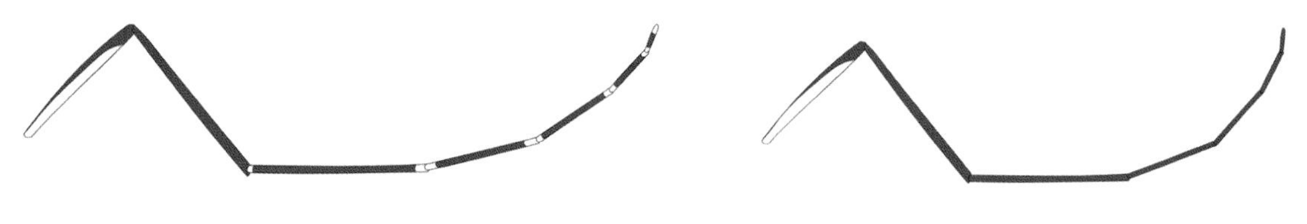

Hind leg of *Cx. coronator*

Hind leg of *Cx. pipiens*

4a Proboscis with distinct pale band near middle; tibia and tarsal segment 1 of hind legs with thin white stripes. ***Cx. tarsalis*** (p. 146).
4b Proboscis without pale band, but often with patch of pale scales on underside near middle; legs without thin white stripes. ***Cx. coronator*** (p. 129).

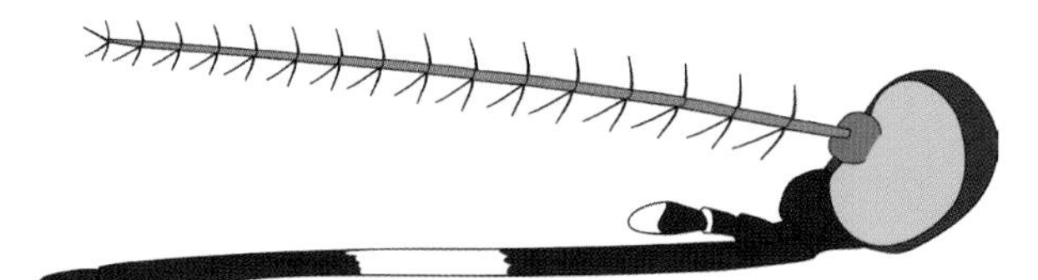

Head of *Cx. tarsalis* female

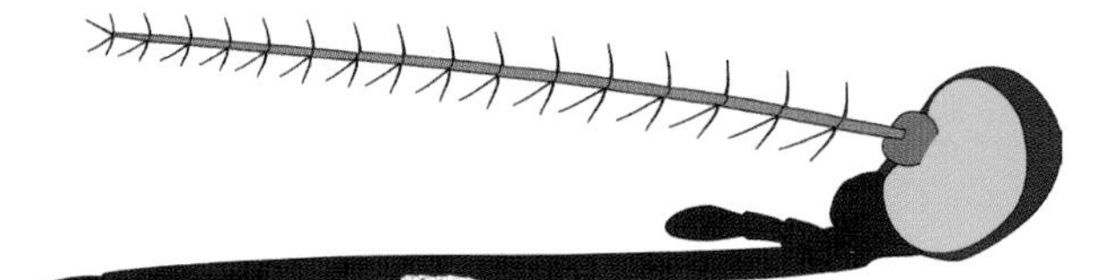

Head of *Cx. coronator* female

Hind leg of *Cx. tarsalis*

Hind leg of *Cx. coronator*

5a Abdominal tergites without pale bands or with very narrow pale basal bands. Go to **6.**
5b Abdominal tergites with conspicuous pale basal bands. Go to **7.**

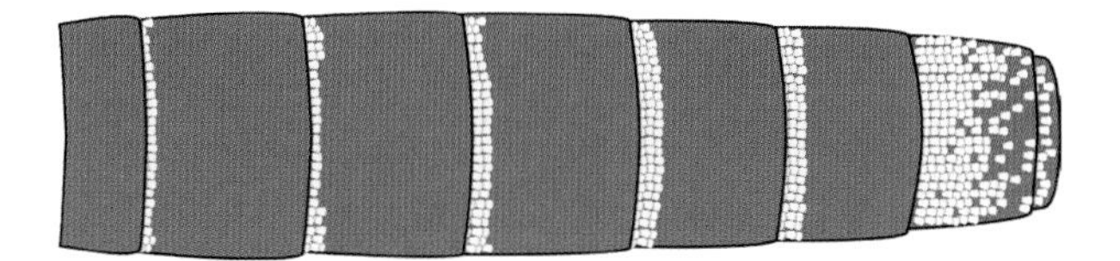

Abdomen of *Cx. salinarius* female

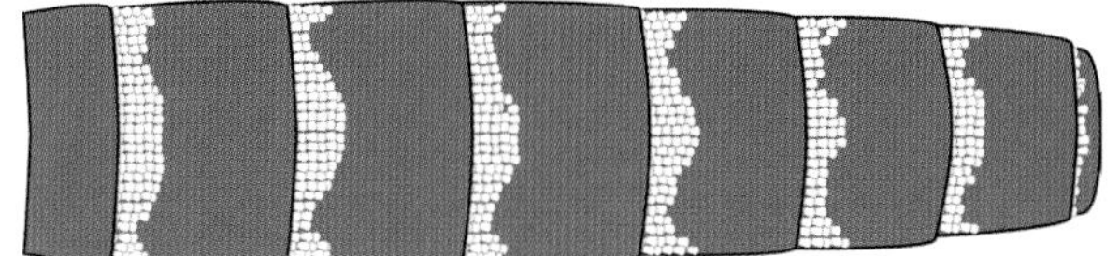

Abdomen of *Cx. pipiens* female

6a Scale patches of lateral thorax absent or in groups of fewer than 6 scales; abdominal tergites often without pale basal bands, segment 7 mostly dark scaled. ***Cx. nigripalpus*** (p. 134).
6b Lateral thorax with several patches of pale scales, each with more than 6 scales; abdominal tergites 2–6 usually with narrow basal bands of yellowish scales, segment 7 mostly covered with yellowish scales. ***Cx. salinarius*** (p. 144).

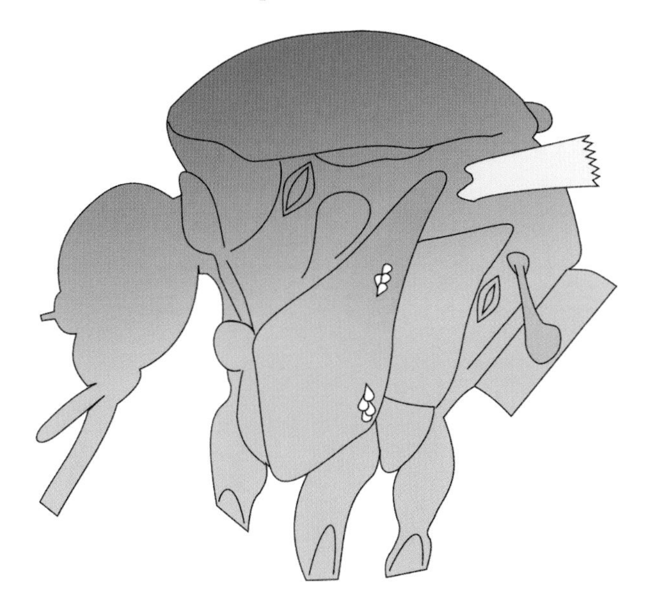

Thorax of *Cx. nigripalpus* female

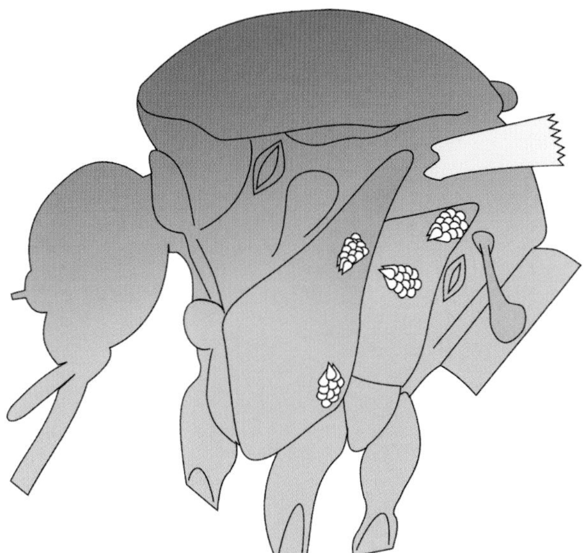

Thorax of *Cx. salinarius* female

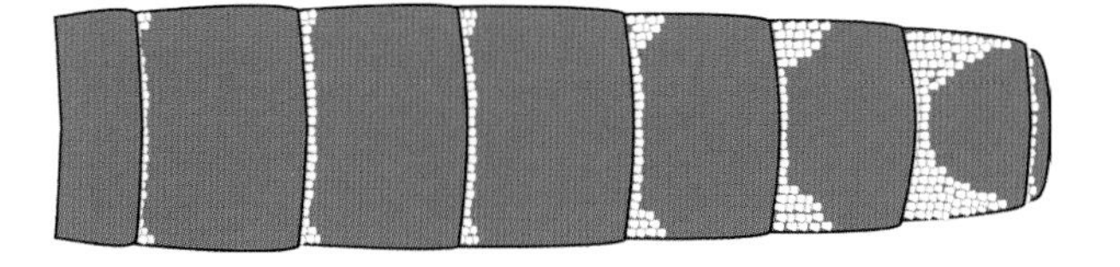

Abdomen of *Cx. nigripalpus* female

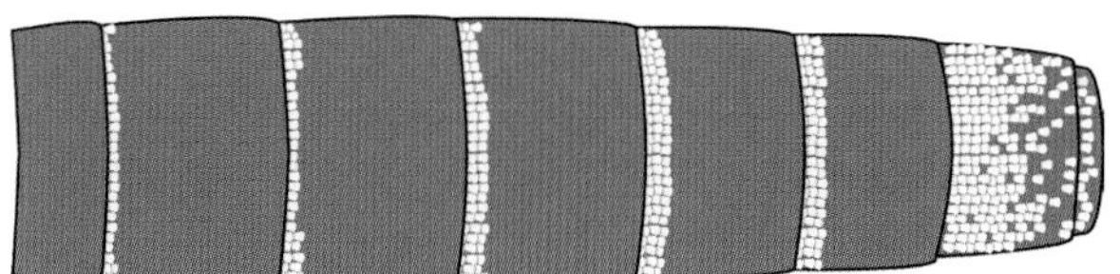

Abdomen of *Cx. salinarius* female

7a Pale bands of abdomen with 2 distinct constrictions, creating a 3-lobed appearance; scutum without pale-scaled spots. ***Cx. pipiens/quinquefasciatus*** (p. 140).

7b Pale bands of abdomen without or with only slight constrictions; scutum with a pair of small pale spots near middle of posterior half. ***Cx. restuans*** (p. 142).

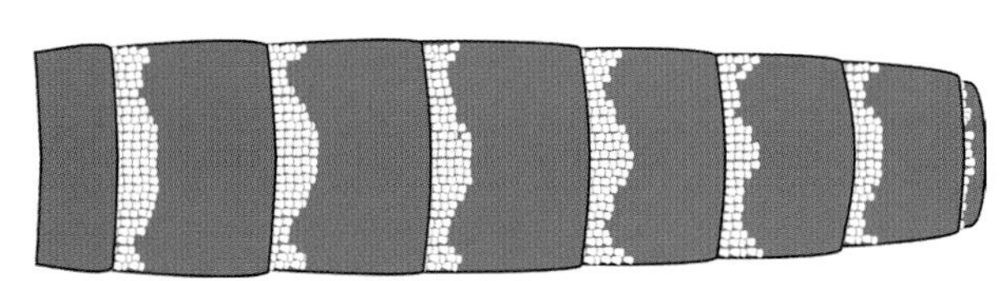

Abdomen of *Cx. pipiens* female

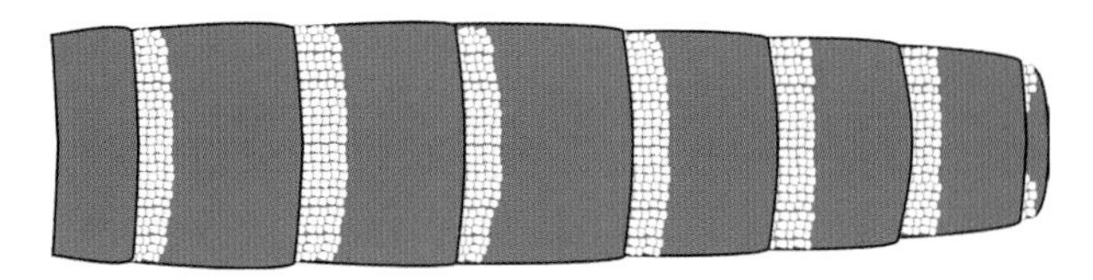

Abdomen of *Cx. restuans* female

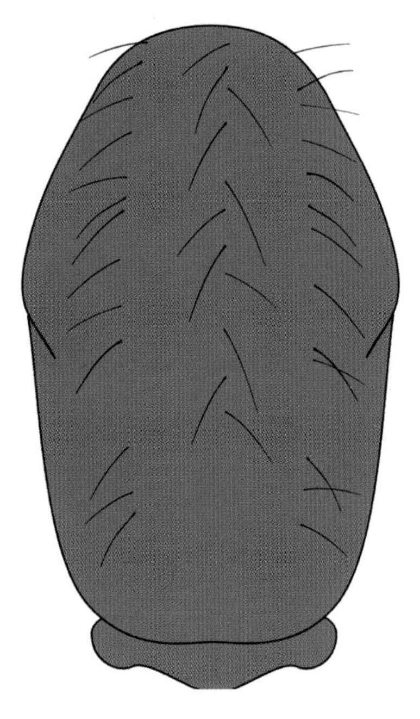

Scutum of *Cx. pipiens*

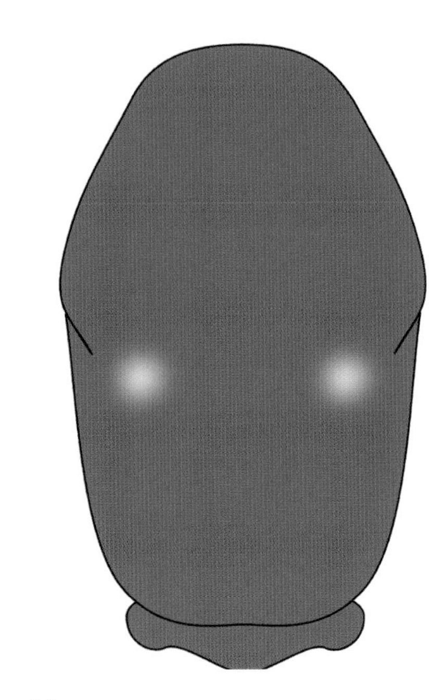

Scutum of *Cx. restuans*

8a Lateral thoracic plate below wing base (mesepimeron) with median patch of white scales. ***Cx. erraticus*** (p. 131).

8b Lateral thoracic plate below wing base (mesepimeron) without median patch of white scales. Go to **9.**

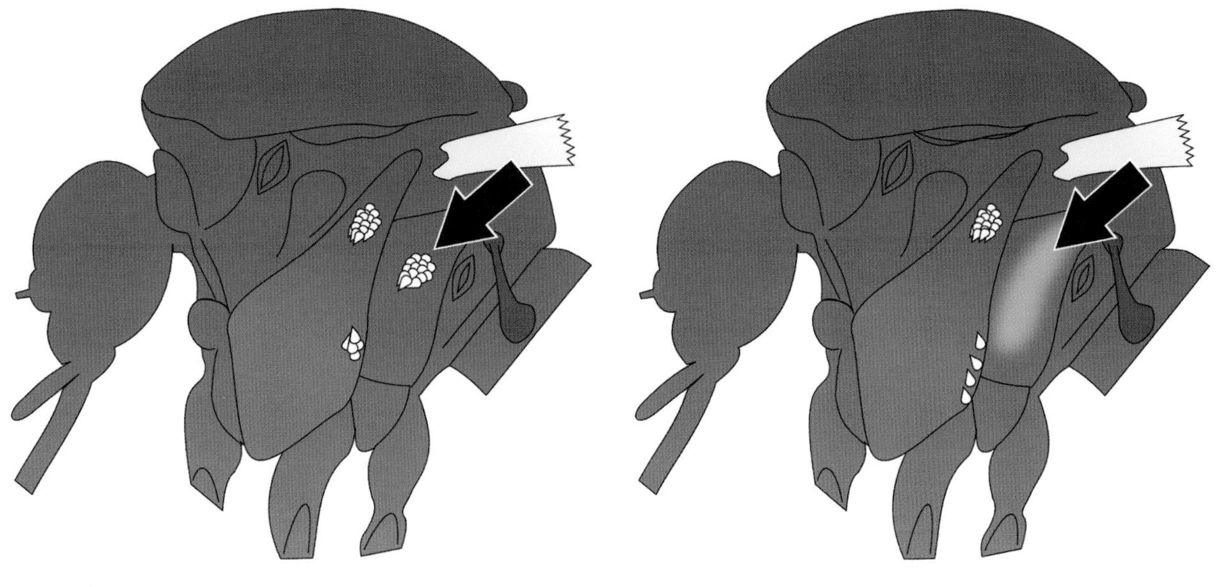

Thorax of *Cx. erraticus* female

Thorax of *Cx. peccator* female

9a Upper mesokatepisternum with patch of 6 or more pale scales; abdominal sternites mostly pale, slightly darker on apical edge. ***Cx. peccator*** (p. 136).

9b Upper mesokatepisternum without scales or with fewer than 6; abdominal sternites with distinct white basal bands and dark apical bands. ***Cx. pilosus*** (p. 138).

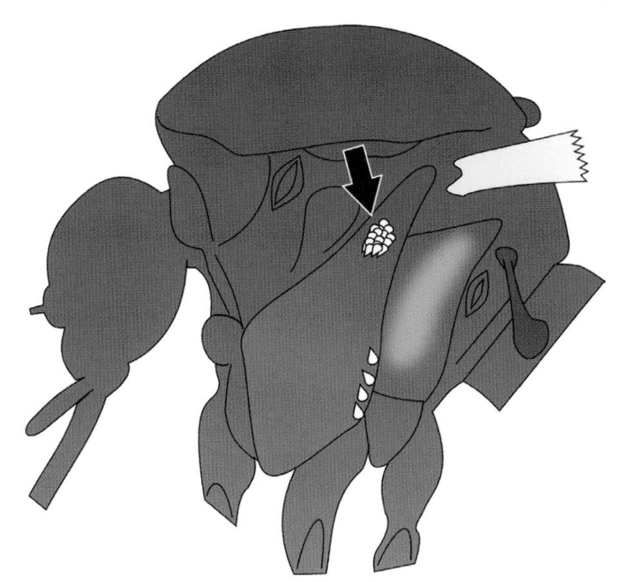

Thorax of *Cx. peccator* female

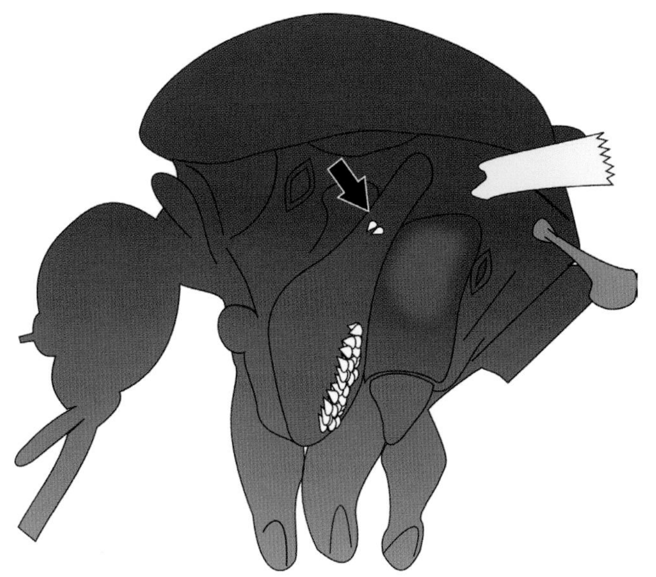

Thorax of *Cx. pilosus* female

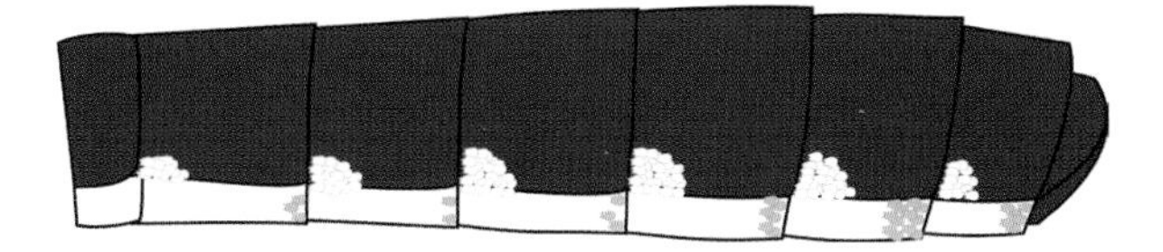

Abdomen of *Cx. peccator* (lateral view)

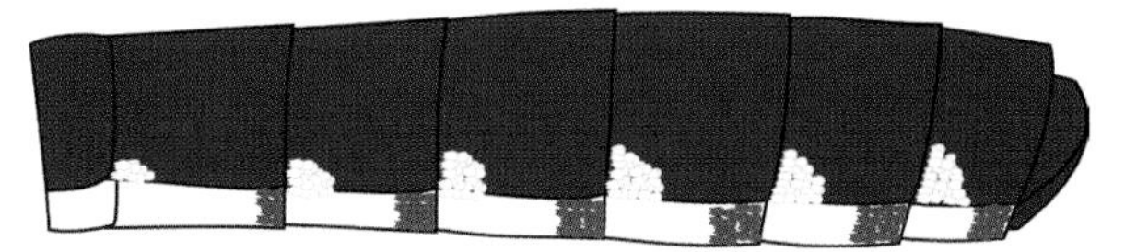

Abdomen of *Cx. pilosus* (lateral view)

Culiseta (Cs.)

1a Abdominal segments with extensive pale scaling; wings with dark and pale scales intermixed on anterior veins. ***Cs. inornata*** (p. 150).

1b Abdominal segments mostly dark scaled, sometimes with narrow pale basal bands; wings with only dark scales. ***Cs. melanura*** (p. 152).

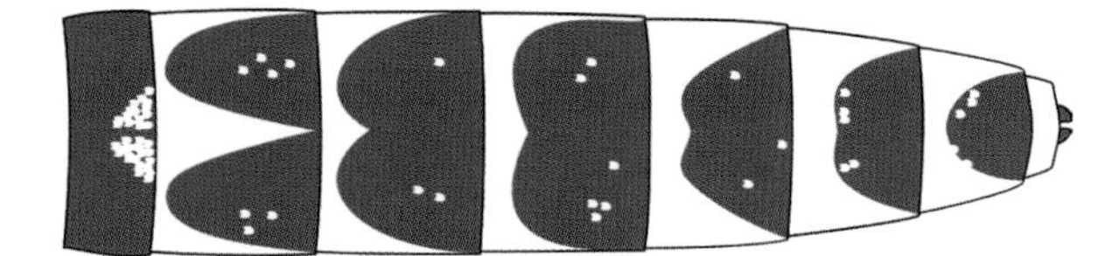

Abdomen of *Cs. inornata* female

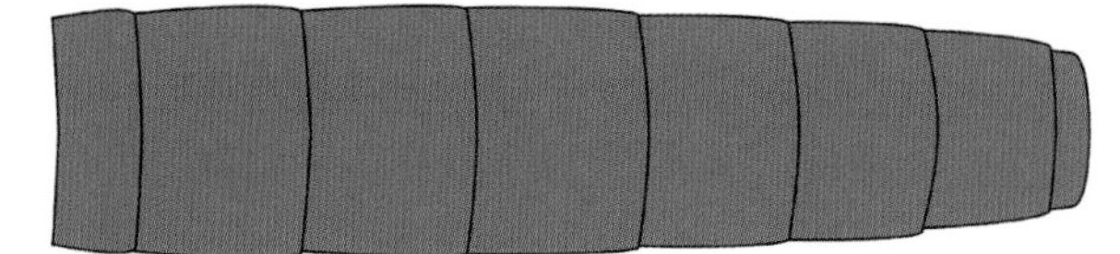

Abdomen of *Cs. melanura* female

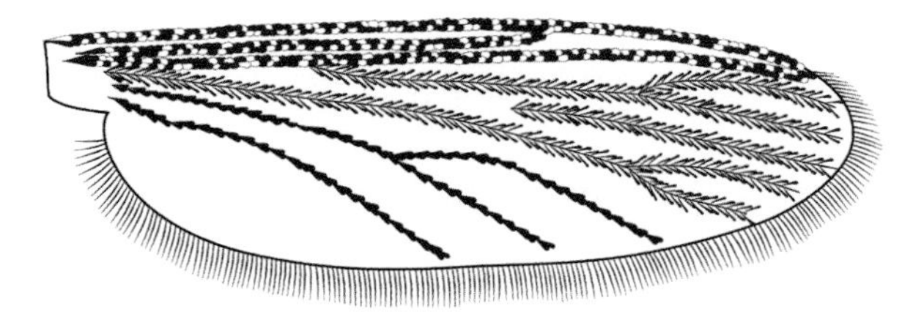

Wing of *Cs. inornata*

Wing of *Cs. melanura*

Mansonia (Mn.)

1a Apical edge of abdominal segment 7 with row of short dark spines; palps one-third or more the length of the proboscis. ***Mn. titillans*** (p. 156).

1b Apical edge of abdominal segment 7 without row of short dark spines; palps less than one-third the length of the proboscis. ***Mn. dyari*** (p. 154).

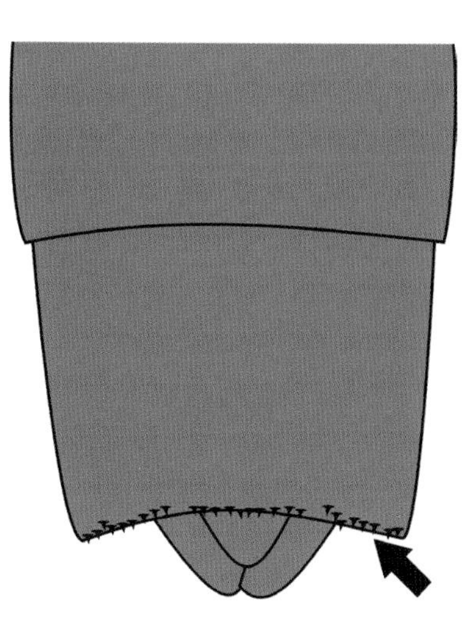

Abdominal segment 7 of *Mn. titillans* female

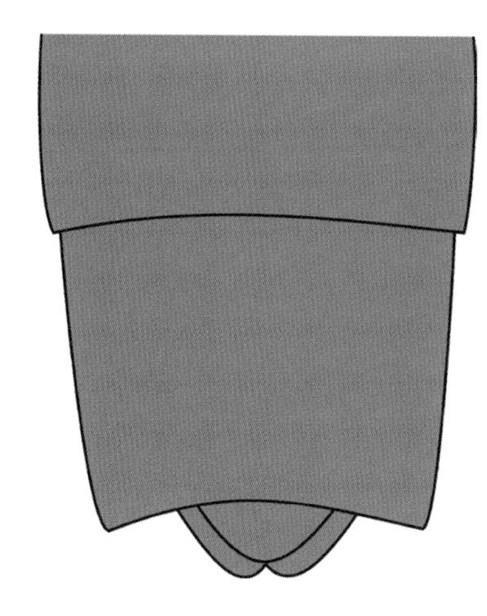

Abdominal segment 7 of *Mn. dyari* female

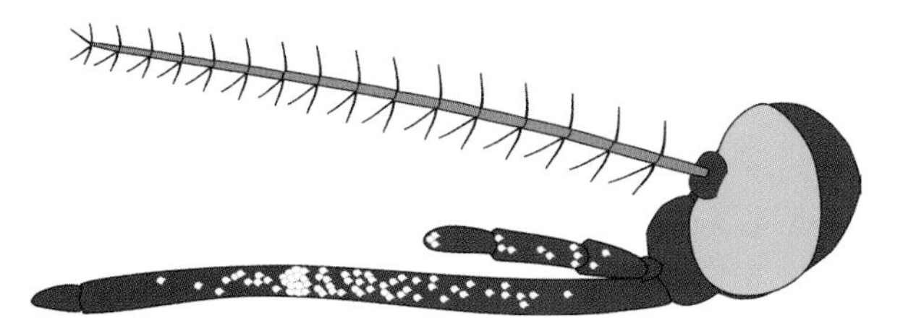

Head of *Mn. titillans* female

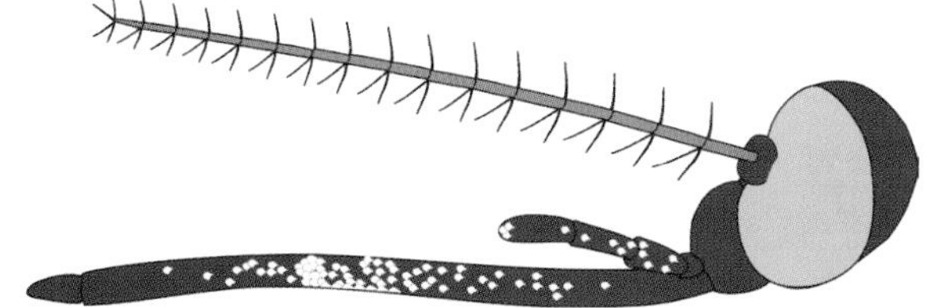

Head of *Mn. dyari* female

Orthopodomyia (Or.)

1a Pale bands at junction of tarsal segments of hind legs slightly broader before junction of segments than after. ***Or. signifera*** (p. 158).

1b Pale bands at junction of tarsal segments of hind legs equally broad before and after junction of segments. ***Or. alba*** (p. 158).

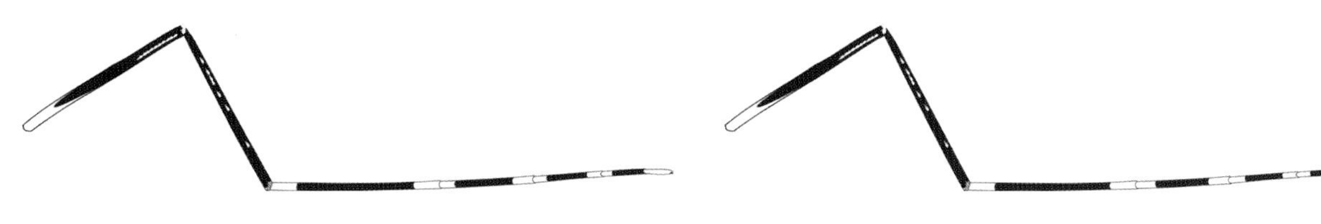

Hind leg of *Or. signifera*

Hind leg of *Or. alba*

Psorophora (Ps.)

1a Wing with dark and pale scales on all veins; hind femur with distinct pale band just before "knee." Go to **2.**

1b Wing scales entirely dark or with only a few pale scales on veins near wing base; hind femur without pale band just before "knee." Go to **3.**

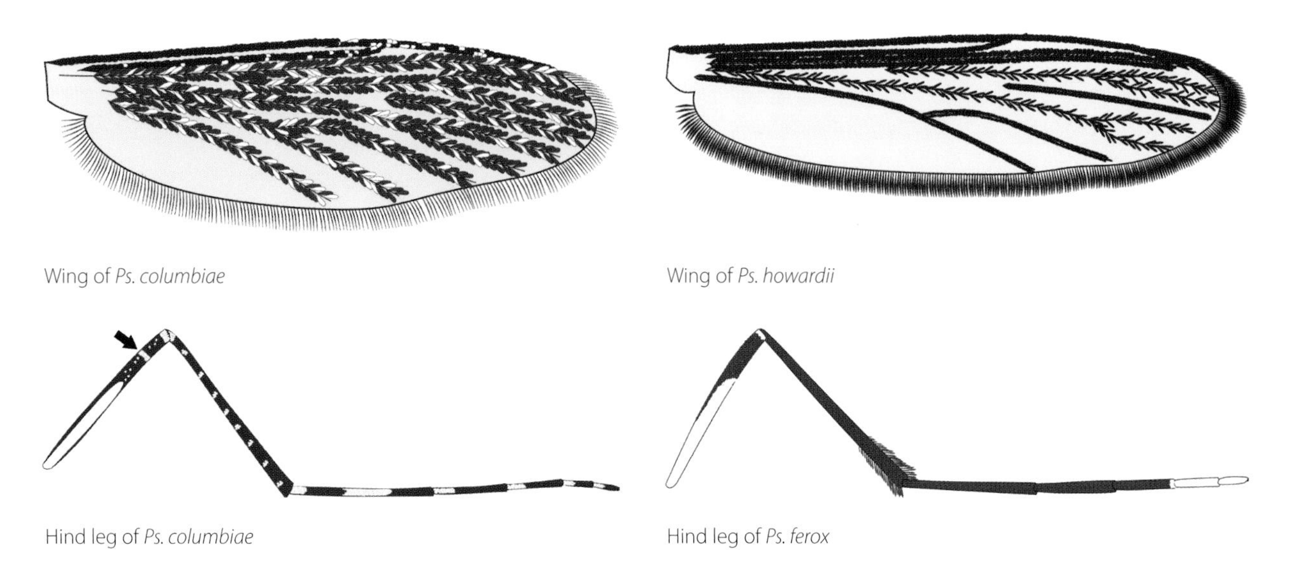

Wing of *Ps. columbiae*

Wing of *Ps. howardii*

Hind leg of *Ps. columbiae*

Hind leg of *Ps. ferox*

2a Tarsomere 1 of hind leg with pale band at base and middle, otherwise dark; dark and pale wing scales intermixed, in no definite pattern. ***Ps. columbiae*** (p. 163).

2b Tarsomere 1 of hind leg mostly pale; wings with definite areas of dark and pale scales. ***Ps. discolor*** (p. 167).

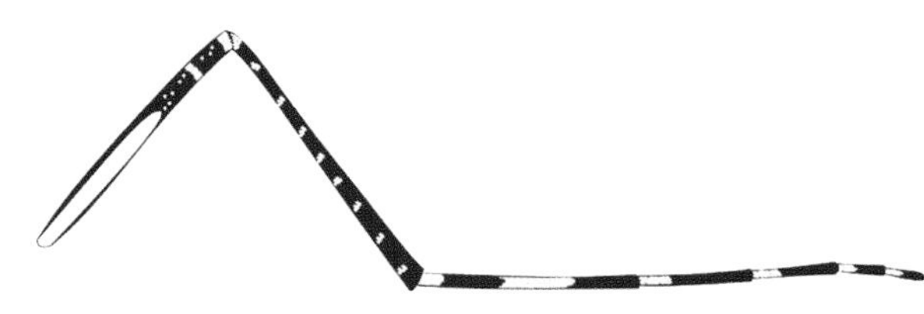

Hind leg of *Ps. columbiae*

Hind leg of *Ps. discolor*

Wing of *Ps. columbiae*

Wing of *Ps. discolor*

3a Palps very long, about one-third to one-half the length of the proboscis; tarsi of forelegs and middle legs with pale band on at least 1 segment; very large mosquitoes. Go to **4.**

3b Palps not so long, less than one-third the length of the proboscis; tarsi of forelegs and middle legs without pale bands; small to large mosquitoes. Go to **5.**

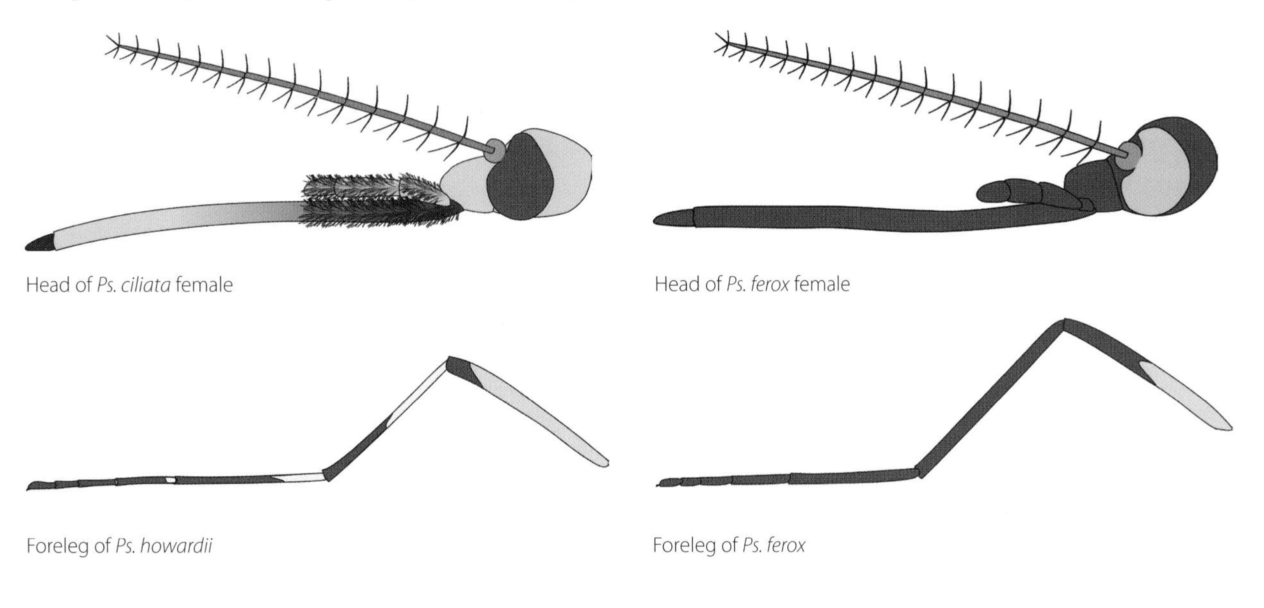

Head of *Ps. ciliata* female

Head of *Ps. ferox* female

Foreleg of *Ps. howardii*

Foreleg of *Ps. ferox*

4a Scutum with narrow median stripe of golden scales; proboscis yellow in distal half, except for labellum. ***Ps. ciliata*** (p. 161).

4b Scutum with median longitudinal stripe of dark scales; proboscis dark. ***Ps. howardii*** (p. 171).

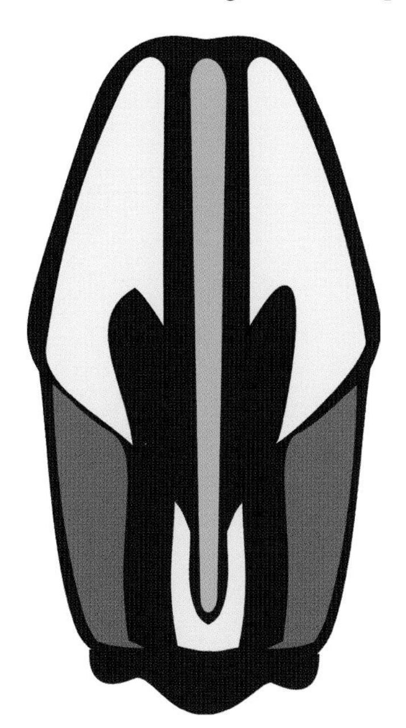

Scutum of *Ps. ciliata*

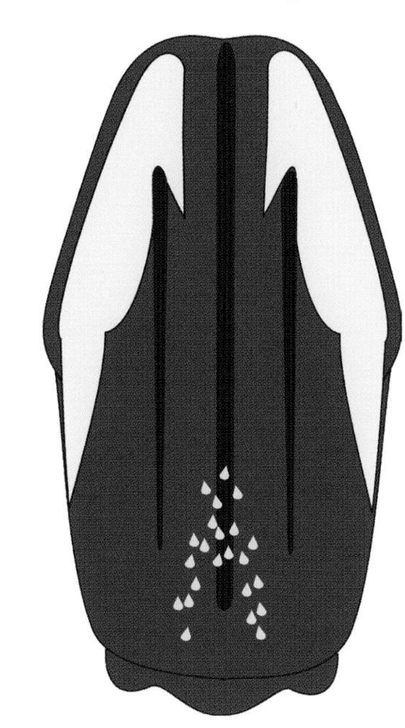

Scutum of *Ps. howardii*

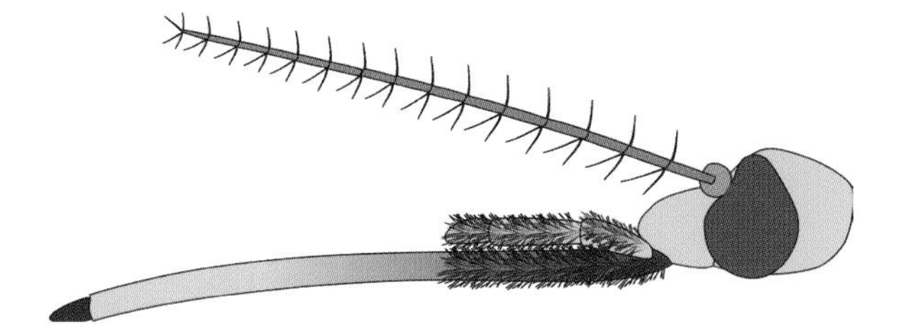

Head of *Ps. ciliata* female

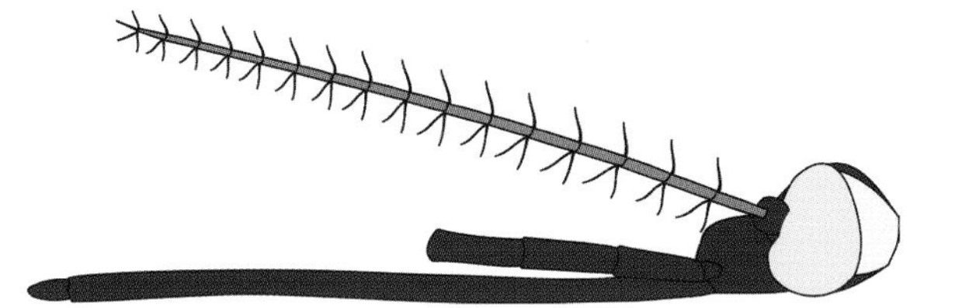

Head of *Ps. howardii* female

5a Hind tarsomeres dark, without pale scales; abdominal tergites with median apical patches of pale scales. ***Ps. cyanescens*** (p. 165).

5b Hind tarsomeres with at least some pale scales; abdominal tergites with only lateral apical patches of pale scales. Go to **6.**

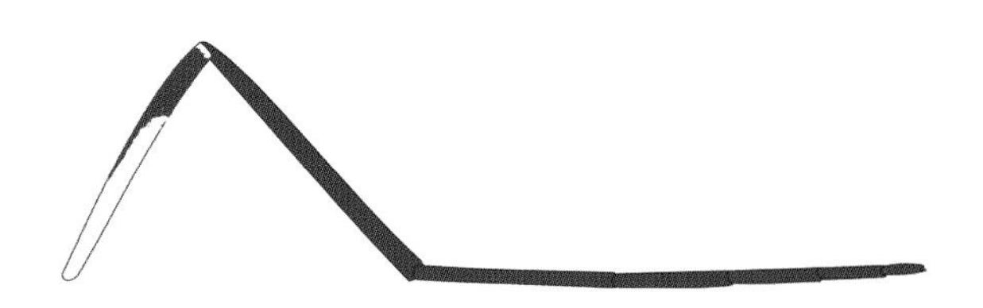

Hind leg of *Ps. cyanescens*

Hind leg of *Ps. ferox*

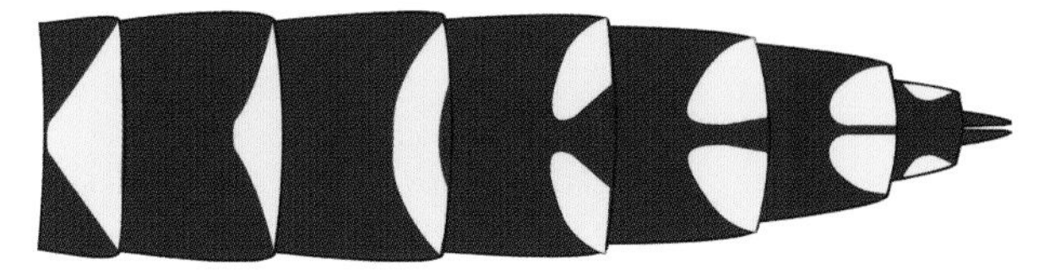

Abdomen of *Ps. cyanescens* female

Abdomen of *Ps. ferox* female

6a Only tarsomere 4 of hind leg with pale scales, other tarsomeres dark. ***Ps. mathesoni*** (p. 173).

6b At least tarsomeres 4 and 5 of hind leg pale. Go to **7.**

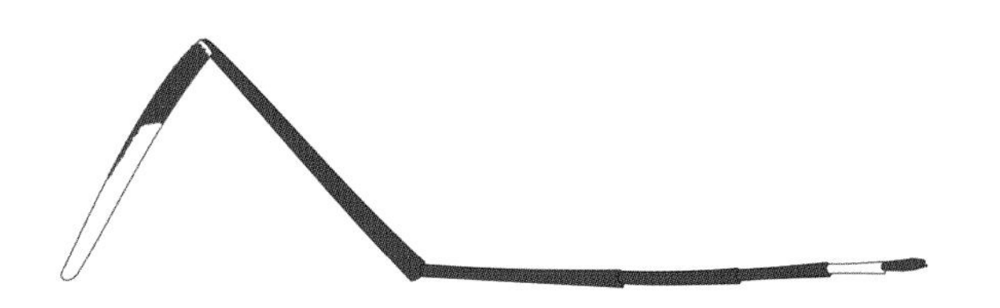

Hind leg of *Ps. mathesoni*

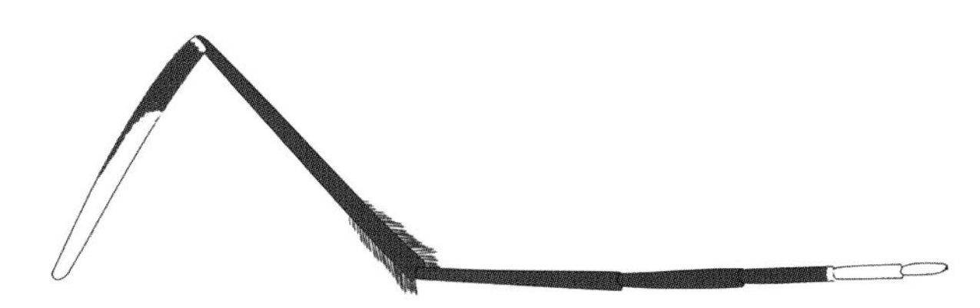

Hind leg of *Ps. ferox*

7a Scutum dark, with scattered golden scales in no definite pattern. ***Ps. ferox*** (p. 168).

7b Scutum with broad median stripe of dark scales, pale scales laterally. ***Ps. horrida*** (p. 170).

Scutum of *Ps. ferox*

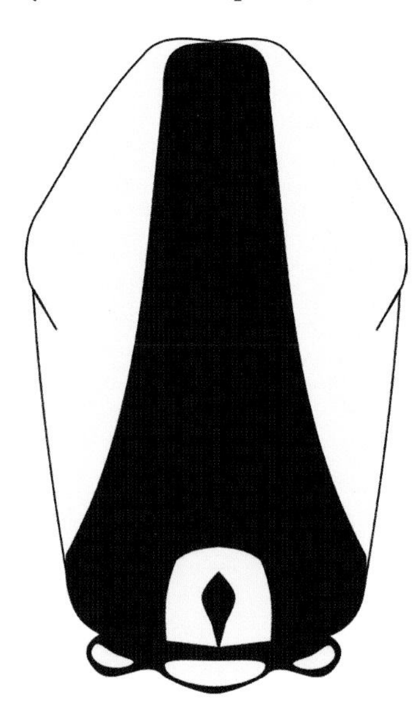

Scutum of *Ps. horrida*

Uranotaenia (Ur.)

1a Hind tarsomeres 4, 5, and part of 3 pale; scutum lacking mid-dorsal row of blue scales. ***Ur. lowii*** (p. 177).
1b Hind tarsomeres all dark; scutum with mid-dorsal stripe of blue scales. ***Ur. sapphirina*** (p. 179).

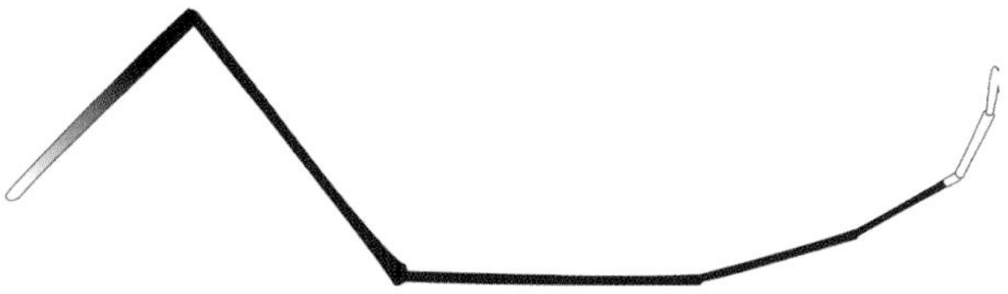

Hind leg of *Ur. lowii*

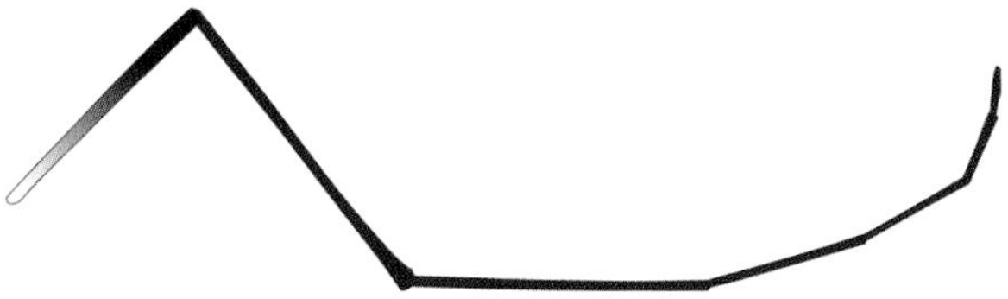

Hind leg of *Ur. sapphirina*

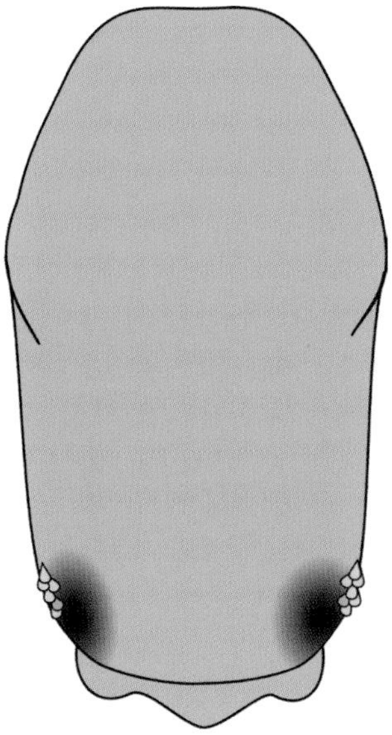

Scutum of *Ur. lowii* female

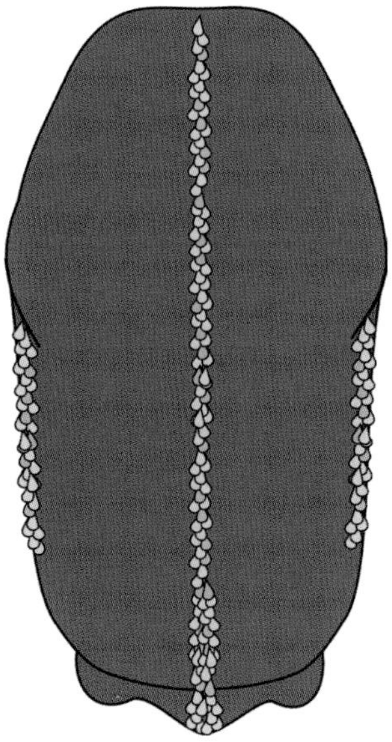

Scutum of *Ur. sapphirina* female

Key to Genera of Fourth-Instar Mosquito Larvae of the Southeastern United States

1a Respiratory siphon absent. ***Anopheles*** (p. 59).
1b Respiratory siphon present. Go to **2.**

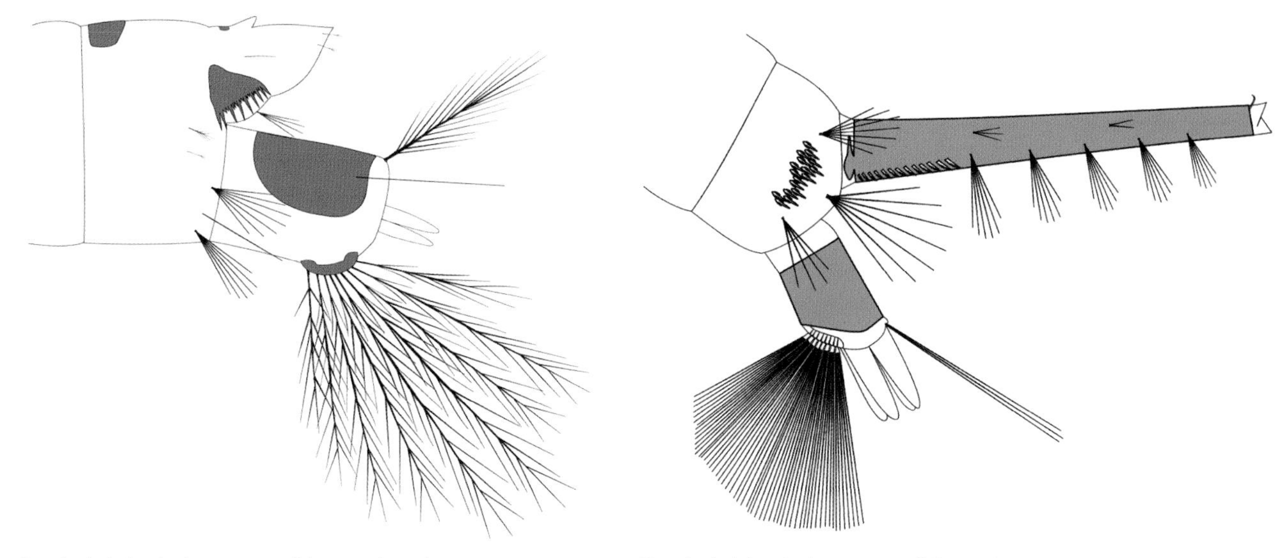

Terminal abdominal segments of *An. punctipennis*

Terminal abdominal segments of *Cx. erraticus*

2a Siphon very short and sharply pointed, with serrations near dorsal tip. Go to **3.**
2b Siphon tubelike, not sharply pointed at its tip. Go to **4.**

Terminal abdominal segments of *Cq. perturbans*

Terminal abdominal segments of *Ur. sapphirina*

3a Antenna with a very long branched seta (half the length of antenna) arising near midpoint; saddle of anal segment pierced along ventral edge with 3–4 setae. ***Mansonia*** (p. 66).

3b Branched seta at midpoint of antenna short; ventral midline of saddle with 1–2 small setae. ***Coquillettidia perturbans*** (p. 127).

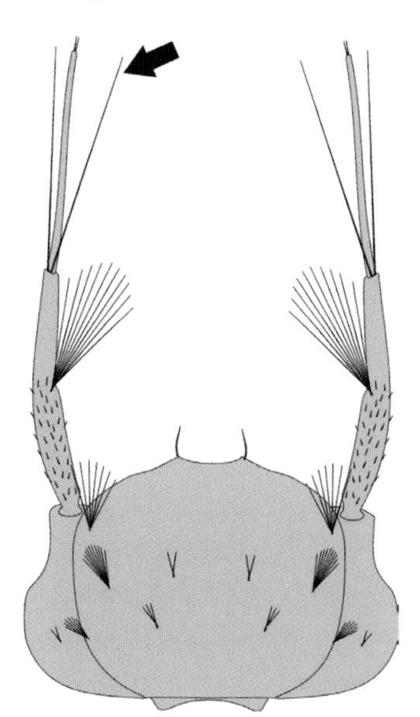

Head of *Mn. dyari*

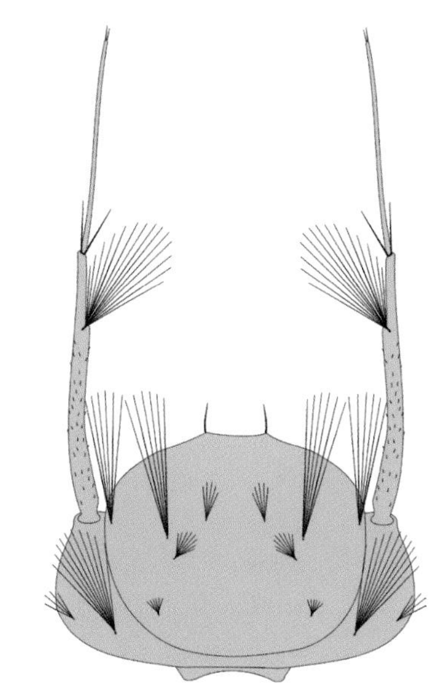

Head of *Cq. perturbans*

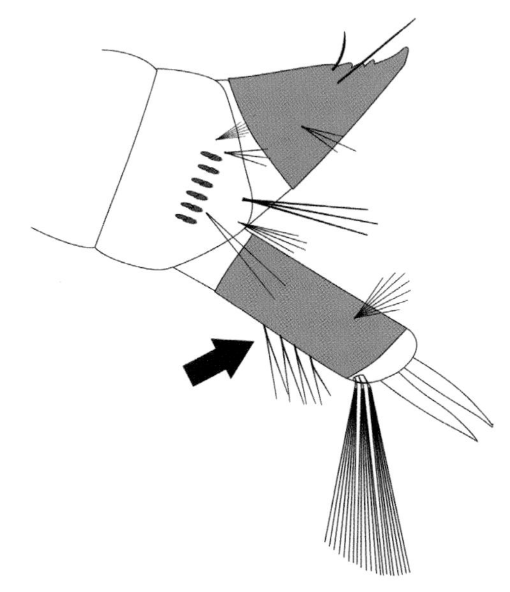

Terminal abdominal segments of *Mn. dyari*

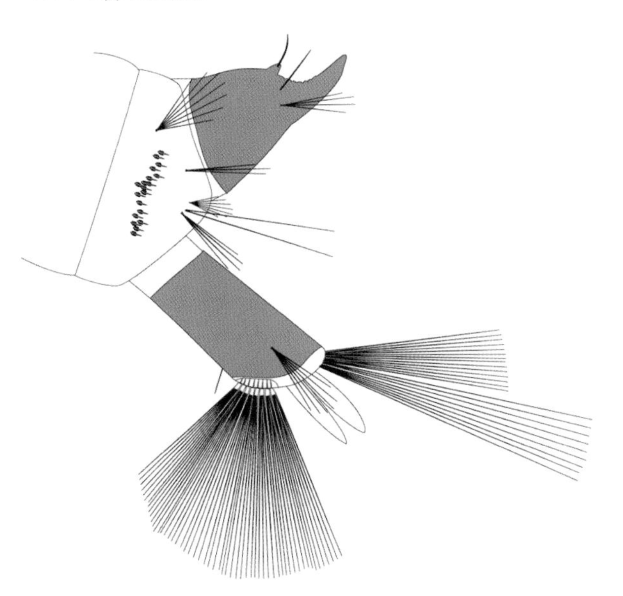

Terminal abdominal segments of *Cq. perturbans*

4a Siphon without pecten spines. Go to **5.**
4b Siphon with pecten spines. Go to **7.**

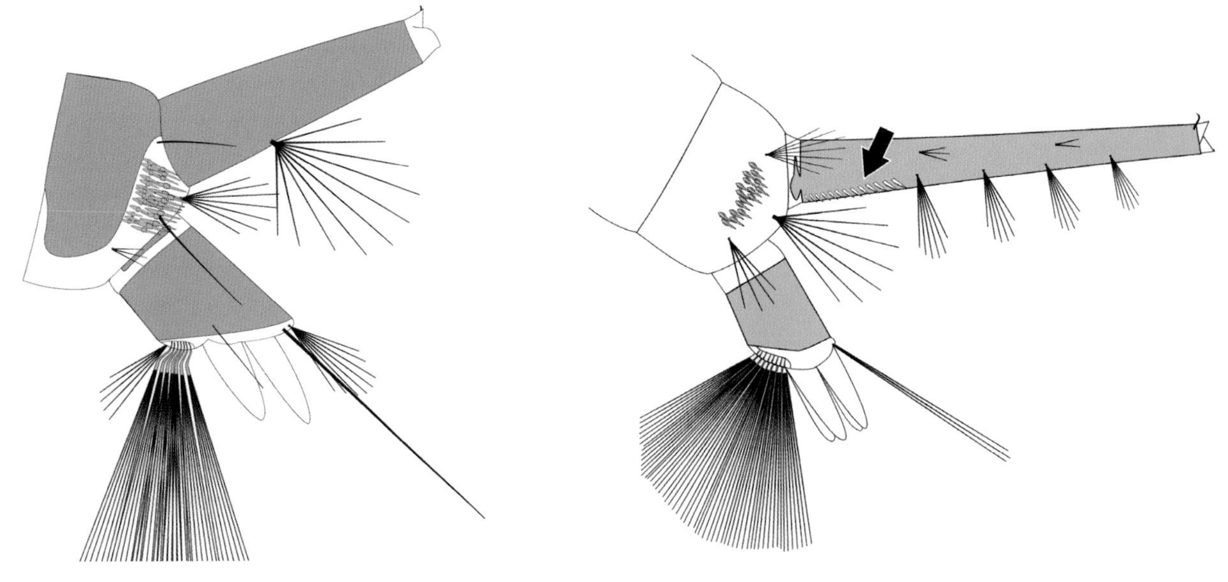

Terminal abdominal segments of *Or. signifera*

Terminal abdominal segments of *Cx. erraticus*

5a Abdominal segment 8 without comb scales. ***Toxorhynchites rutilus*** (p. 175).
5b Abdominal segment 8 with comb scales. Go to **6.**

Terminal abdominal segments of *Tx. rutilus*

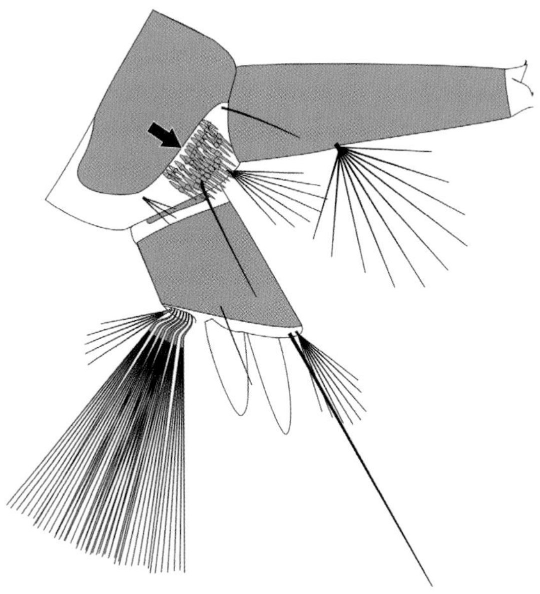

Terminal abdominal segments of *Or. signifera*

6a Saddle completely encircling anal segment of abdomen. ***Orthopodomyia*** (p. 67).

6b Saddle not completely encircling anal segment of abdomen. ***Wyeomyia smithii*** (p. 181).

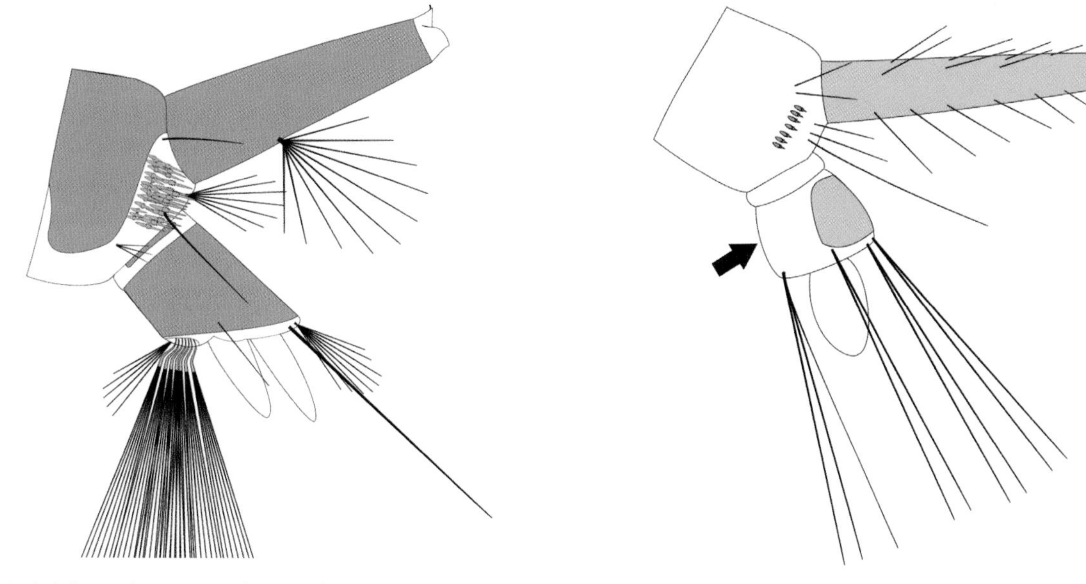

Terminal abdominal segments of *Or. signifera*

Terminal abdominal segments of *Wy. smithii*

7a Abdominal segment 8 with large lateral comb plate; head longer than wide. ***Uranotaenia*** (p. 71).

7b Abdominal segment 8 without comb plate or with only small plate; head wider than long. Go to **8.**

Terminal abdominal segments of *Ur. sapphirina*

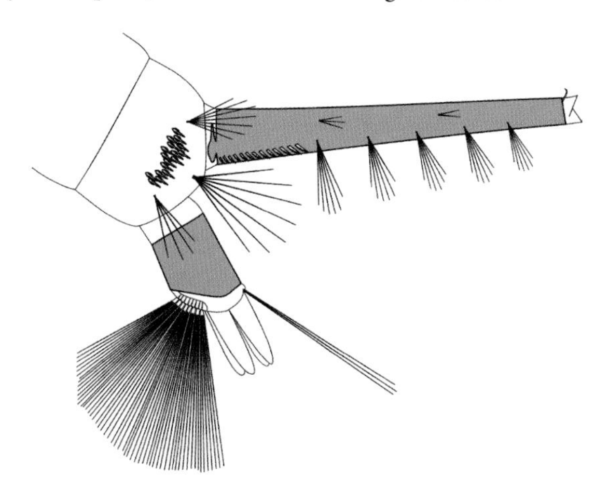

Terminal abdominal segments of *Cx. erraticus*

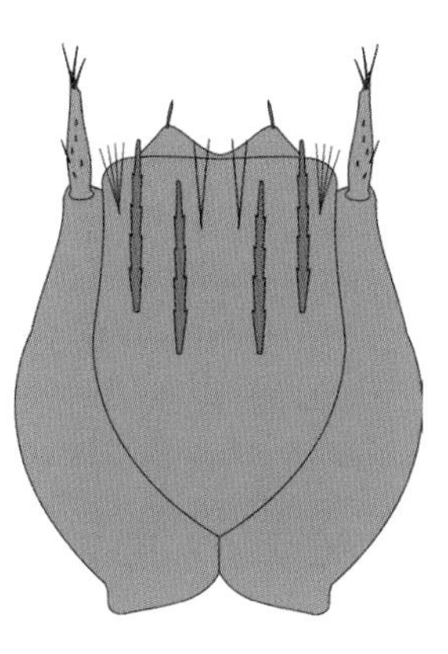

Head of *Ur. sapphirina*

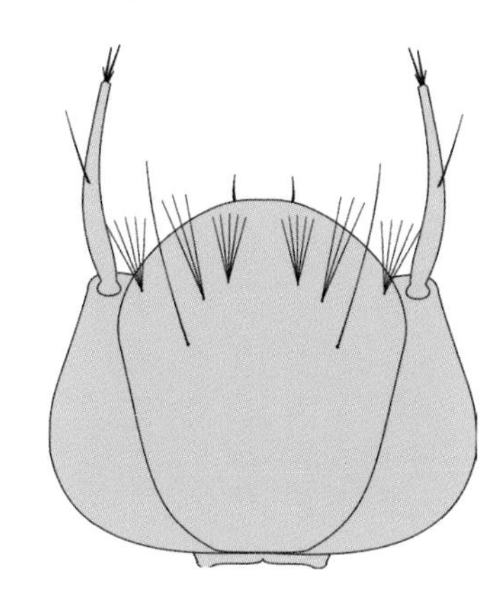

Head of *Ae. triseriatus*

8a Siphon with 3 or more pairs of setae (other than the seta at the dorsal tip). Go to **9.**
8b Siphon with only 1 pair (rarely 2 pairs) of setae (other than the seta at the dorsal tip). Go to **10.**

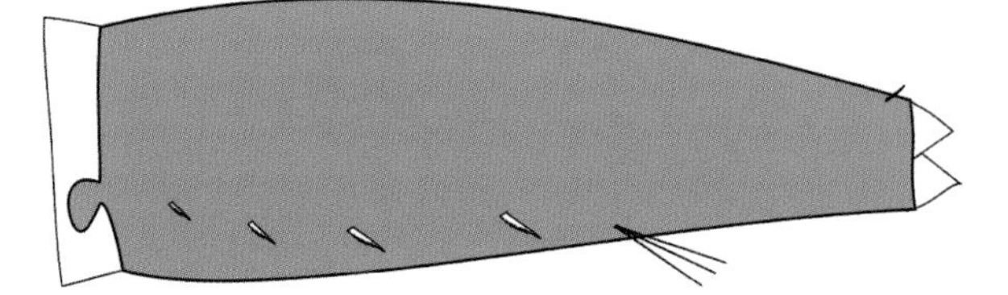

Siphon of *Cs. melanura*

Siphon of *Ps. columbiae*

9a Siphon with a branched seta arising from its base. ***Culiseta*** (p. 66).
9b Siphon without a branched seta arising from its base. ***Culex*** (p. 63).

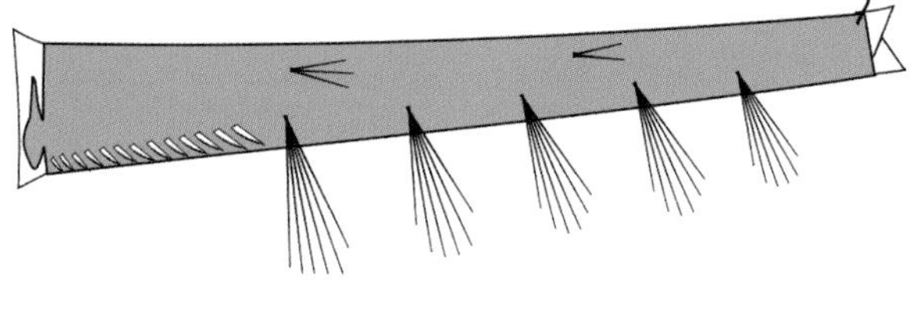

Siphon of *Cs. melanura*

Siphon of *Cx. erraticus*

10a Anal segment without setae along ventral midline. ***Aedes*** (p. 49).
10b Anal segment with a row of setae along the ventral midline. ***Psorophora*** (p. 68).

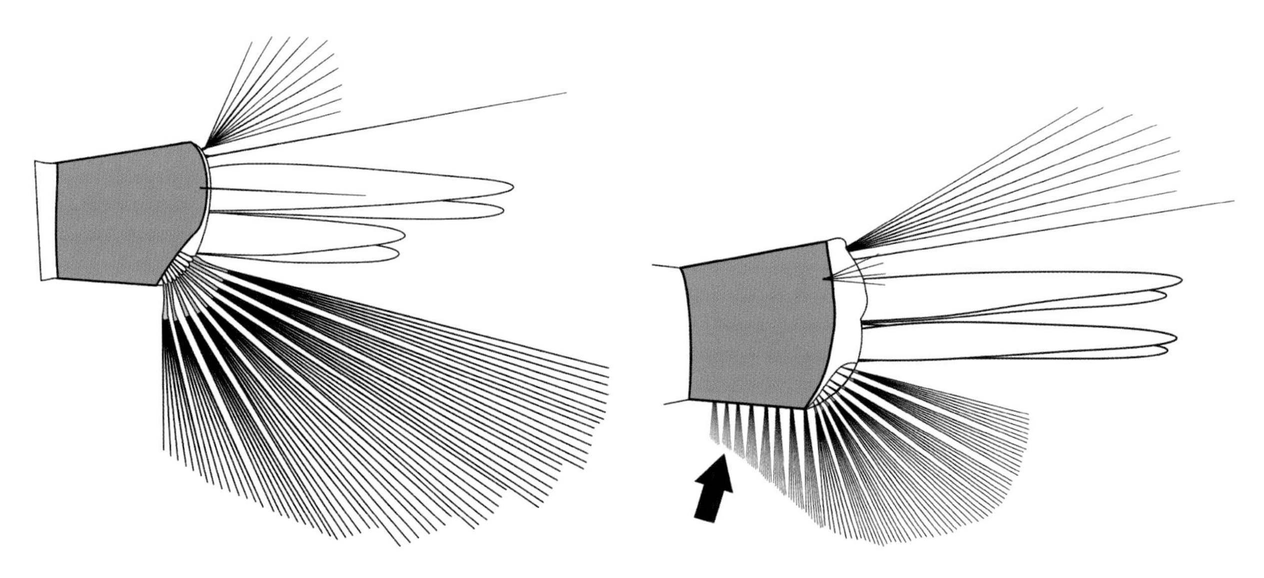

Anal segment of *Ae. atlanticus*

Anal segment of *Ps. ciliata*

Aedes (Ae.)

1a Saddle completely encircling anal segment of abdomen. Go to **2.**
1b Saddle not completely encircling anal segment of abdomen. Go to **10.**

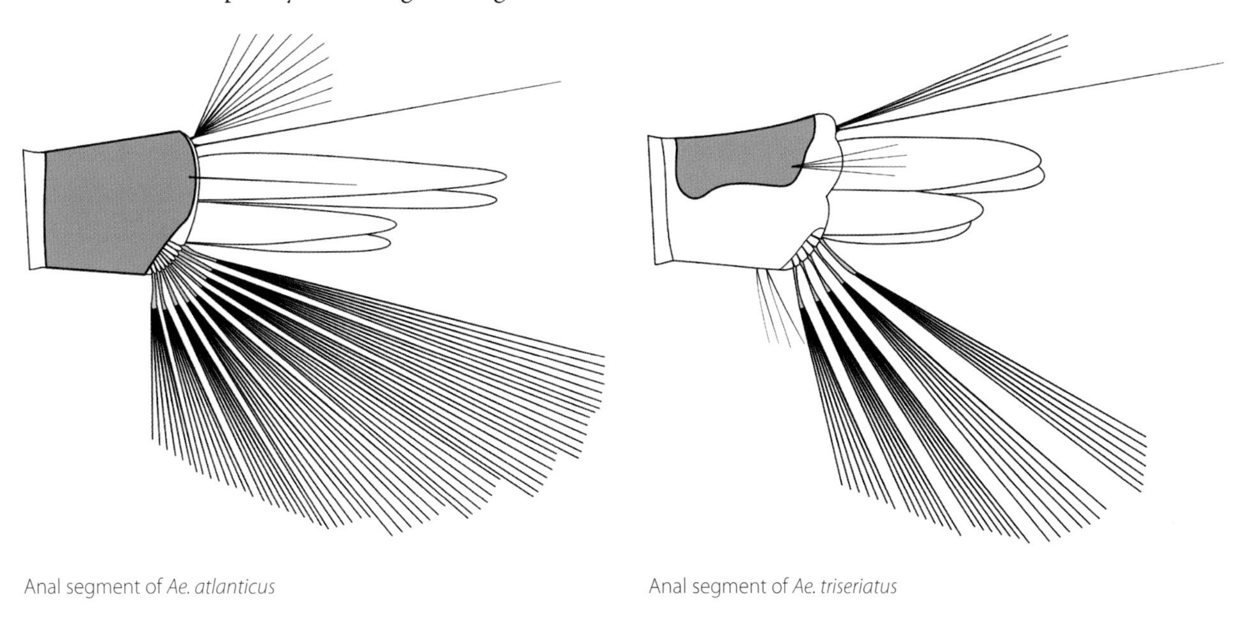

Anal segment of *Ae. atlanticus*

Anal segment of *Ae. triseriatus*

2a Pecten with 2 distal spines separated from other spines. ***Ae. fulvus pallens*** (p. 92).
2b Pecten with all spines more or less evenly spaced. Go to **3.**

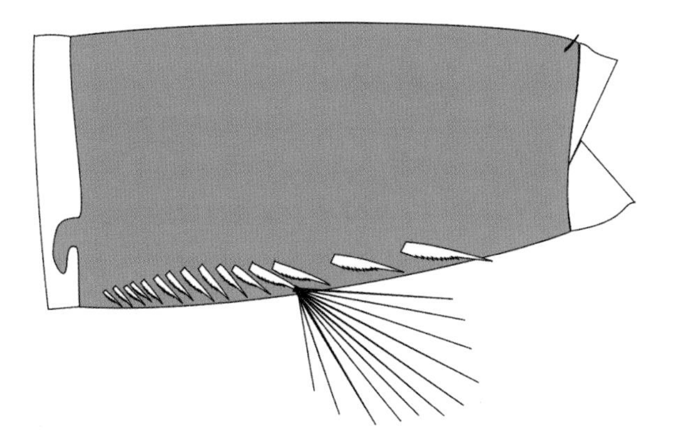

Siphon of *Ae. fulvus pallens*

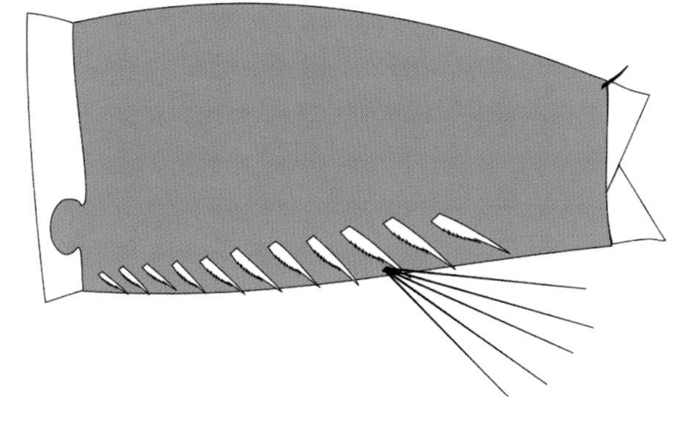

Siphon of *Ae. tormentor*

3a Branched seta of siphon arising within pecten. ***Ae. tormentor*** (p. 79).
3b Branched seta of siphon not arising within pecten. Go to **4.**

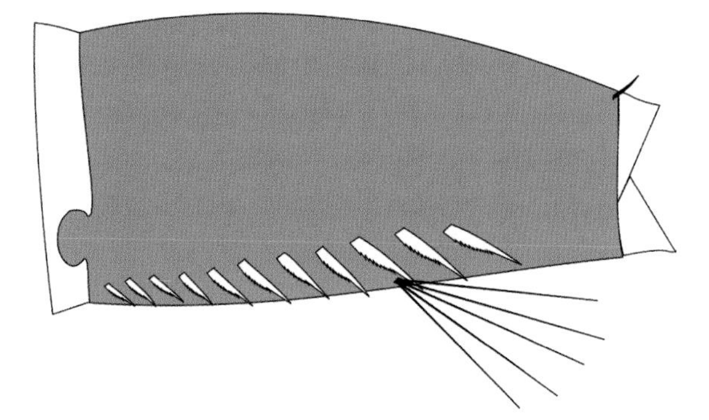

Siphon of *Ae. tormentor*

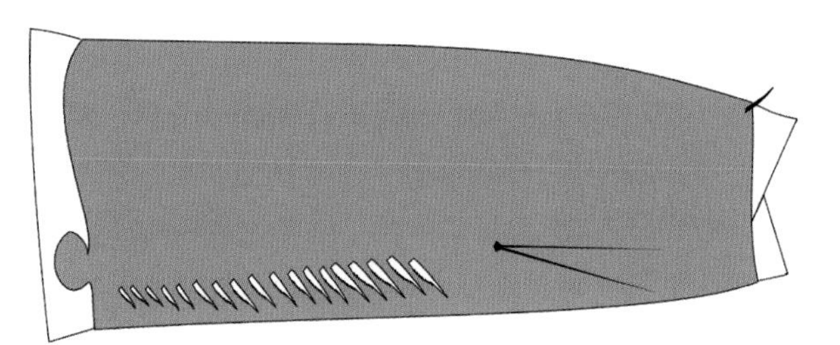

Siphon of *Ae. triseriatus*

4a Comb scales with median spine 5 or more times longer than other spines; thorax glossy smooth, without small spines. Go to **5.**
4b Comb scales with median spine 3 or less times longer than other spines; thorax with tiny spines giving it a rough appearance. Go to **8.**

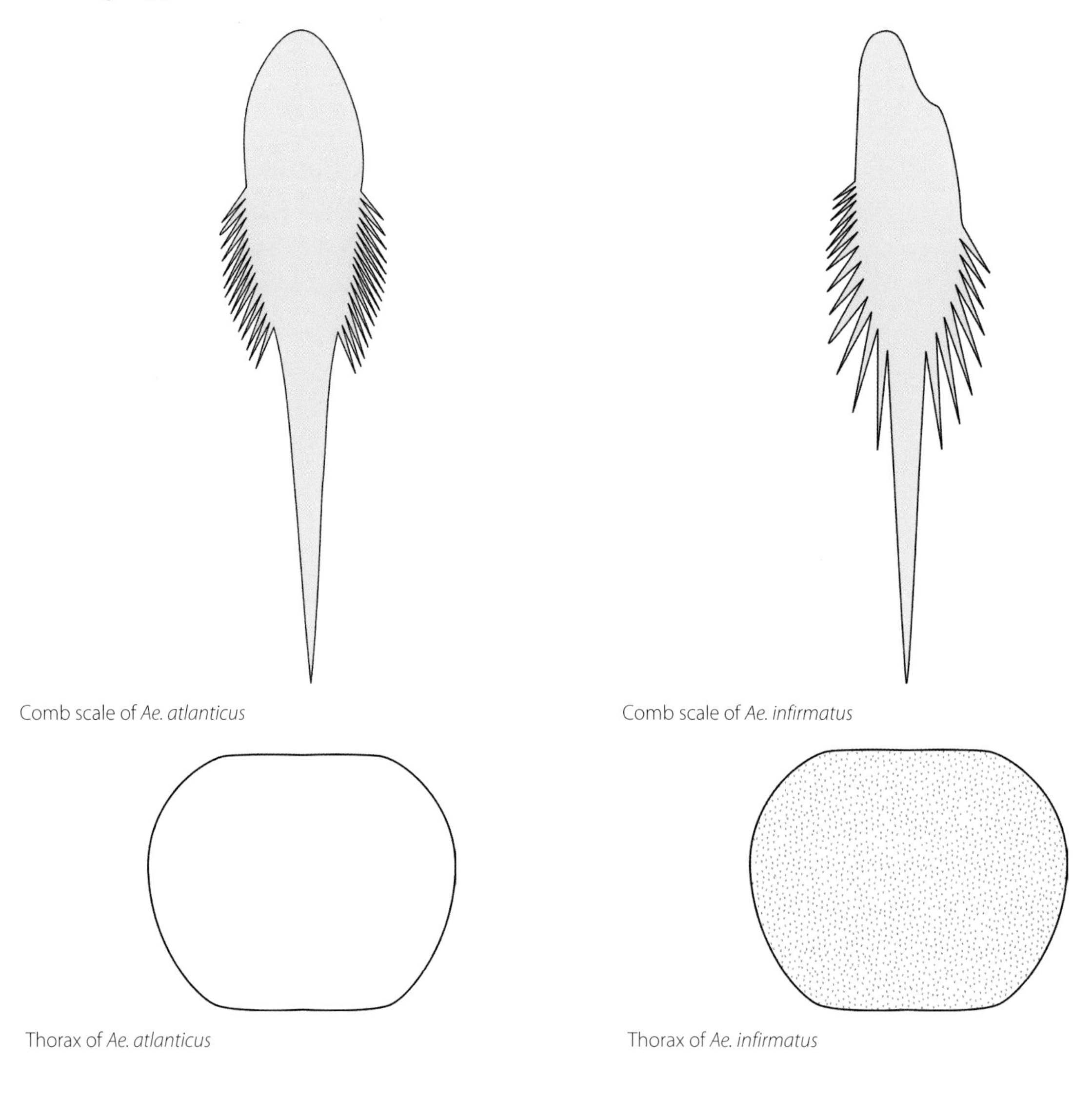

Comb scale of *Ae. atlanticus*

Comb scale of *Ae. infirmatus*

Thorax of *Ae. atlanticus*

Thorax of *Ae. infirmatus*

5a Anal papillae very long, dark, and thin, around 8 times longer than saddle. ***Ae. dupreei*** (p. 90).
5b Anal papillae not dark, not more than 2–3 times longer than saddle. Go to **6.**

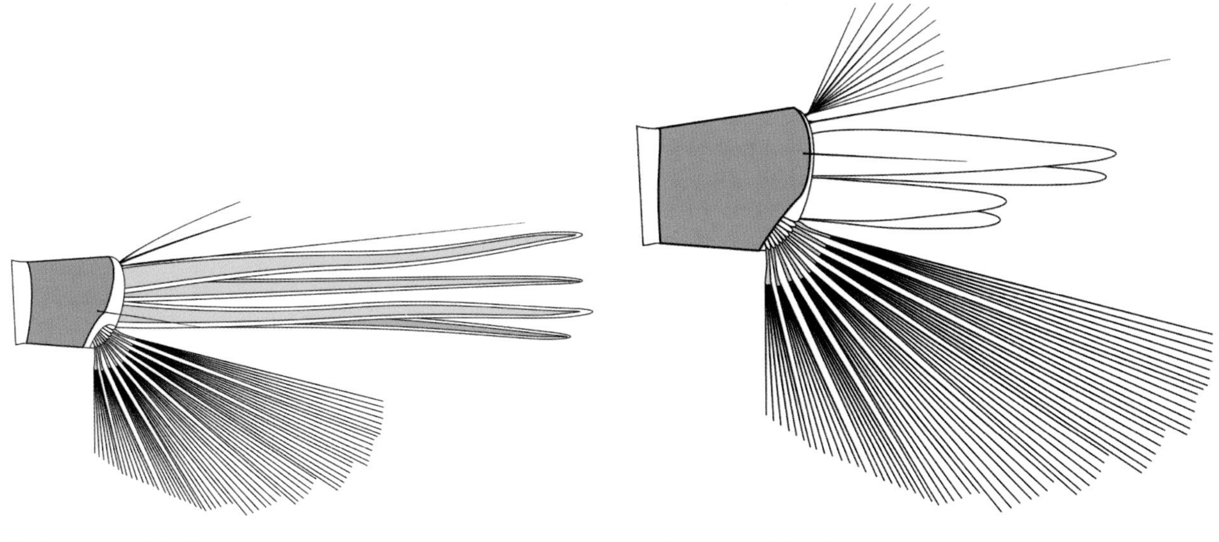

Anal segment of *Ae. dupreei*

Anal segment of *Ae. atlanticus*

6a Comb scales numbering 4–6, large. ***Ae. atlanticus*** (p. 79).
6b Comb scales numbering 10–30, small. Go to **7.**

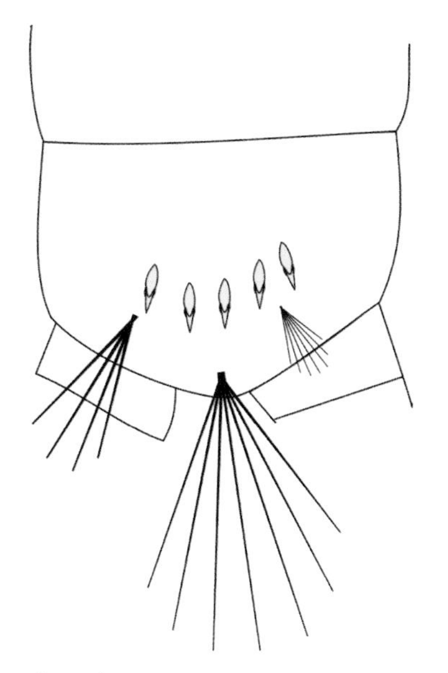

Comb scales of *Ae. atlanticus*

Comb scales of *Ae. sollicitans*

7a Siphon 3–5 times longer than wide at its midpoint; pecten not reaching past midpoint of siphon. ***Ae. mitchellae*** (p. 99).

7b Siphon 2–2.5 times longer than wide at its midpoint; pecten reaching well past midpoint of siphon. ***Ae. sollicitans*** (p. 101).

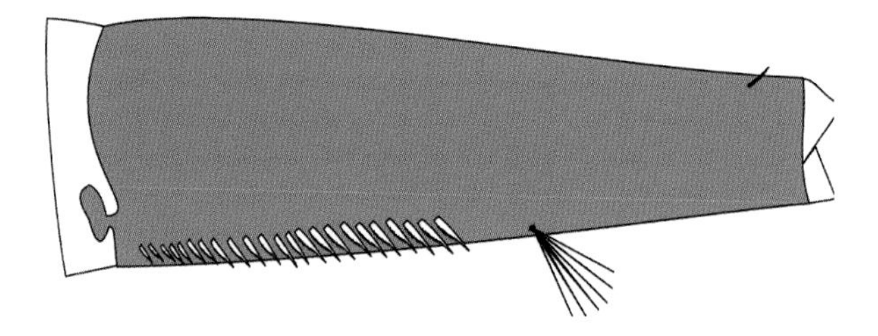

Siphon of *Ae. mitchellae*

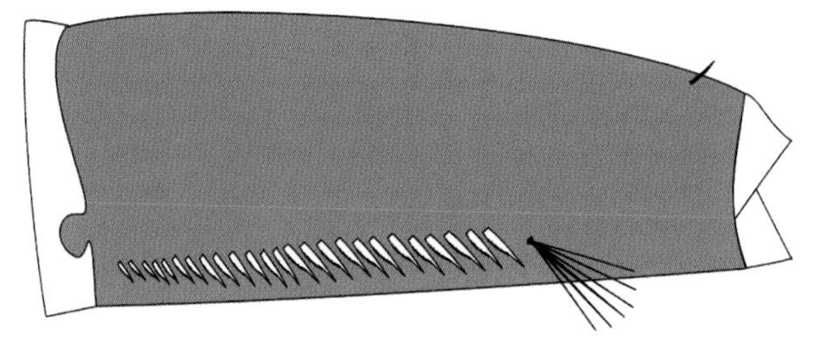

Siphon of *Ae. sollicitans*

8a Comb scales with spines of similar size. ***Ae. taeniorhynchus*** (p. 104).

8b Comb scales with median spine distinctly longer and wider than other spines. Go to **9.**

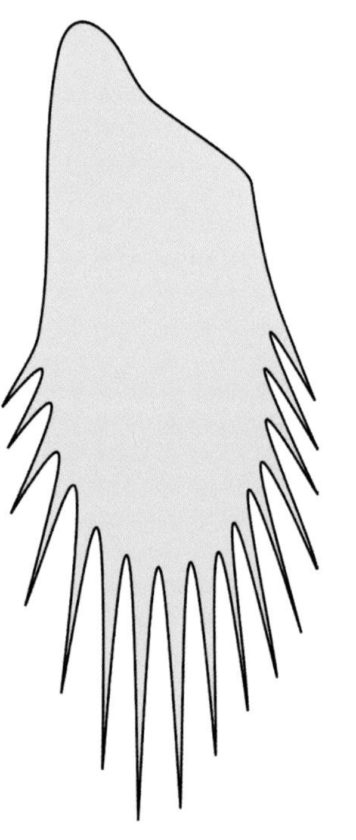

Comb scale of *Ae. taeniorhynchus*

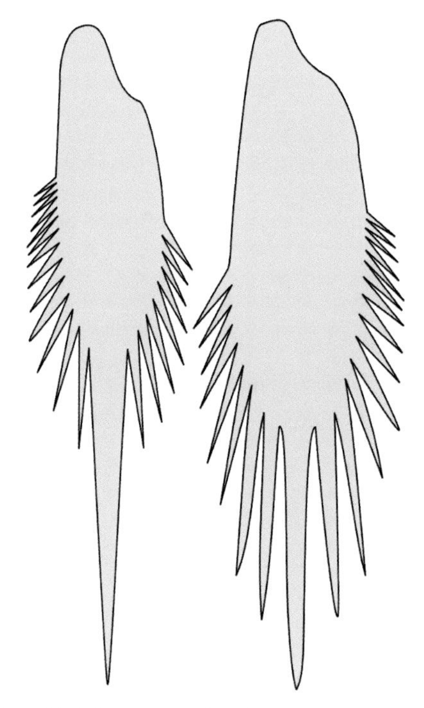

Comb scales of *Ae. infirmatus* and *Ae. trivittatus*

9a Median spine of comb scales about 5–6 times broader at its base than other spines. ***Ae. infirmatus*** (p. 95).

9b Median spine of comb scales about 2–3 times broader at its base than other spines. ***Ae. trivittatus*** (p. 110).

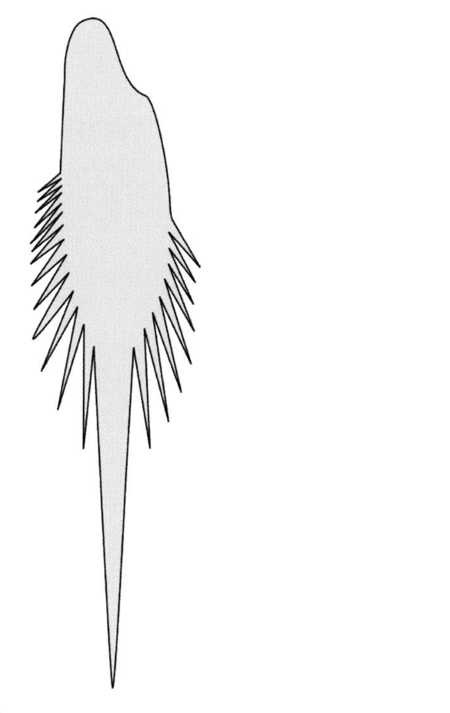

Comb scale of *Ae. infirmatus*

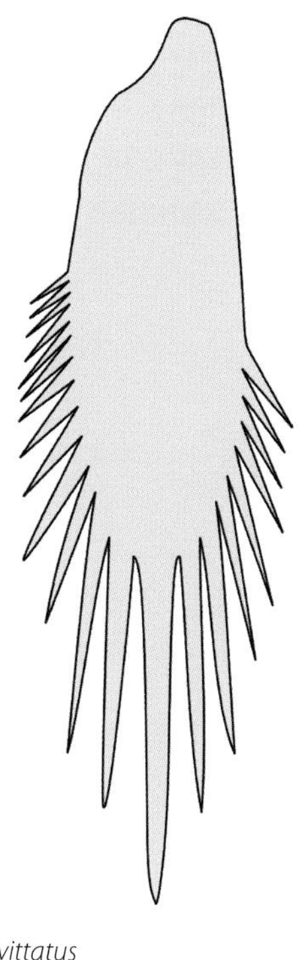

Comb scale of *Ae. trivittatus*

10a Pecten with 1 or more spines separated from other spines. Go to **11.**

10b Pecten with all spines more or less evenly spaced. Go to **15.**

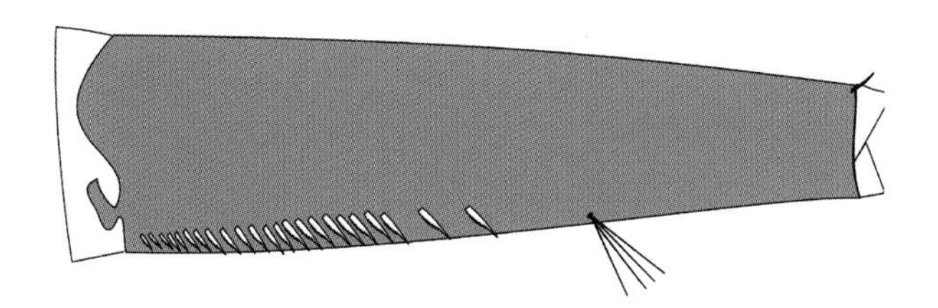

Siphon of *Ae. vexans*

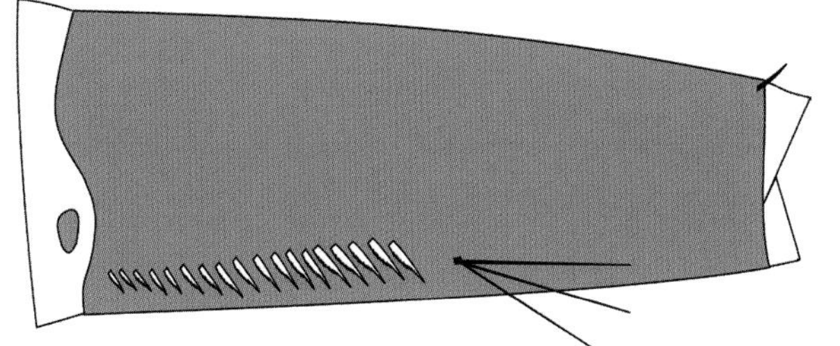

Siphon of *Ae. hendersoni*

11a Branched seta arising within pecten. Go to **12.**
11b Branched seta not arising within pecten. Go to **13.**

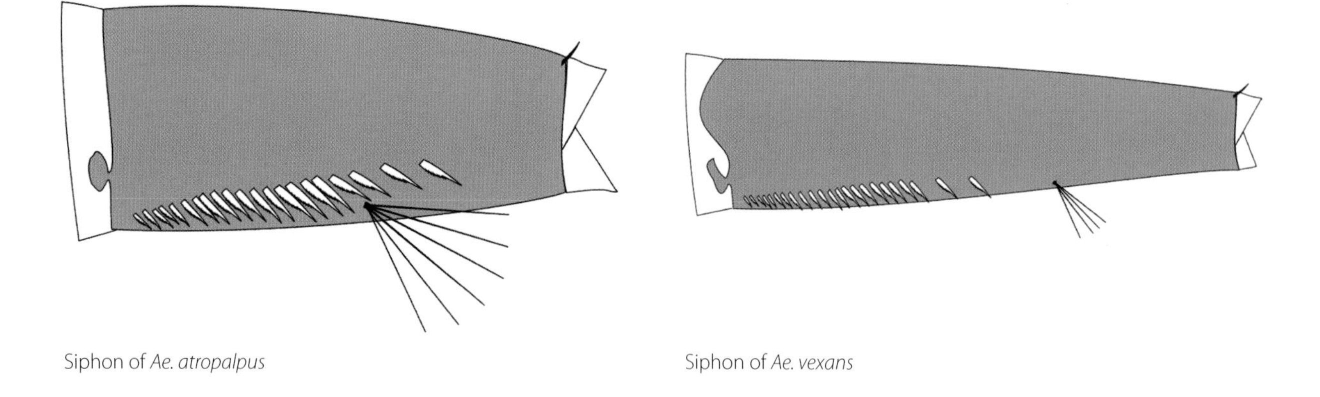

Siphon of *Ae. atropalpus*

Siphon of *Ae. vexans*

12a Setae at front of head (5–7-C) with 5–6 branches ***Ae. japonicus*** (p. 97).
12b Setae at front of head (5–7-C) unbranched or with 2 branches. ***Ae. atropalpus*** (p. 82).

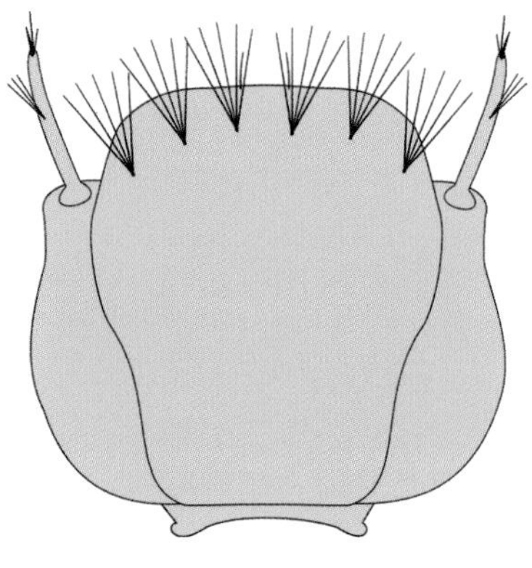

Head of *Ae. japonicus*

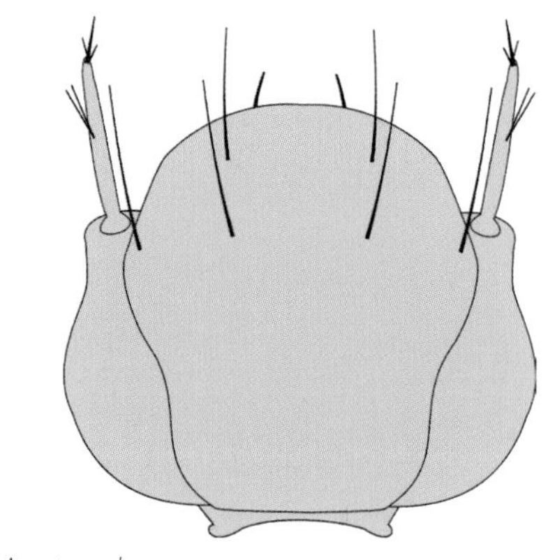

Head of *Ae. atropalpus*

13a Antennae about as long as head, or only slightly shorter. ***Ae. aurifer*** (p. 84).
13b Antennae distinctly shorter than head. Go to **14.**

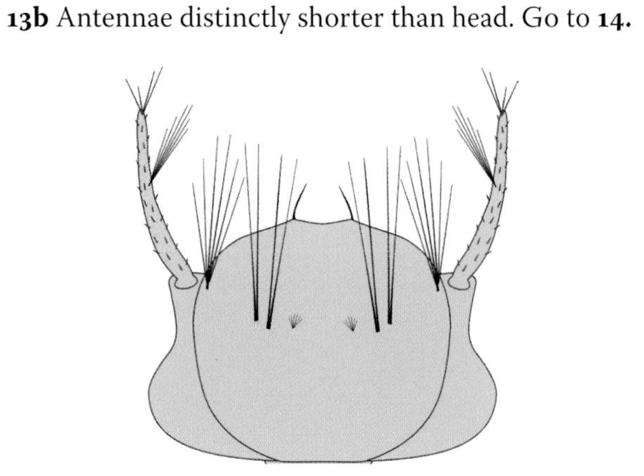

Head of *Ae. aurifer*

Head of *Ae. vexans*

14a Bases of branched setae of head (5–7-C) not in a straight line. ***Ae. vexans*** (p. 111).
14b Bases of branched setae of head (5–7-C) in a straight line. ***Ae. cinereus*** (p. 88).

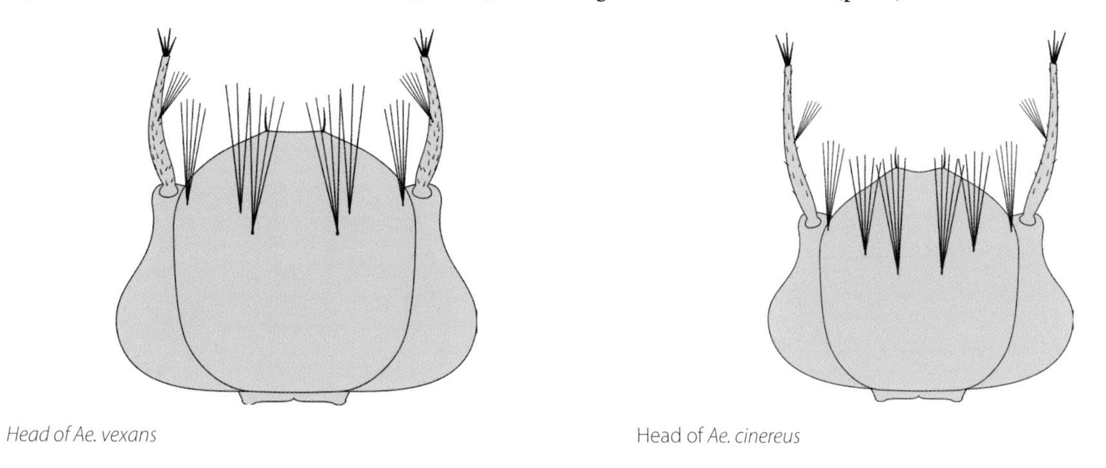

Head of Ae. vexans

Head of *Ae. cinereus*

15a Seta at midpoint of antenna (1-A) with 2 branches or unbranched; antennae without small spines.
Go to **16.**
15b Seta at midpoint of antenna (1-A) with 4 or more branches; antennae with many small spines.
Go to **19.**

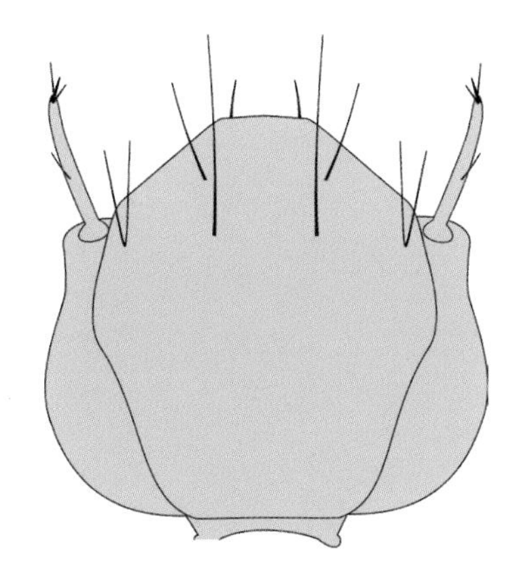

Head of *Ae. albopictus*

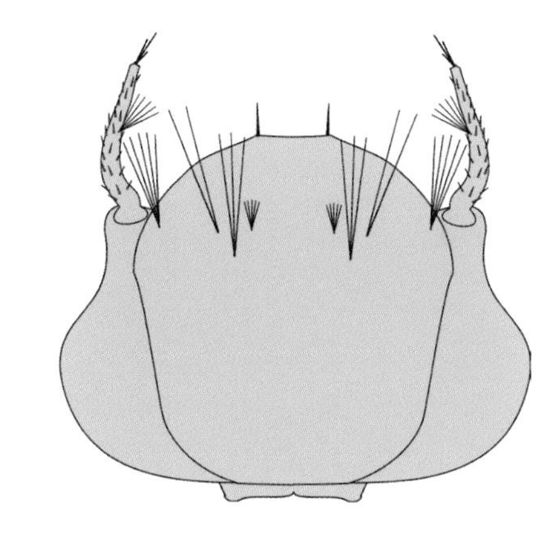

Head of *Ae. grossbecki*

16a Comb scales with stout median spine, sharply and narrowly pointed at their tips. Go to **17.**
16b Comb scales broadly pointed at their tips and fringed with many small spines. Go to **18.**

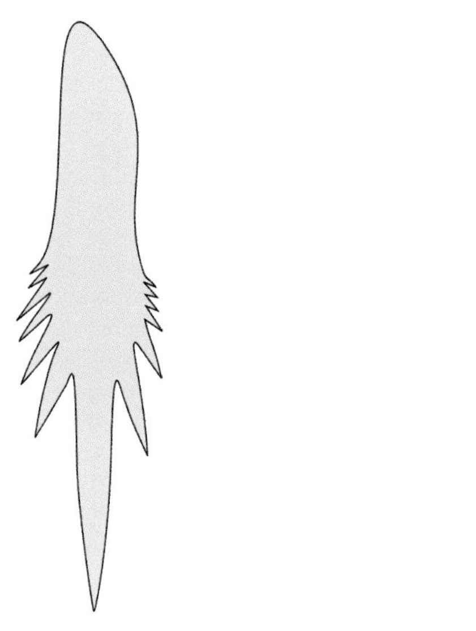

Comb scale of *Ae. aegypti*

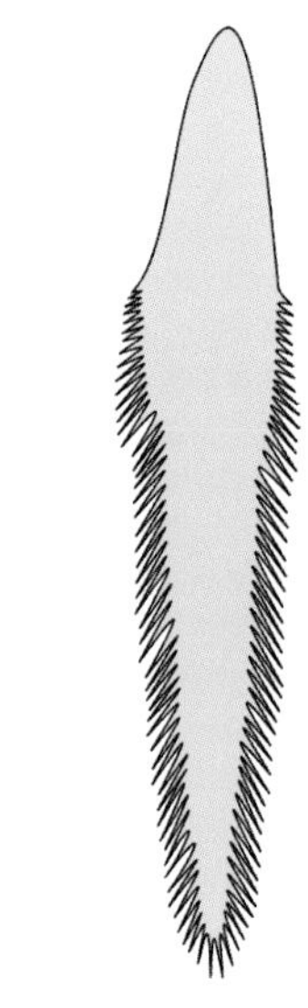

Comb scale of *Ae. triseriatus*

17a Comb scales with 1 stout median spine and several short yet stout spines surrounding it; seta at base of antenna (7-C) unbranched. ***Ae. aegypti*** (p. 75).
17b Comb scales with 1 stout median spine and a fringe of tiny spines at its base; seta at base of antenna (7-C) with 2 branches. ***Ae. albopictus*** (p. 77).

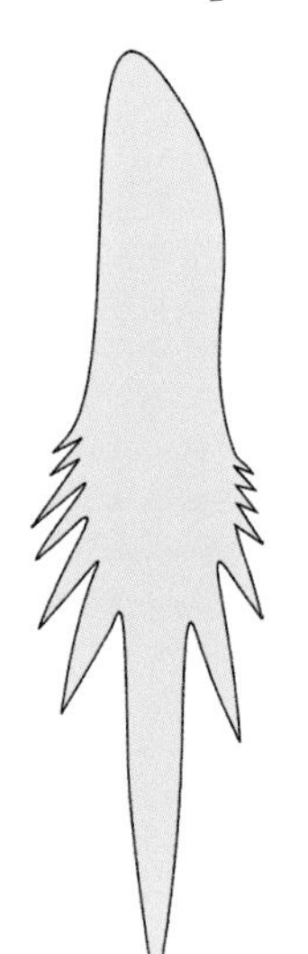

Comb scale of *Ae. aegypti*

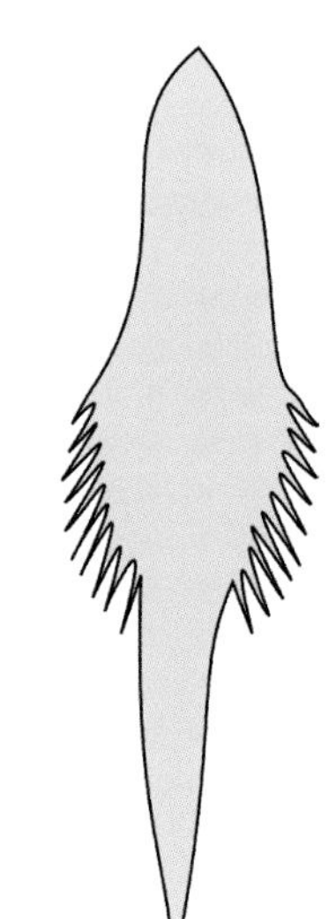

Comb scale of *Ae. albopictus*

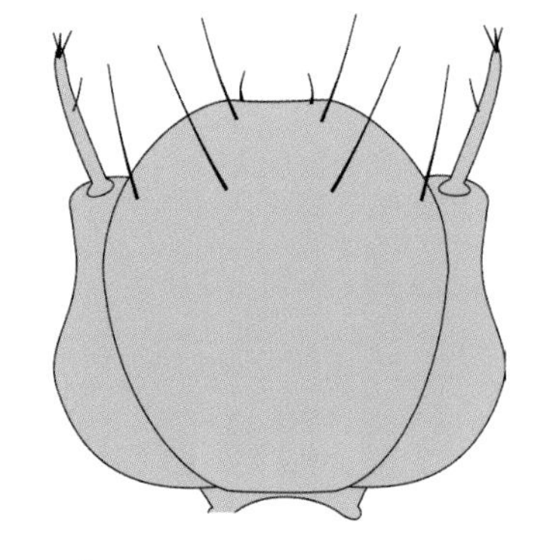

Head of *Ae. aegypti*

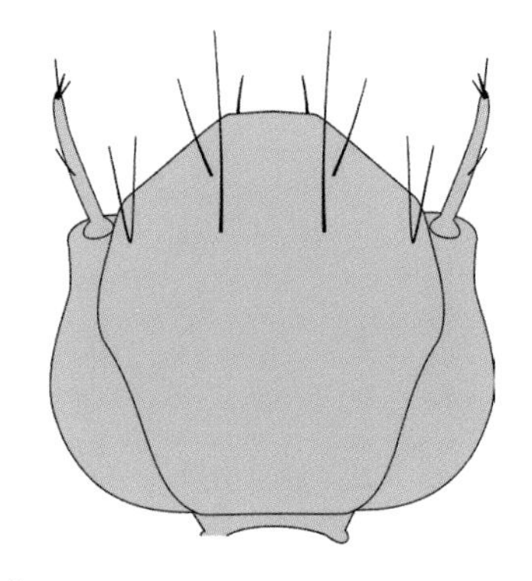

Head of *Ae. albopictus*

18a Anal papillae not bulbous, dorsal pair longer than ventral pair; siphon with small plate (acus) attached to its base. ***Ae. triseriatus*** (p. 108).

18b Anal papillae bulbous, dorsal and ventral pair of about equal length; siphon with small detached plate (acus) near its base. ***Ae. hendersoni*** (p. 108).

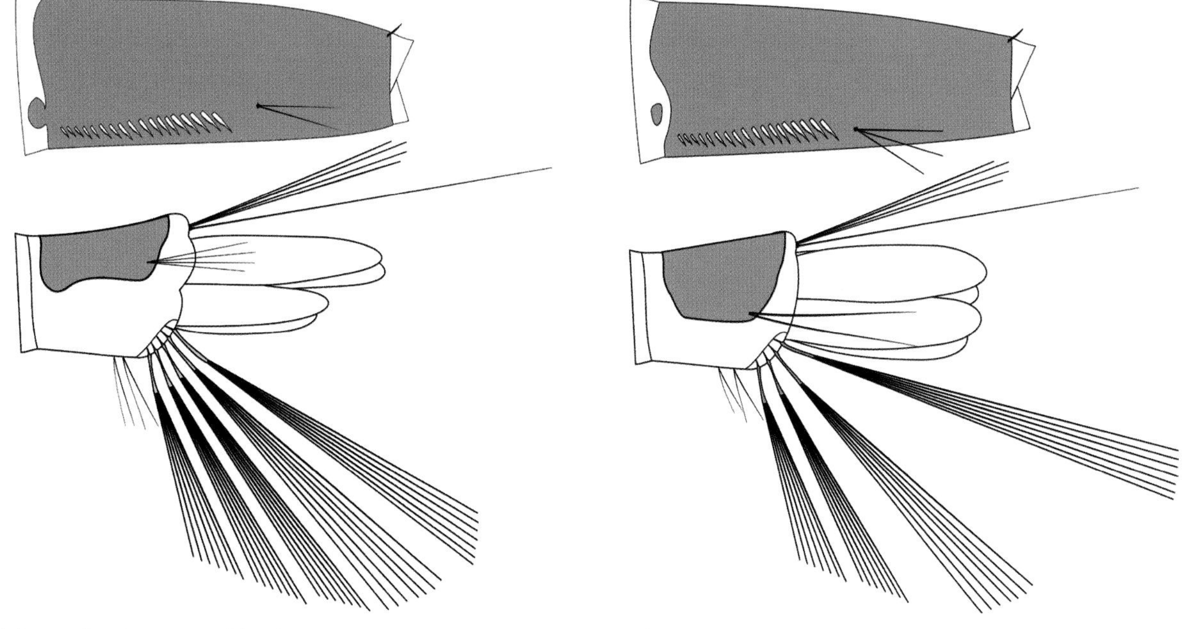

Siphon and anal segment of *Ae. triseriatus*

Siphon and anal segment of *Ae. hendersoni*

19a Comb scales with large median spine 3 or more times longer than other spines. ***Ae. sticticus*** (p. 103).

19b Comb scales with small median spine less than 3 times larger than other spines. Go to **20.**

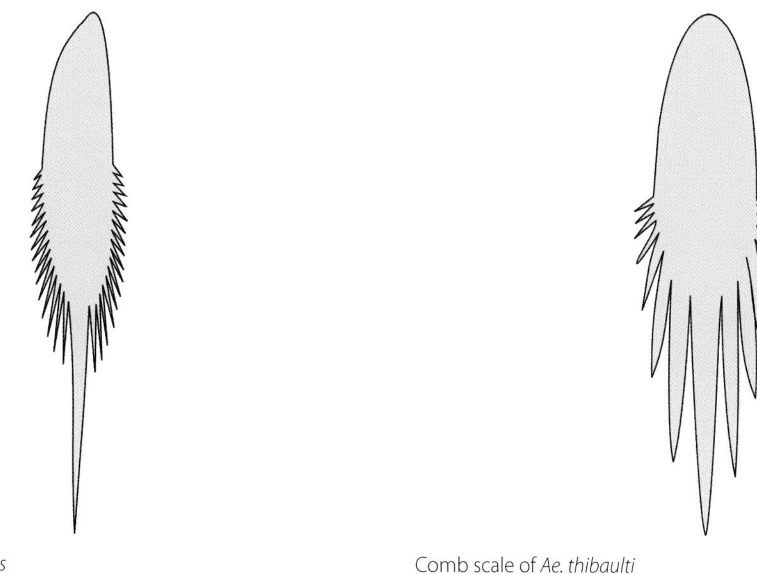

Comb scale of *Ae. sticticus*

Comb scale of *Ae. thibaulti*

20a Siphon long, usually about 5 times longer than wide at its midpoint; antennae about as long as head. ***Ae. thibaulti*** (p. 106).

20b Siphon not so long, 2.5–4 times longer than wide at its midpoint; antennae shorter than head. Go to **21.**

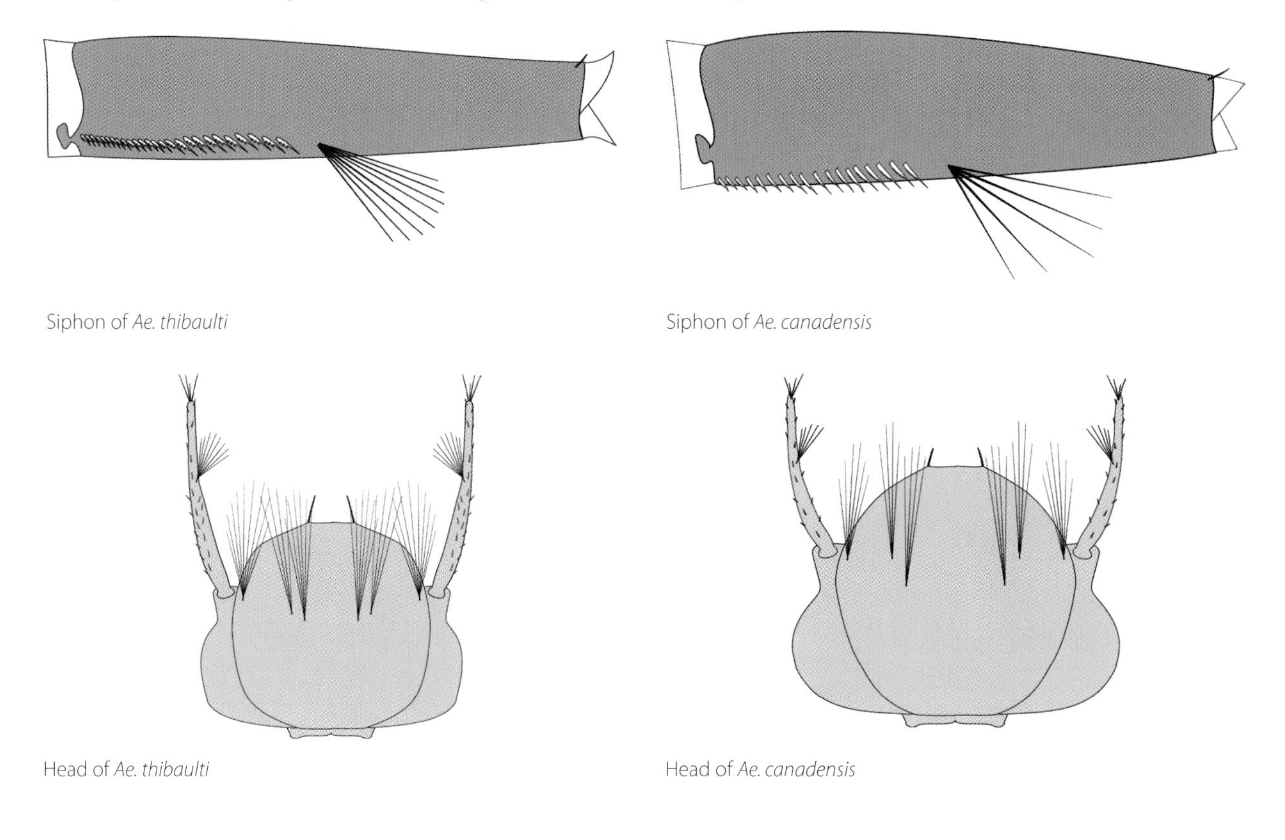

Siphon of *Ae. thibaulti*

Siphon of *Ae. canadensis*

Head of *Ae. thibaulti*

Head of *Ae. canadensis*

21a Anal papillae short, less than half the length of the saddle. ***Ae. cantator*** (p. 87).

21b Anal papillae not so short, about equal in length to saddle or longer. Go to **22.**

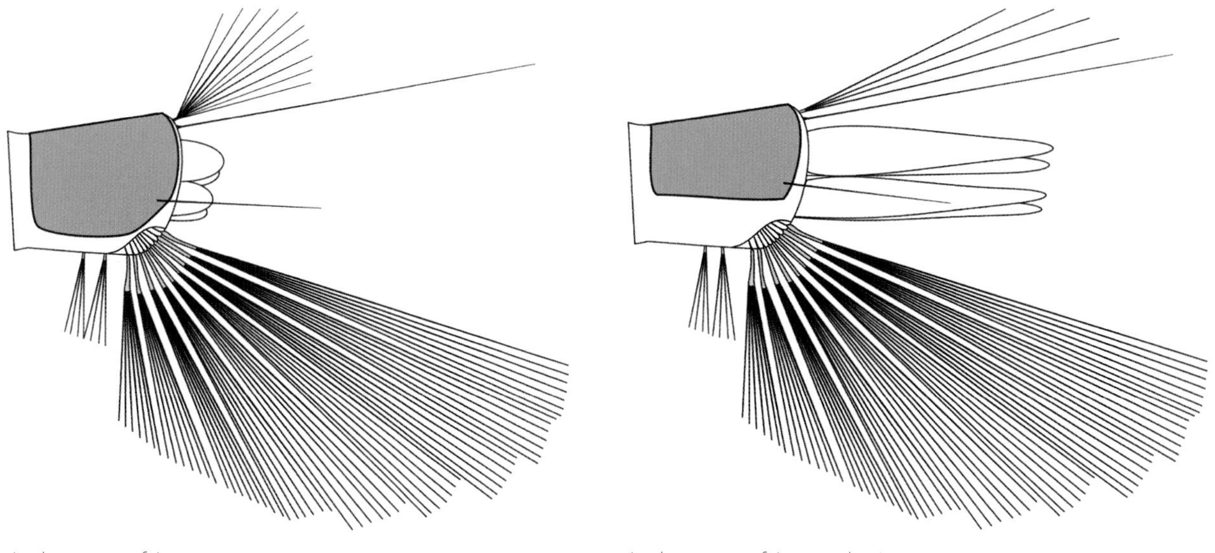

Anal segment of *Ae. cantator*

Anal segment of *Ae. canadensis*

22a Median head setae (5-C and 6-C) usually with 2 or 3 branches. ***Ae. grossbecki*** (p. 93).
22b Median head setae (5-C and 6-C) usually with 4 or more branches. ***Ae. canadensis*** (p. 85).

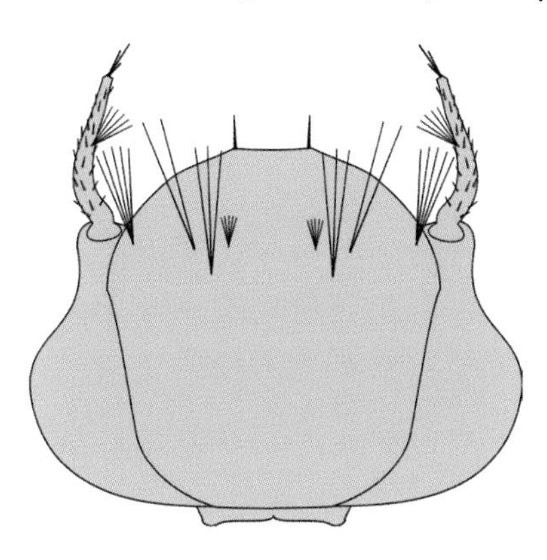
Head of *Ae. grossbecki*

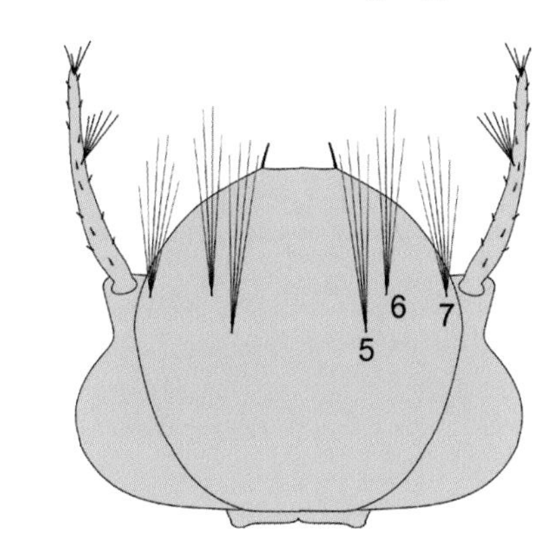

Head of *Ae. canadensis*

Anopheles (An.)

1a Setae of head small, unbranched. ***An. barberi*** (p. 115).
1b Setae of head mostly large, branched. Go to **2.**

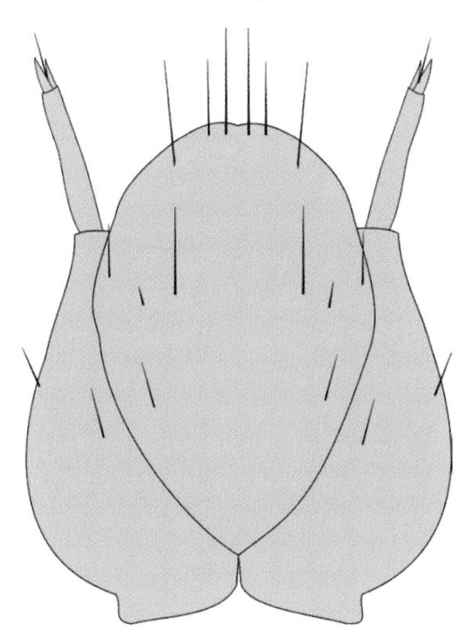
Head of *An. barberi*

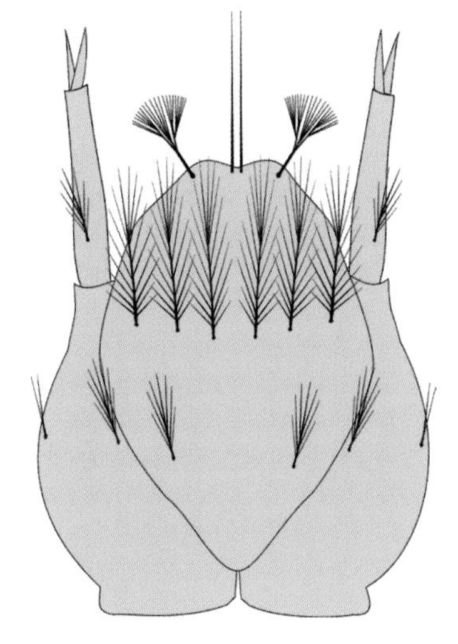
Head of *An. punctipennis*

2a Setae along front edge of head (frontal setae) unbranched. ***An. pseudopunctipennis*** (p. 120).
2b Outer frontal setae (3-C) with 5 or more branches. Go to **3.**

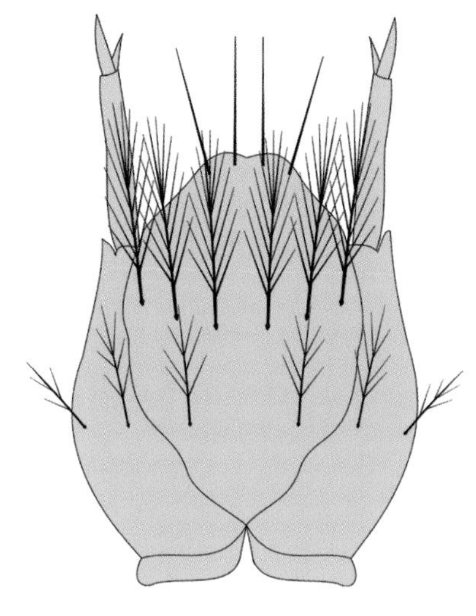

Head of *An. pseudopunctipennis*

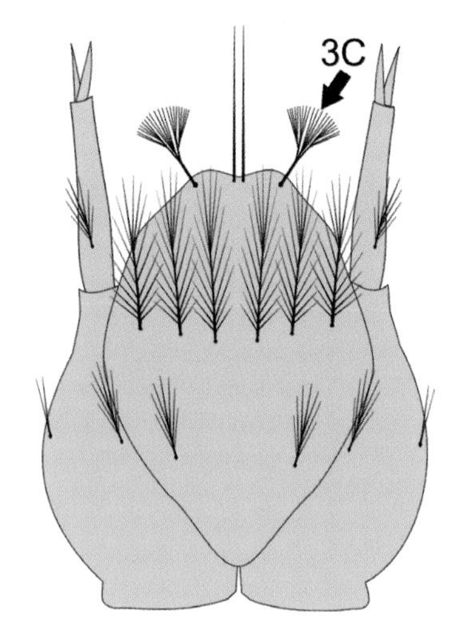

Head of *An. punctipennis*

3a Outer frontal setae (3-C) with 10 or fewer branches. ***An. atropos*** (p. 113).
3b Outer frontal setae with many branches, appearing brushlike. Go to **4.**

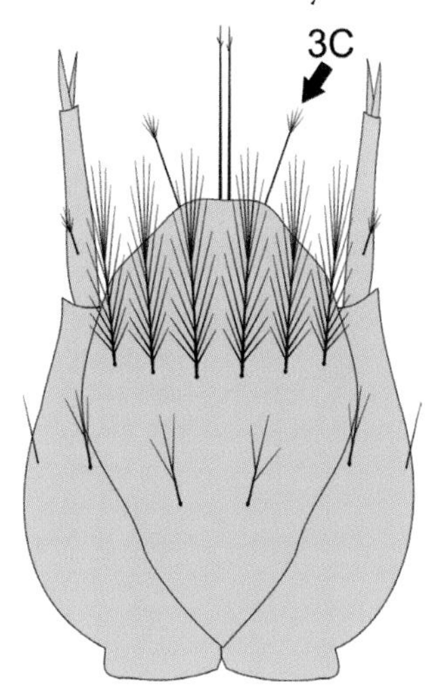

Head of *An. atropos*

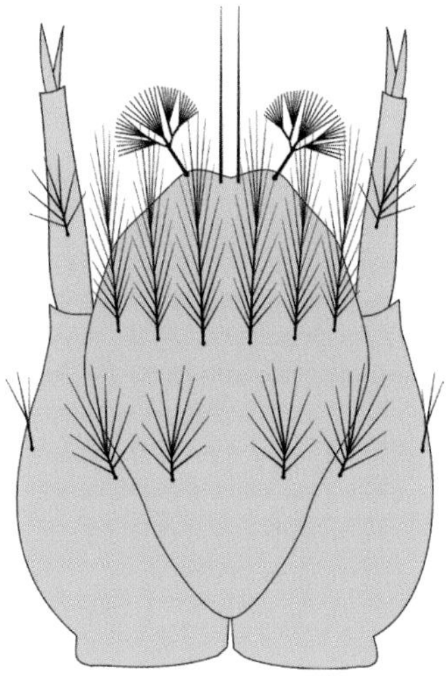

Head of *An. quadrimaculatus*

4a Anterior seta (seta 0) of abdominal segments 4 and 5 with 4 or more branches. ***An. crucians* complex** (p. 116).

4b Anterior seta (seta 0) of abdominal segments 4 and 5 small, with fewer than 4 branches. Go to **5.**

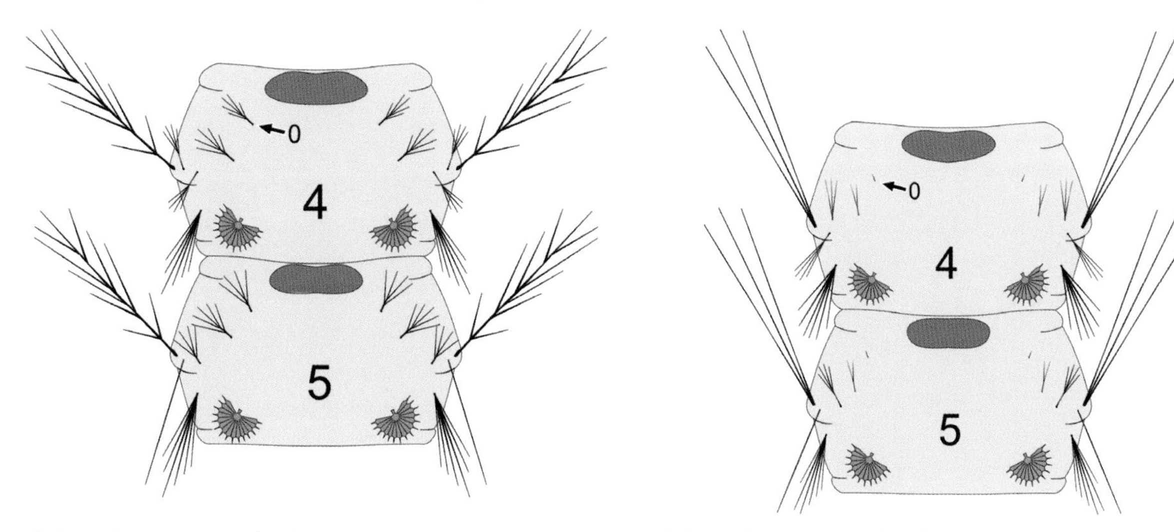

Abdominal segments 4 and 5 of *An. crucians*

Abdominal segments 4 and 5 of *An. walkeri*

5a Inner frontal setae (2C) with sparse aciculae (branches) near the tip. ***An. walkeri*** (p. 125).

5b Inner frontal setae (2C) without small aciculae. Go to **6.**

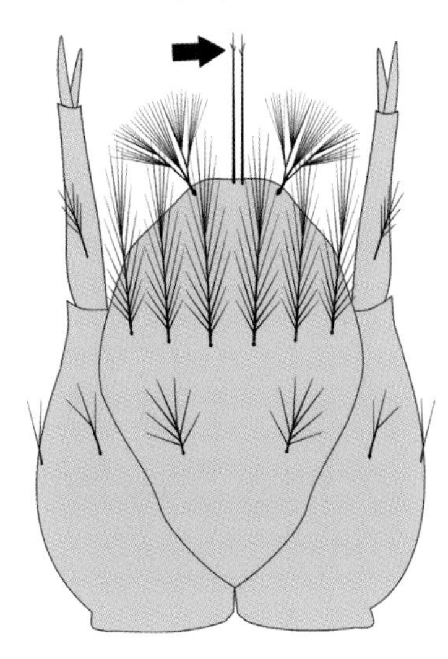

Head of *An. walkeri*

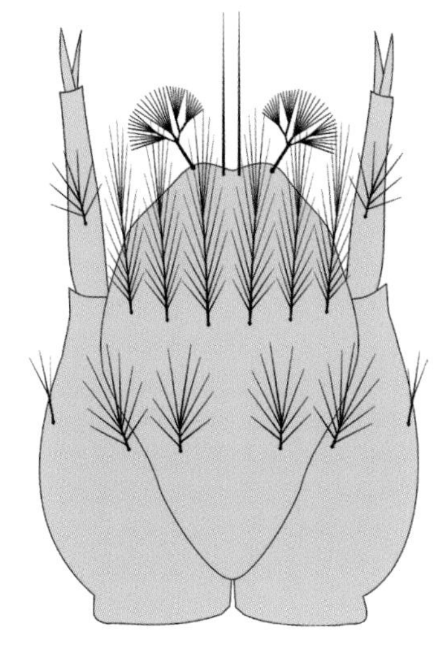

Head of *An. quadrimaculatus*

6a Abdominal segments 4 through 6 with large palmate setae. ***An. bradleyi*** and ***An. georgianus*** (p. 116).
6b Abdominal segments 3 through 7 with large palmate setae. Go to **7.**

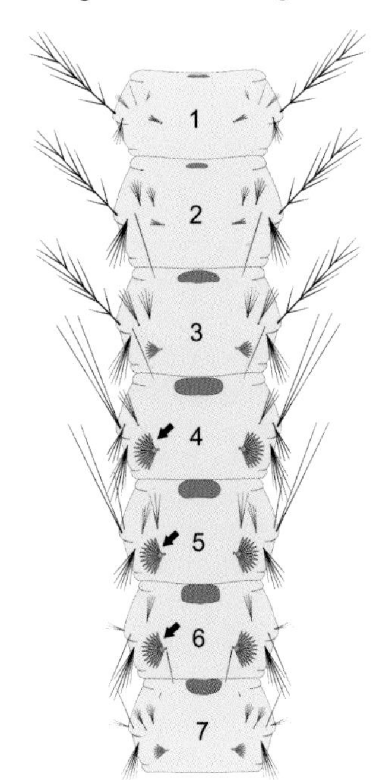

Abdominal segments of *An. bradleyi*

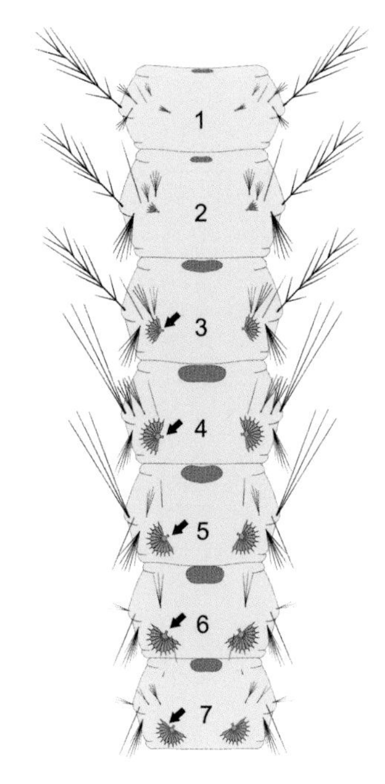

Abdominal segments of *An. quadrimaculatus*

7a Bases (alveoli) of inner frontal setae separated by distance greater than the diameter of 1 base. ***An. quadrimaculatus* group** (p. 123).
7b Bases (alveoli) of inner frontal setae closer than the diameter of 1 base. ***An. perplexens*** (p. 118) and ***An. punctipennis*** (p. 121).

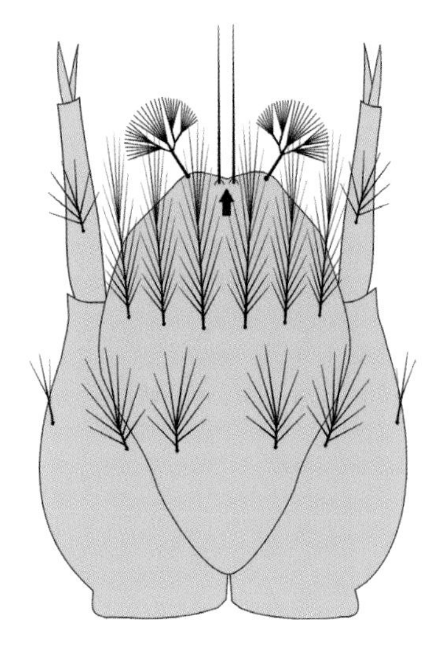
Head of *An. quadrimaculatus*

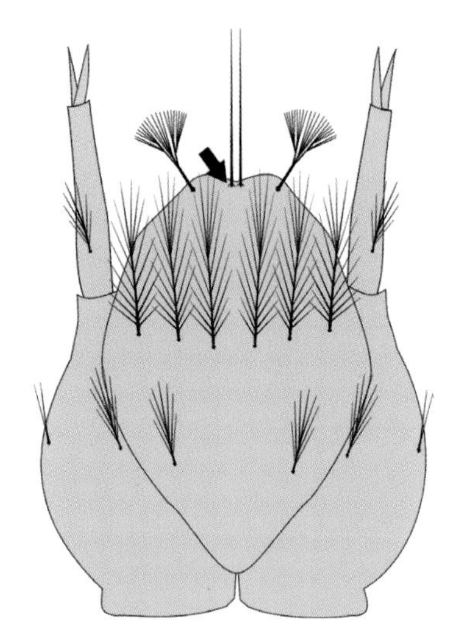
Head of *An. punctipennis*

Culex (Cx.)

1a Median setae of head (6-C and usually 5-C) with 3 or more branches. Go to **2.**
1b Median setae of head (6-C and usually 5-C) unbranched or with 2 branches (rarely 3). Go to **7.**

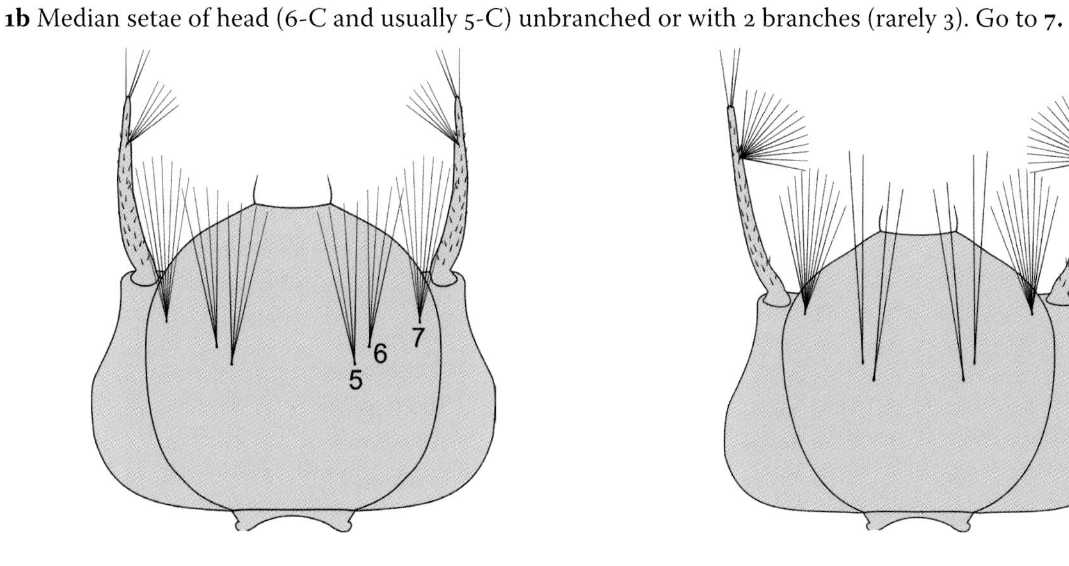

Head of *Cx. pipiens*

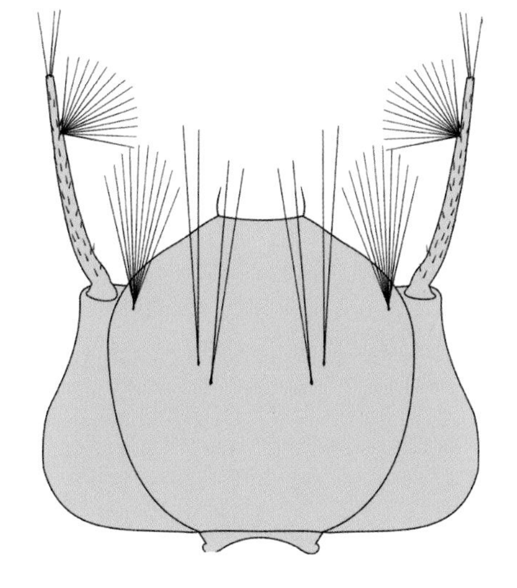

Head of *Cx. territans*

2a Setae of siphon long, mostly unbranched, and irregularly placed. ***Cx. restuans*** (p. 142).
2b Setae of siphon short, mostly branched, and mostly in a line. Go to **3.**

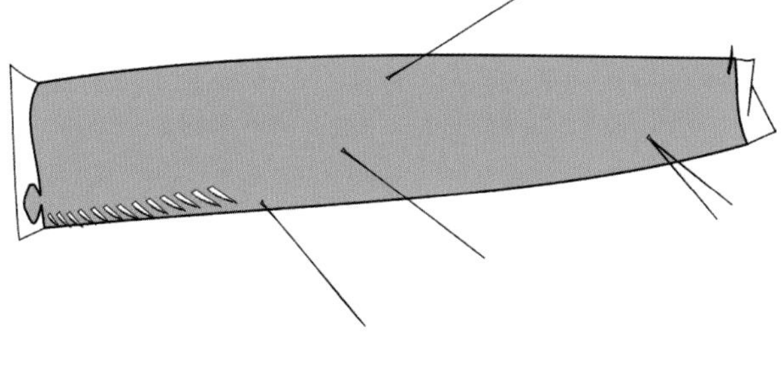

Siphon of *Cx. restuans*

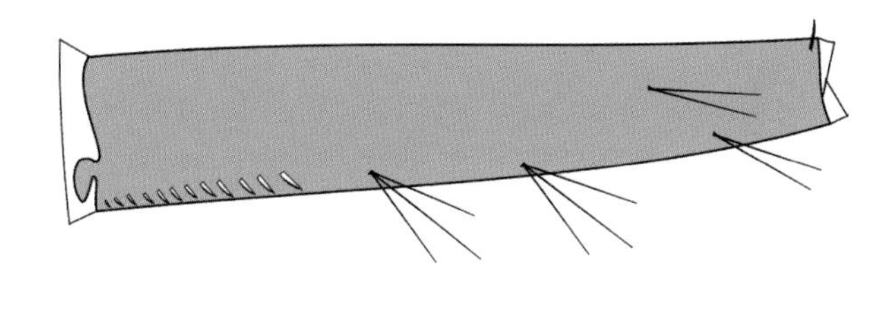

Siphon of *Cx. pipiens*

3a Siphon with several small spines near apex. ***Cx. coronator*** (p. 129).
3b Siphon without spines near apex. Go to **4.**

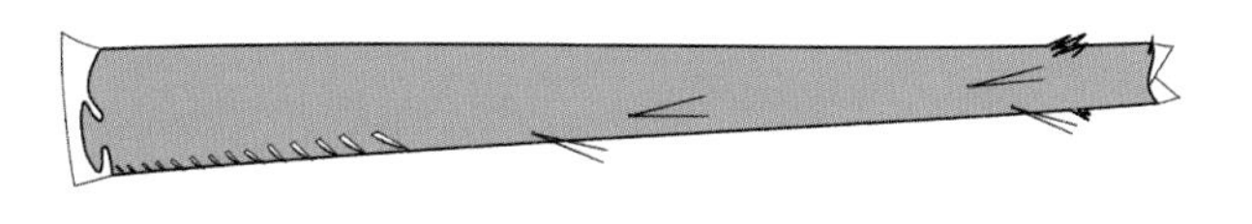

Siphon of *Cx. coronator*

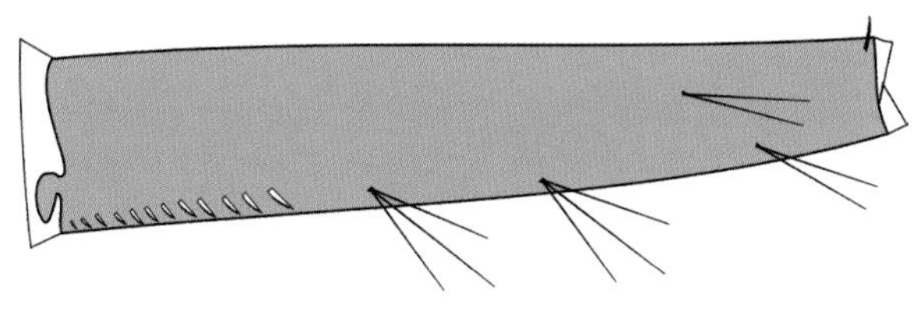

Siphon of *Cx. pipiens*

4a Setae of siphon in a straight ventral line, usually with 5 pairs of branched setae. ***Cx. tarsalis*** (p. 146).
4b Setae of siphon not all in a straight line (1 or 2 setae out of line dorsally). Go to **5.**

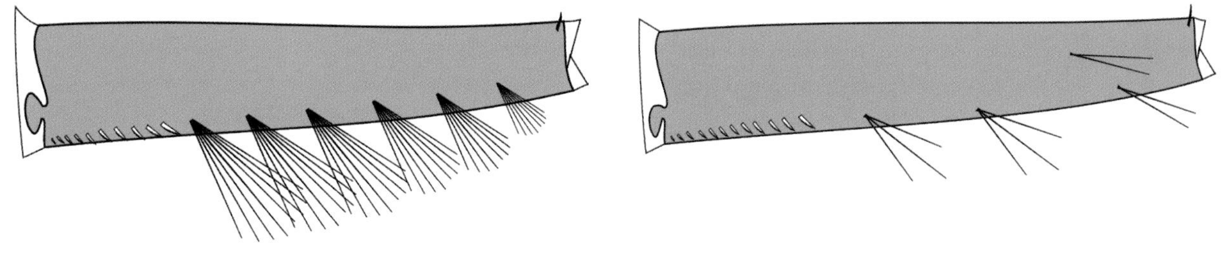

Siphon of *Cx. tarsalis*

Siphon of *Cx. pipiens*

5a Siphon 4–5 times longer than wide at its midpoint. ***Cx. pipiens*** and ***Cx. quinquefasciatus*** (p. 140).
5b Siphon 6–8 times longer than longer than wide at its midpoint. Go to **6.**

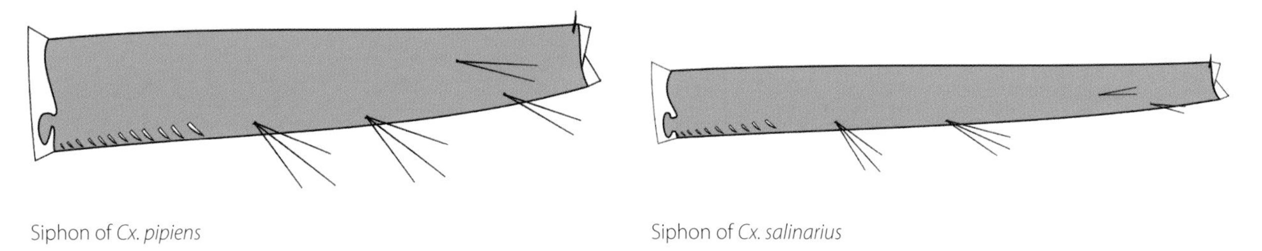

Siphon of *Cx. pipiens*

Siphon of *Cx. salinarius*

6a Thorax covered with fine spines, giving it a rough appearance; seta of saddle usually unbranched. ***Cx. nigripalpus*** (p. 134).
6b Thorax without fine spines, having a glossy appearance; seta of saddle usually with 2 branches. ***Cx. salinarius*** (p. 144).

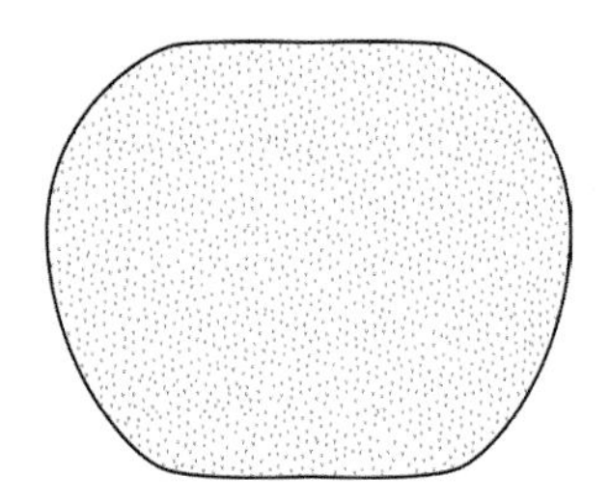

Thorax of *Cx. nigripalpus*

Thorax of *Cx. salinarius*

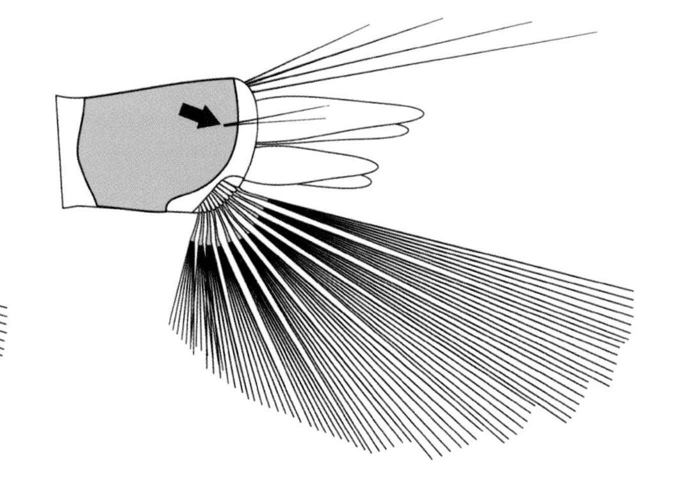

Anal segment of *Cx. nigripalpus*

Anal segment of *Cx. salinarius*

7a Setae of siphon very short, barely longer than the width of the siphon at its midpoint; apical seta of siphon (seta 2-S) not curved. ***Cx. territans*** (p. 148).

7b Setae of siphon not so short, much longer than the width of the siphon at its midpoint; apical seta of siphon (seta 2-S) curved. Go to **8.**

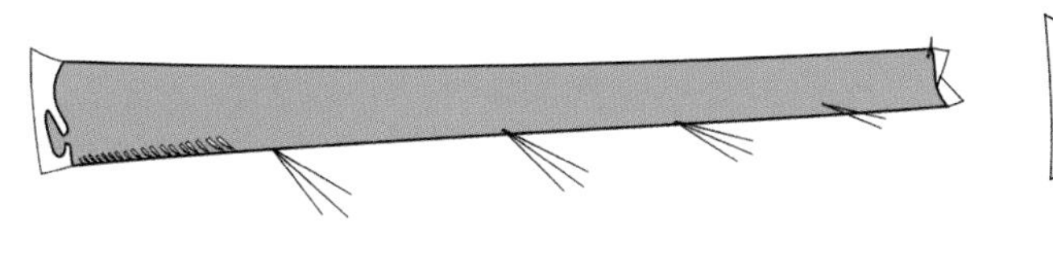

Siphon of *Cx. territans*

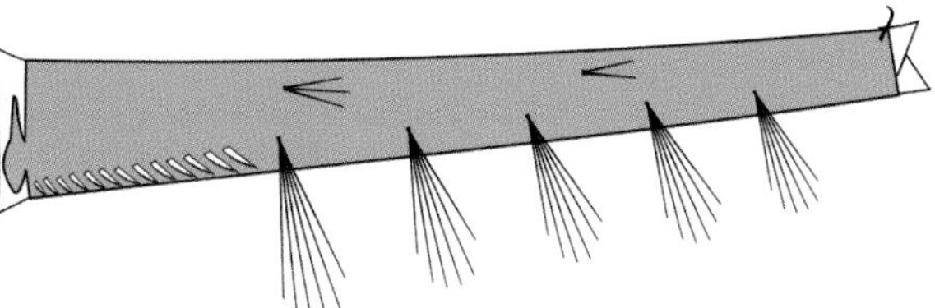

Siphon of *Cx. erraticus*

8a Siphon short and stout, with long branched setae along its entire length. ***Cx. pilosus*** (p. 138).

8b Siphon long and thin, with setae along only three-quarters of its length. Go to **9.**

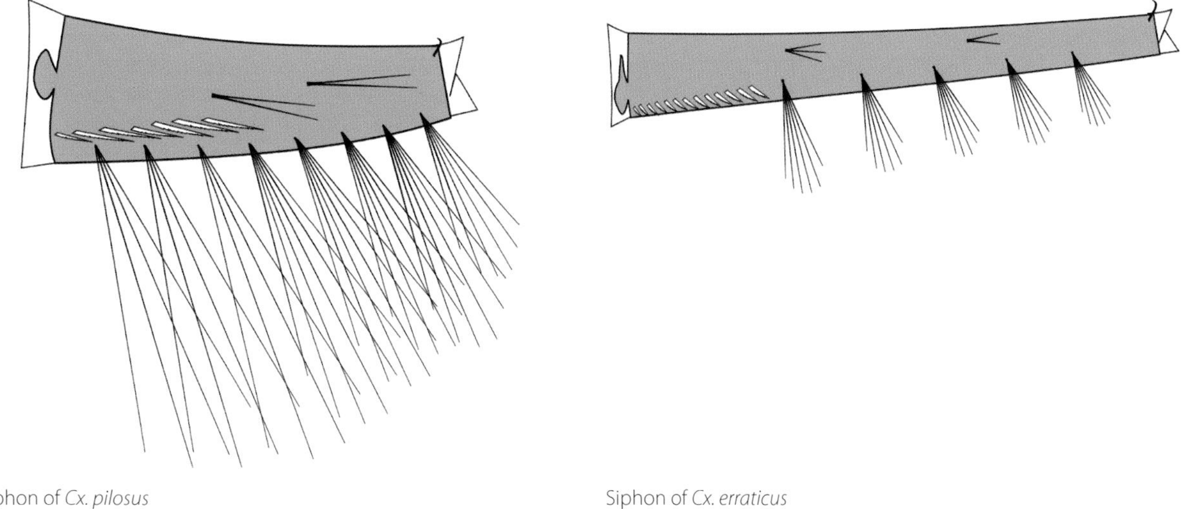

Siphon of *Cx. pilosus*

Siphon of *Cx. erraticus*

9a Siphon with dark band near middle. ***Cx. peccator*** (p. 136).

9b Siphon without dark band near middle. ***Cx. erraticus*** (p. 131).

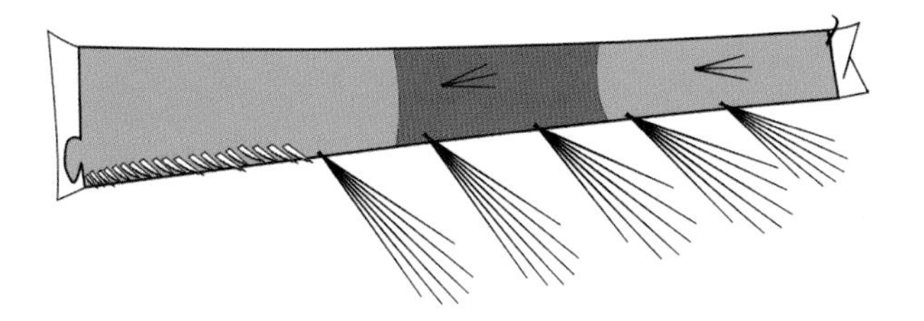

Siphon of *Cx. peccator*

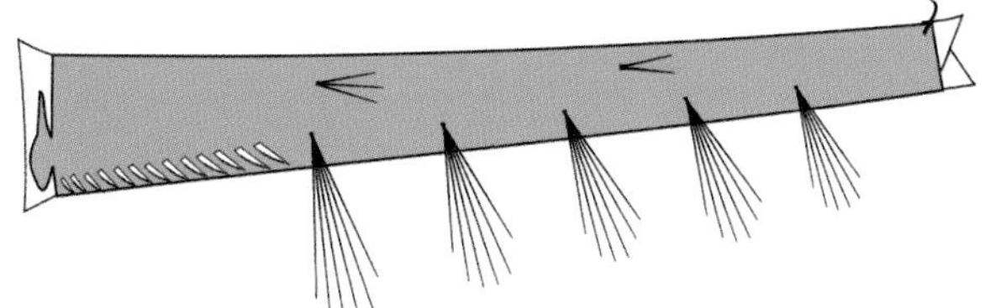

Siphon of *Cx. erraticus*

Culiseta (Cs.)

1a Siphon very long, straight, and tubelike. ***Cs. melanura*** (p. 152).
1b Siphon short and stout. ***Cs. inornata*** (p. 150).

Siphon of *Cs. melanura*

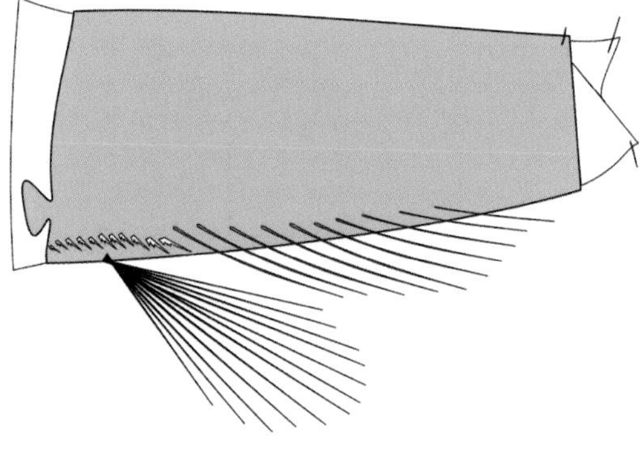

Siphon of *Cs. inornata*

Mansonia (Mn.)

1a Comb scales slender, with a single long spine. ***Mn. titillans*** (p. 156).
1b Comb scales broad, with several stout spines. ***Mn. dyari*** (p. 154).

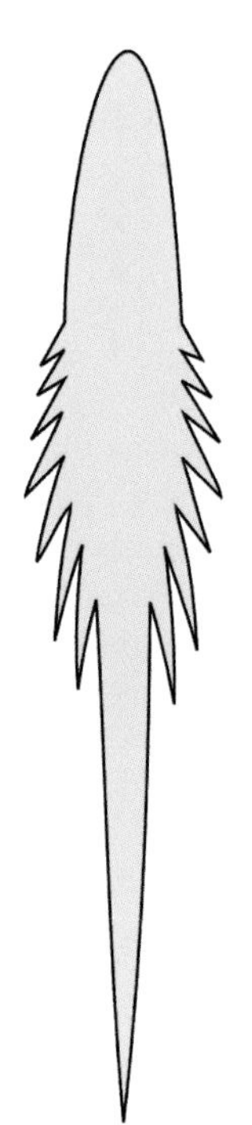

Comb scale of *Mn. titillans*

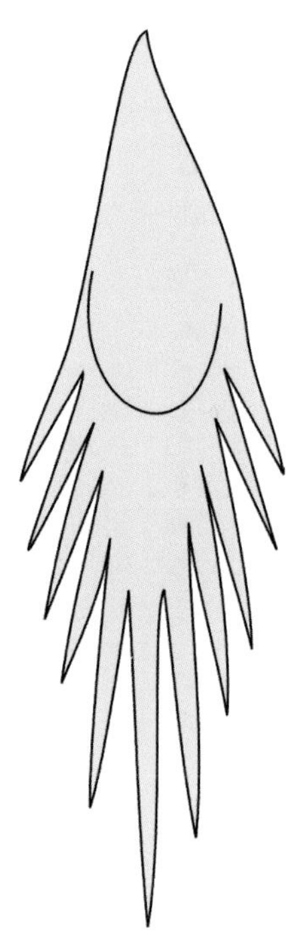

Comb scale of *Mn. dyari*

Orthopodomyia (Or.)

1a Abdominal segment 8 with large dorsal plate. ***Or. signifera*** (p. 158).
1b Abdominal segment 8 without dorsal plate. ***Or. alba*** (p. 158).

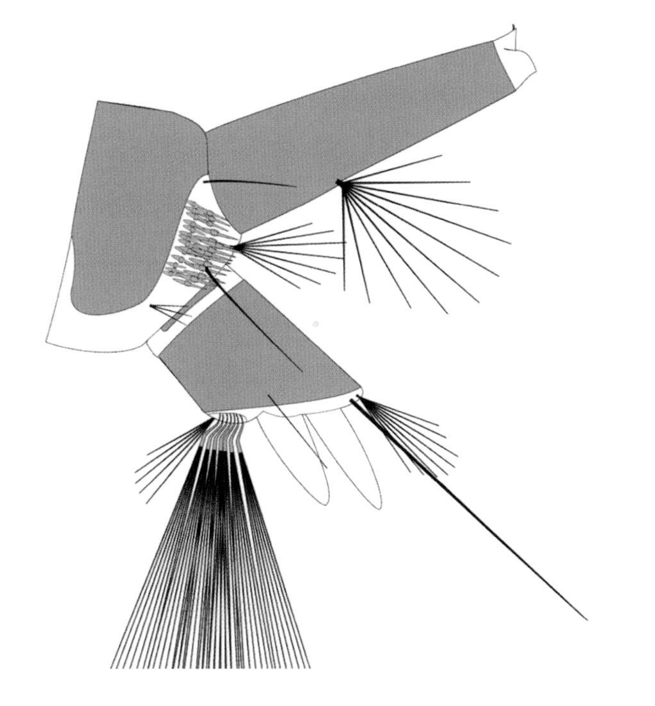

Terminal abdominal segments of *Or. signifera*

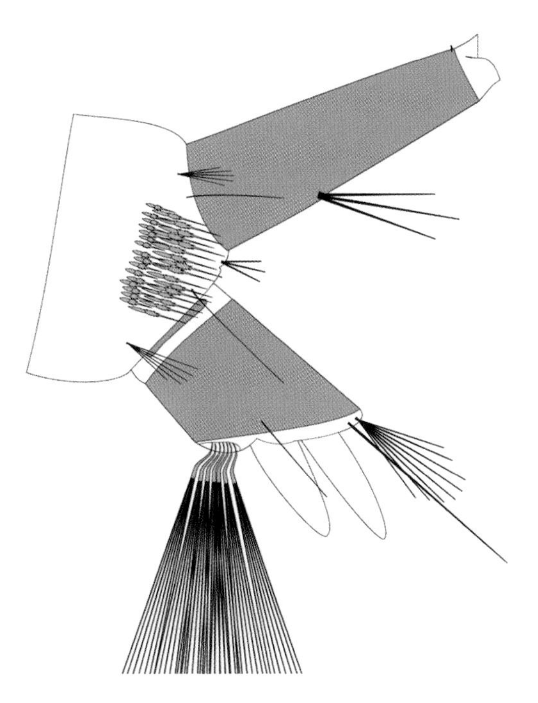

Terminal abdominal segments of *Or. alba*

Psorophora (Ps.)

1a Antennae small, barely reaching beyond head; pecten with 12 or more filamentous spines. Go to **2.**
1b Antennae long, reaching well beyond head; pecten with fewer than 10 spines. Go to **3.**

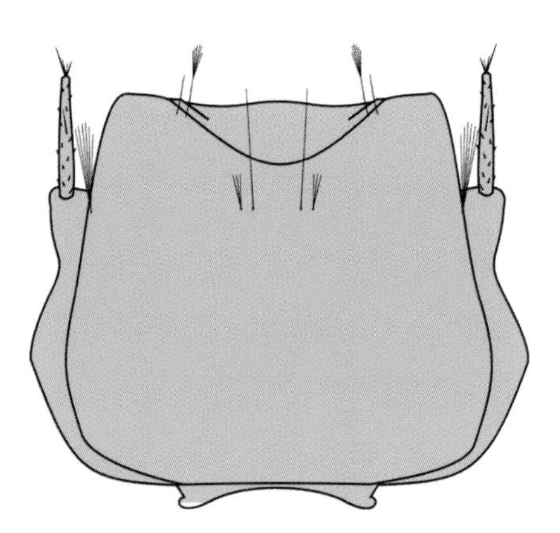

Head of *Ps. ciliata*

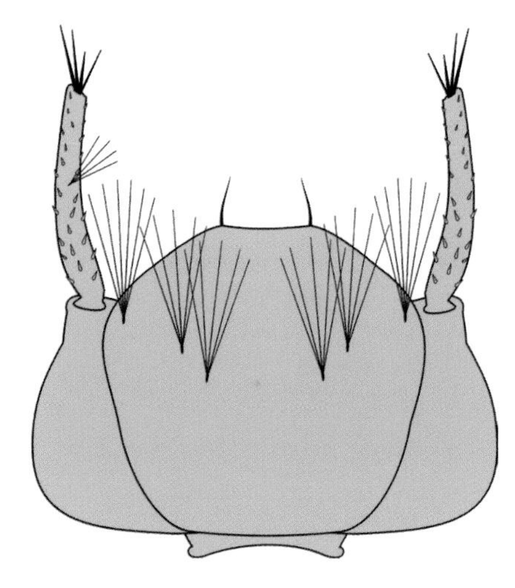

Head of *Ps. columbiae*

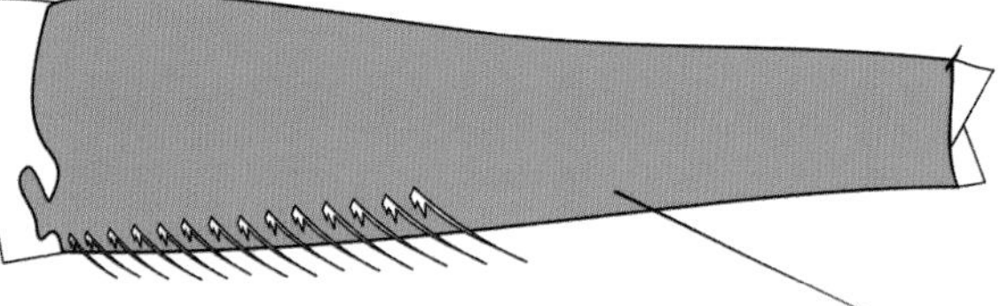
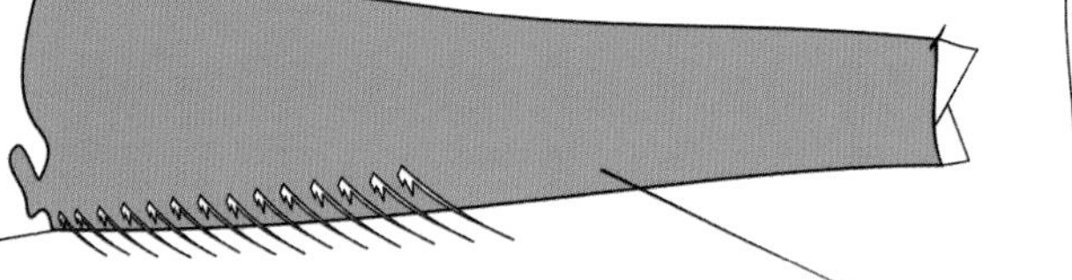

Siphon of *Ps. howardii*

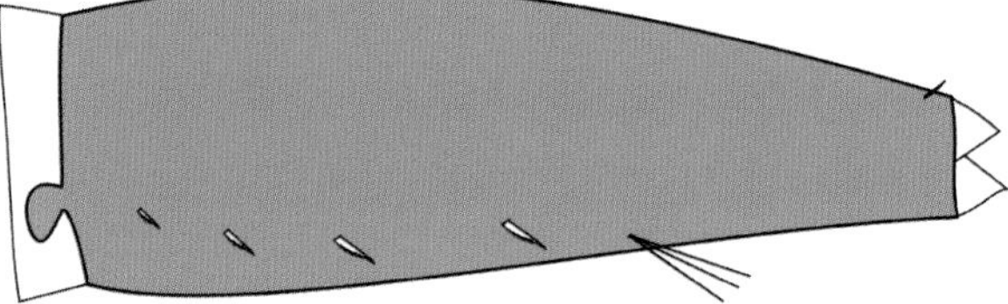

Siphon of *Ps. columbiae*

2a Lateral seta of saddle with 3–4 branches. ***Ps. ciliata*** (p. 161).
2b Lateral seta of saddle unbranched or branching near apex. ***Ps. howardii*** (p. 171).

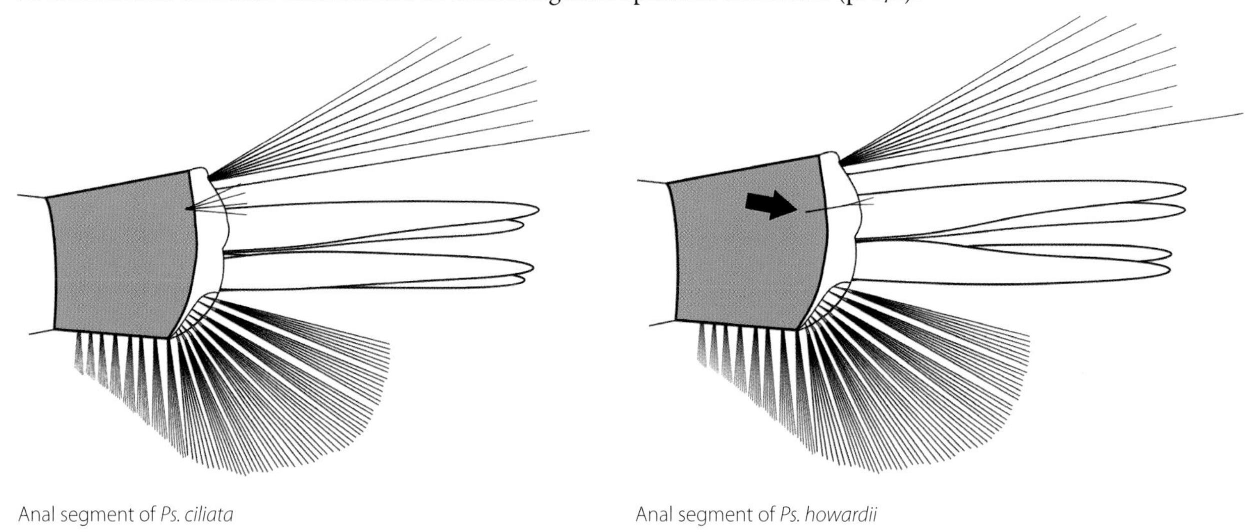

Anal segment of *Ps. ciliata*

Anal segment of *Ps. howardii*

3a Siphon small, with very large multibranched seta arising near tip. ***Ps. discolor*** (p. 167).
3b Siphon large, with small seta with 3–4 branches. Go to **4.**

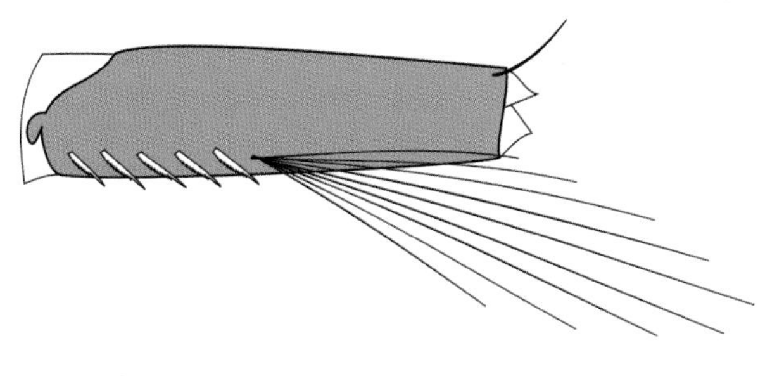

Siphon of *Ps. discolor*

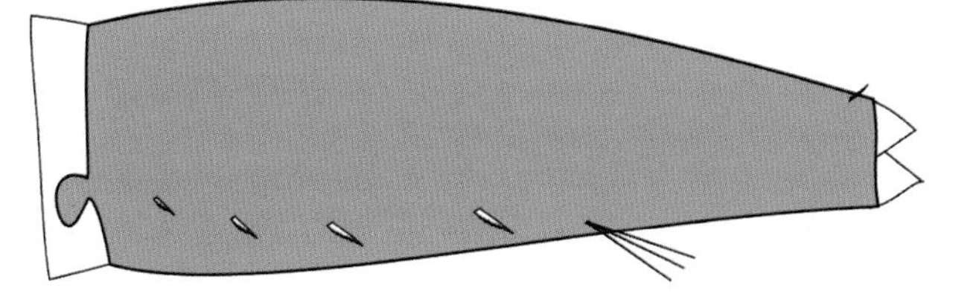

Siphon of *Ps. columbiae*

4a Antennae shorter than head. Go to **5.**
4b Antennae longer than or equal to head. Go to **6.**

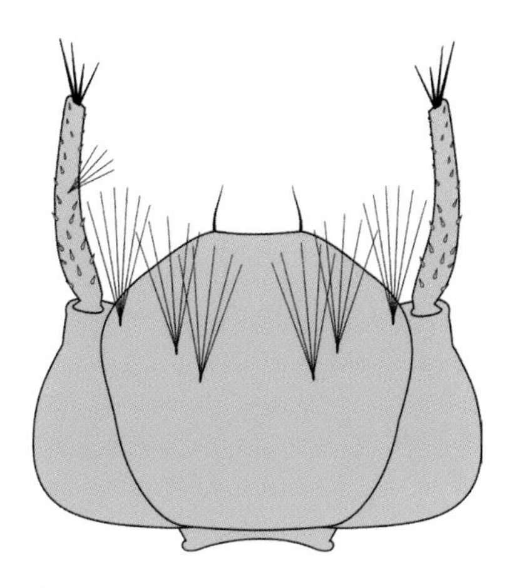

Head of *Ps. columbiae*

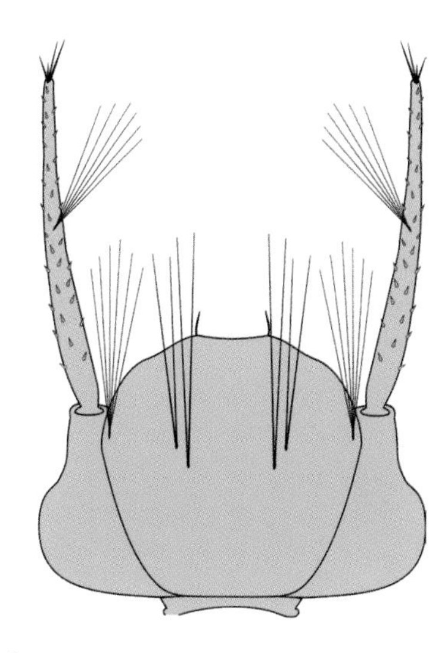

Head of *Ps. ferox*

5a Setae of head with 4 or more branches. ***Ps. columbiae*** (p. 163).
5b Setae of head unbranched or with 2 branches. ***Ps. cyanescens*** (p. 165).

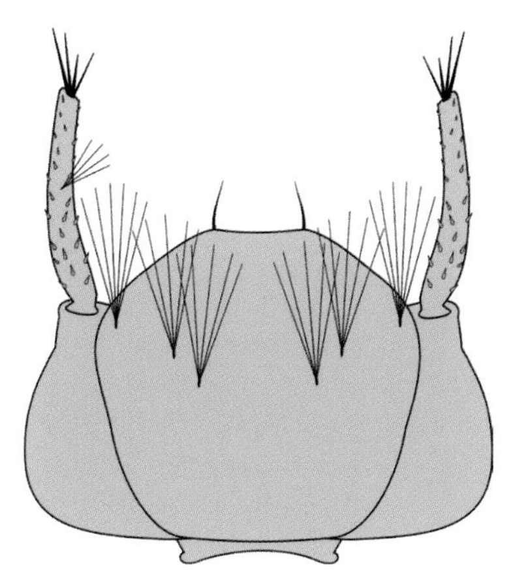

Head of *Ps. columbiae*

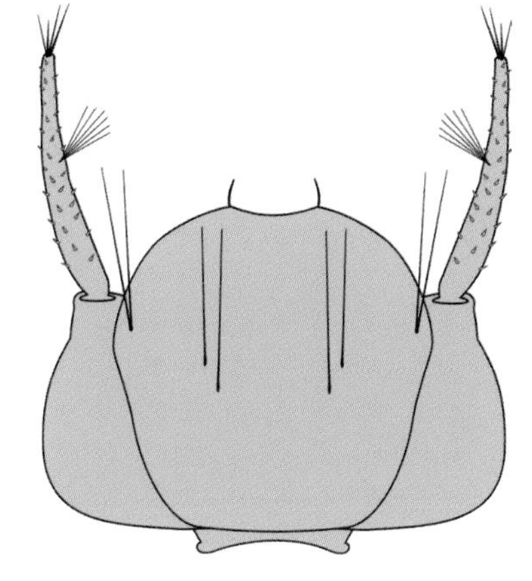

Head of *Ps. cyanescens*

6a Antennae distinctly longer than head. ***Ps. ferox*** (p. 168).
6b Antennae barely longer than or equal to head. Go to **7.**

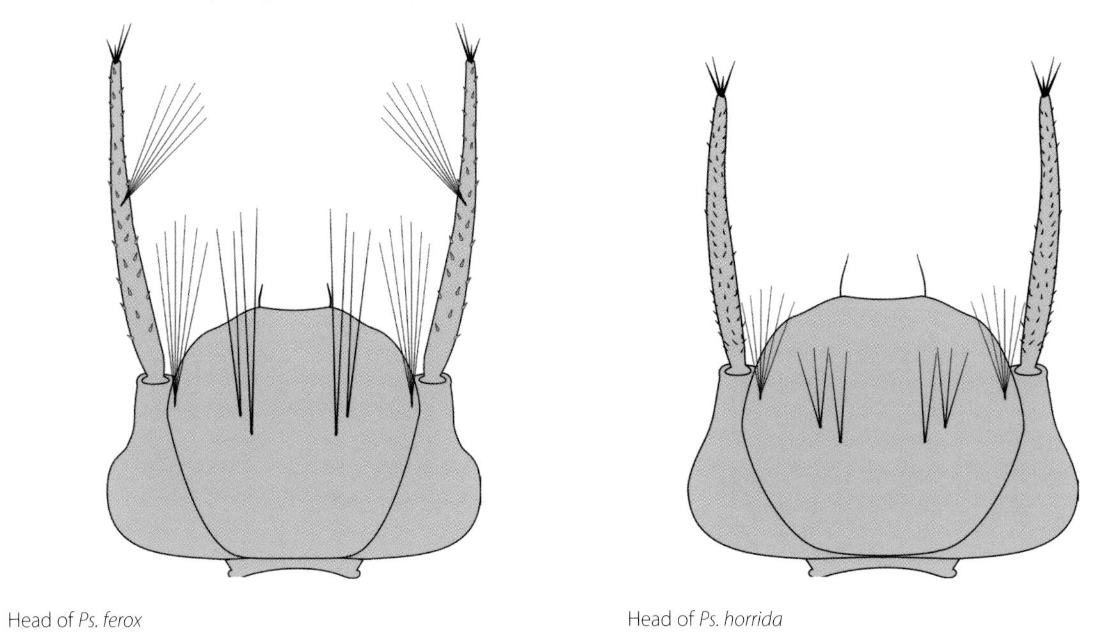

Head of *Ps. ferox* Head of *Ps. horrida*

7a Lateral setae of abdominal segments 4, 5, and 6 shorter than abdominal segments, and with 2–3 branches.
Ps. horrida (p. 170).
7b Lateral setae of abdominal segments 4, 5, and 6 longer than abdominal segments, and unbranched.
Ps. mathesoni (p. 173).

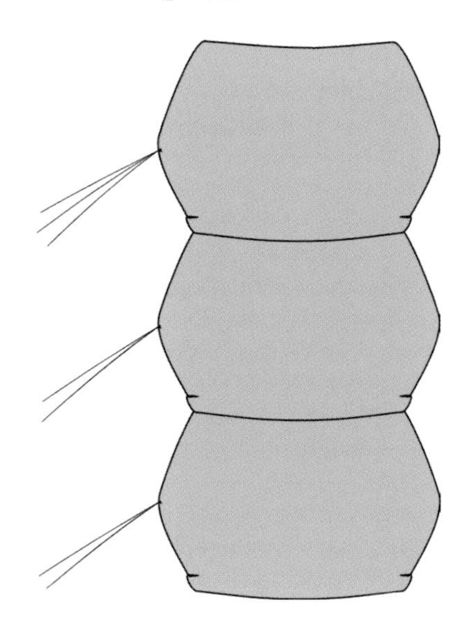

Abdominal segments 4–6 of *Ps. horrida*

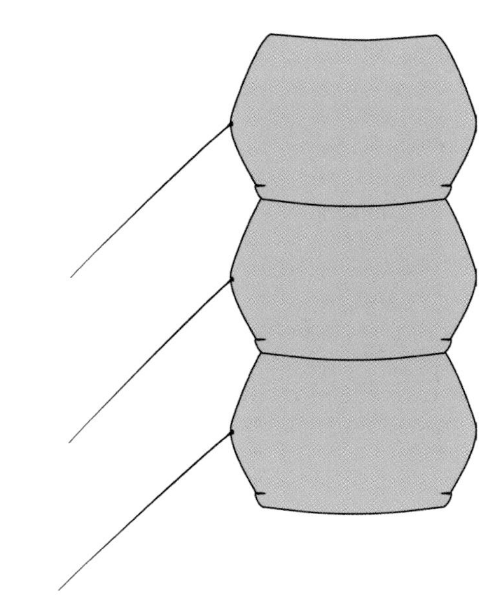

Abdominal segments 4–6 of *Ps. mathesoni*

Uranotaenia (Ur.)

1a Branched seta near anterior midline of thorax (seta 3-P) long, with 4–8 branches and more than half the length of the neighboring seta (seta 1-P). ***Ur. lowii*** (p. 177)

1b Branched seta near anterior midline of thorax (seta 3-P) short, with 8–10 branches and much less than half the length of the neighboring seta (seta 1-P). ***Ur. sapphirina*** (p. 179)

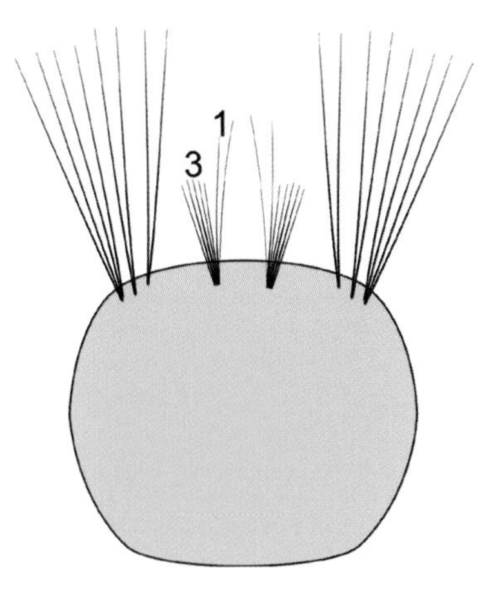

Thorax of *Ur. lowii*

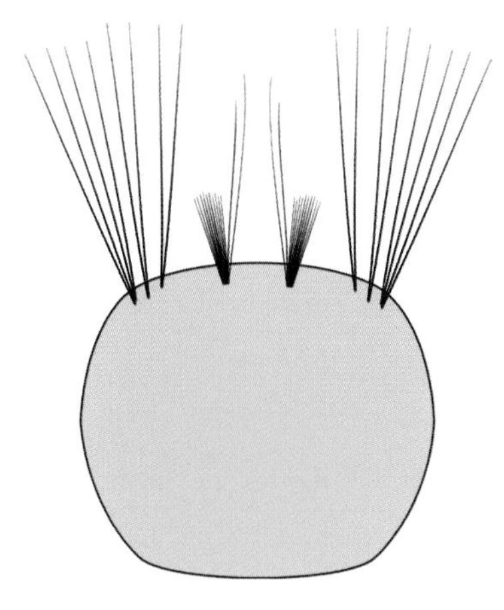

Thorax of *Ur. sapphirina*

Part Two

Species Accounts

The following section contains biological information, distribution maps, and images of the mosquito species that occur in Alabama, Georgia, Mississippi, North Carolina, South Carolina, and Tennessee. For each species a summary is provided of the physical description and habitats of larvae and adults, animals fed upon, and human and animal diseases associated with each species. The information provided here is by no means a complete account of all of the information available for each species but is a short summary of our basic biological knowledge for each mosquito. A wealth of detailed information is available for many of the mosquitoes of the southeastern United States, especially those of medical importance or common pest species.

A Warning against "Picture-Booking" to Identify Mosquitoes

Users of this book may be tempted to compare the photographs in the following section against their own unidentified specimens in order to circumvent using the key (part I). Most mosquito workers would probably consider this unwise. Many species appear quite similar and can easily be misidentified. Females of *Culiseta melanura*, for example, superficially resemble those of some *Culex* species and can sometimes confuse even experienced researchers. Rather than "picture-booking," the better option is to first use the key to make a positive identification and then confirm the identification with the photographs.

There are several mosquito species that have been reported from southeastern states in the past but are extremely rare or not likely to actually inhabit the area. Several other species are found in areas bordering the Southeast and thus have the potential to be encountered in southeastern states. Those species for which records are considered doubtful and those that occur in adjacent states have not been included in the species accounts in this book. For reference, they are briefly described here.

Aedes dorsalis is a northern and western species that has been reported from Kentucky, Louisiana, and Mississippi, although these reports are considered outside the normal range of this mosquito. The hind legs are largely pale scaled and the tarsi have pale basal and apical bands. The hind femur and tibia of adults are mostly pale scaled, with scattered dark scales. The proboscis and palps are dark, with scattered pale scales. The scutum has a broad dark longitudinal stripe bordered by pale-scaled areas. The wings have dark and pale scales intermixed, although pale scales usually predominate. The abdomen has extensive areas of pale scales, including a mid-dorsal pale-scaled stripe.

Aedes nigromaculis is primarily a western species that also occurs in eastern Louisiana and Arkansas; its range thus borders western portions of Mississippi and Tennessee and it could potentially be encountered in these states. Adults have hind tarsi with broad pale bands, and the proboscis has a pale ring. The wings are lightly speckled with pale scales. This species is easily confused with *Aedes sollicitans*, which has areas of pale yellowish scales on the abdomen, whereas *Aedes nigromaculis* has only white-scaled pale areas on the abdomen.

Aedes stimulans is a northern species that has also been collected from Mississippi (although not since 1920), far from its normally accepted range. Adults have hind tarsi with broad pale bands, and the proboscis and palps are speckled with pale scales. The abdominal sternites are usually pale scaled, with small patches of dark scales.

Aedes zoosophus is found in the south-central United States, including portions of eastern Louisiana and Arkansas; thus it has the potential to be collected in western portions of Mississippi. The basal half of the hind femora of adults is pale scaled, while the apical half is dark scaled. The proboscis and palps are dark scaled.

Uranotaenia anhydor syntheta is reported from southeastern Arkansas and thus has the potential to be encountered in northern Louisiana and western portions of Mississippi. The hind tarsi of adults are dark scaled, and the scutum lacks a median stripe of blue scales.

Mosquito Classification

The classification used here follows that of the Systematic Catalog of Culicidae, compiled and maintained by the Walter Reed Biosystematics Unit. The common names that follow the scientific names are a mixture of official common names (those approved by the Entomological Society of America), common names found in other mosquito literature, and names generated by the author.

The scientific name of each mosquito is followed by the name of the person(s) who first described the species in a scientific publication, and the year of that publication. An author name and year that appear in parentheses indicate that the genus has changed since the original description.

Aedes

A series of scientific papers published between 2000 and 2009 proposed splitting the genus *Aedes* into a number of "new" genera. This proposed reclassification was met with immediate criticism and has since become a matter of debate among mosquito taxonomists (the scientists that study the classification of mosquito species). The proposed reclassification would necessitate changing the scientific names of hundreds of mosquito species. Due to the confusion surrounding the reclassification, and the need for more research to fully resolve the issue, the *Journal of Medical Entomology*, a highly respected scientific journal, has encouraged authors to use the traditional (pre-reclassification) names. In this guide I follow this suggestion and treat *Aedes* as a single group. Note that the subgenus is included in the species accounts, under the species name.

Aedes is a very diverse group of mosquitoes, both biologically and morphologically. More than 900 species occur worldwide, 21 of which are found in our area. The adults range from small to large and feed on a wide variety of hosts (although most species feed primarily on mammals). The larvae occur in various temporary larval habitats, including rock pools, tree holes, man-made containers, and floodwater pools. However, most are not usually found in large permanent bodies of water such as open marshes, lakes, and ponds. In general, females lay single eggs (not in clusters) in places that will one day be flooded with water. Many species of *Aedes* are important in the transmission of human pathogens.

Aedes aegypti (Linnaeus, 1762)

Subgenus *Stegomyia*

Yellow fever mosquito

Adults Adults of *Aedes aegypti* are small to medium sized, with striking black-and-white patterns on body and legs. Their most distinguishing feature is the silver "lyre-shaped" pattern on the thorax. The abdomen is black, with horizontal white basal bands. The legs have broad white bands at the bases of the tarsal segments, and the last tarsal segment of the hind leg is solid white. The palps are also white at their tips, and the clypeus has a patch of silver scales. Wings have only dark scales. Adults will often rest on the walls of shaded buildings during daylight hours. Females feed mostly on mammals, and humans are one of the primary hosts. Hungry females will fly indoors in search of a blood meal and usually bite below the knee. Cats, dogs, and rodents are also commonly fed upon. *Aedes aegypti* females transmit the viruses of yellow fever, dengue, and chikungunya. Yellow fever was common in the southeastern United States and along the Atlan-

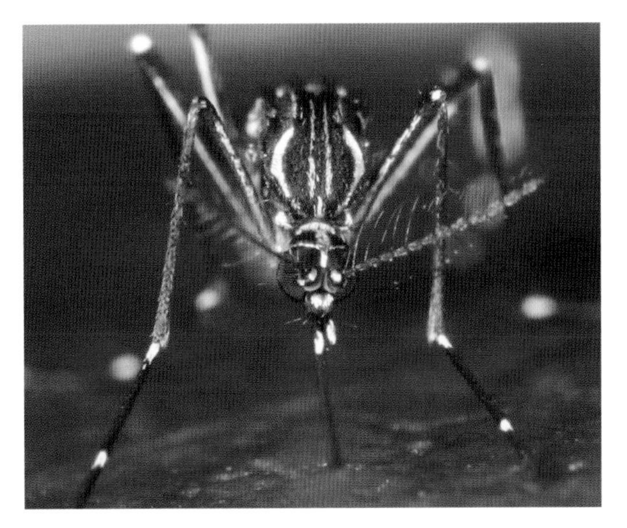

Aedes aegypti female

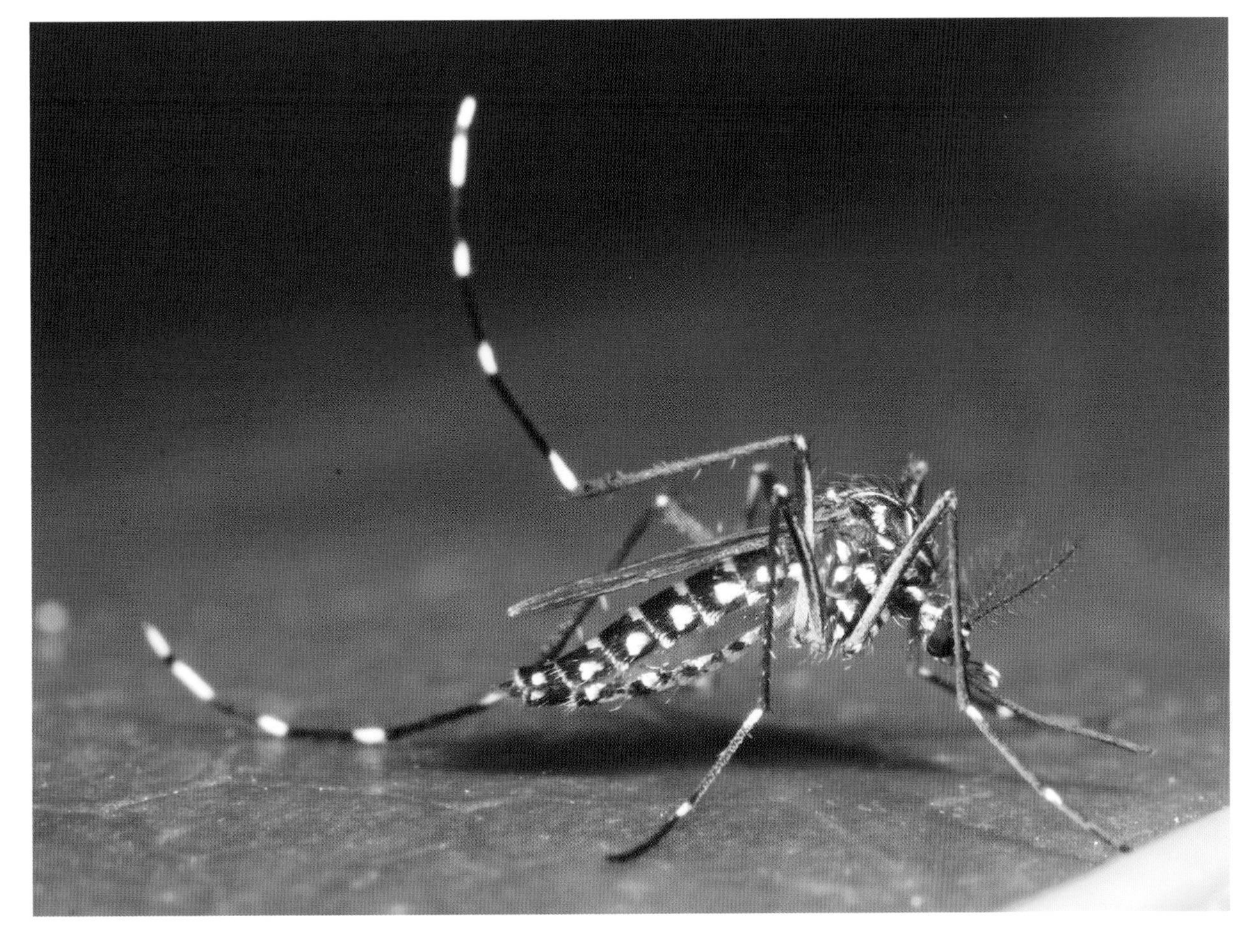

Aedes aegypti female

Aedes aegypti larva

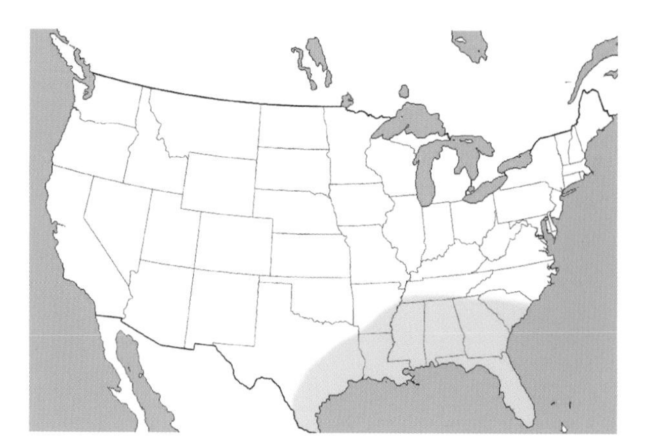

Distribution of *Aedes aegypti* in the U.S.

tic Coast in the 1800s and early 1900s. It claimed the lives of many Americans, until scientists discovered that mosquitoes transmitted the virus. This discovery, which occurred at the beginning of the twentieth century, prompted the initiation of mosquito control in the United States. While yellow fever has been eradicated from the United States, it persists in many tropical parts of Africa and the Americas.

Larvae Larvae of *Aedes aegypti* are medium sized. Setae of the head and antennae are mostly unbranched. The siphon is short and slightly tapered, with a pecten that reaches to its midpoint and a pair of large setae. The saddle nearly surrounds the anal segment but is "broken" along the ventral midline. The comb is a single curved row of 7–12 scales. Larvae are most commonly found in man-made objects that hold water and organic debris, especially fallen leaves. Discarded automobile tires are common breeding sites. Larvae are also occasionally encountered in natural containers, such as tree cavities that hold water.

Distribution *Aedes aegypti* is native to Africa but was accidentally brought to the Americas aboard ships traveling from Africa to the New World. This mosquito was once common and well established in much of the eastern United States. In the 1980s another mosquito, *Aedes albopictus,* was introduced into the United States and has since displaced *Aedes aegypti* throughout most of its range. Today, *Aedes aegypti* is found in southern Florida, the Rio Grande Valley of Texas, and occasionally elsewhere.

Aedes albopictus (Skuse 1894)

Subgenus *Stegomyia*

Asian tiger mosquito

Adults The Asian tiger mosquito is one species that is easy to recognize with the naked eye. Adults are small to medium sized, with striking black-and-white coloration on body and legs. This is the only mosquito species in the southeastern United States with a single bright white stripe down the center of the scutum and broad white bands on the hind legs. Wings and proboscis are dark scaled. *Aedes albopictus* is a very pestiferous daytime biter and will even enter homes in search of human blood. Females will bite in direct sunlight, an uncommon behavior for most mosquito species. This mosquito seems to prefer human blood but also feeds on dogs, cats, and a variety of small mammals, including chipmunks, squirrels, and other rodents. *Aedes albopictus* transmits the viruses causing dengue (also known as breakbone fever), chikungunya, and yellow fever in the tropics.

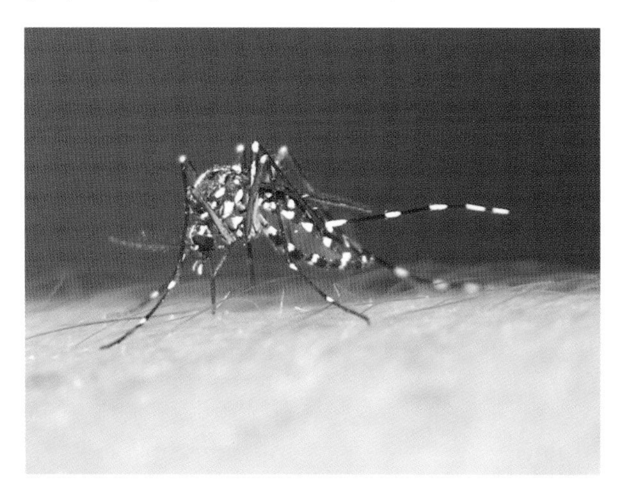

Aedes albopictus female

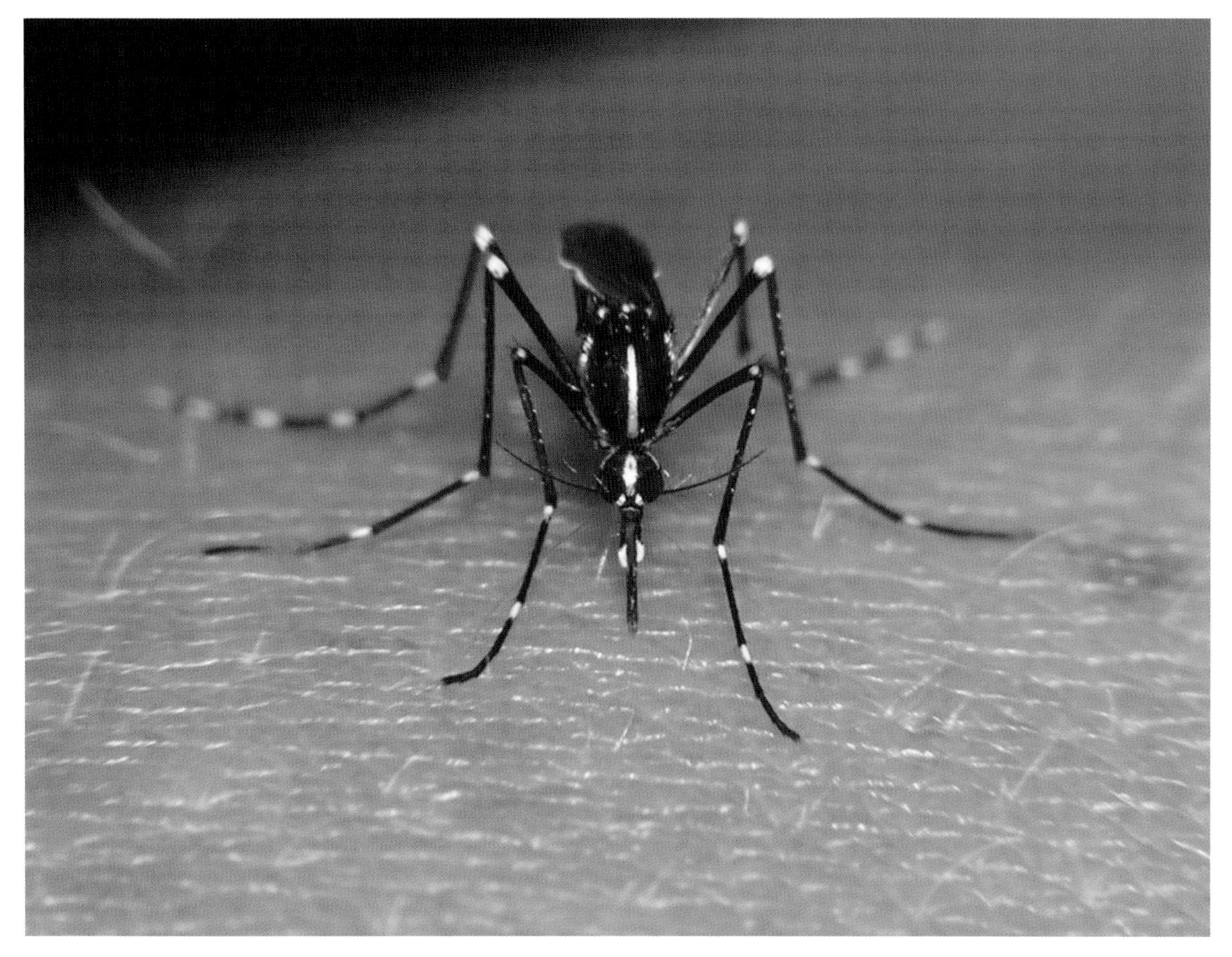

Aedes albopictus female

Aedes albopictus larva

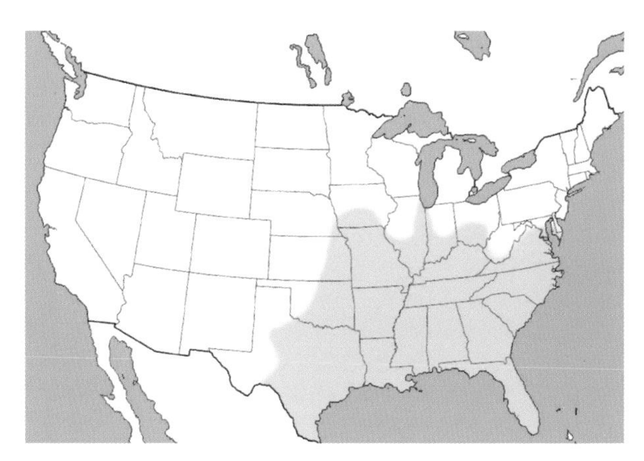

Distribution of *Aedes albopictus* in the U.S.

Larvae Larvae of *Aedes albopictus* are small to medium sized. Setae of the head and antennae are mostly unbranched. The siphon is short and appears slightly swollen near its midpoint. The pecten reaches to the midpoint of the siphon. The saddle nearly completely surrounds the anal segment but is "broken" along the ventral midline. The comb is a single curved row of 8–12 scales. *Aedes albopictus* is one of the mosquitoes most commonly found in water-filled man-made containers in the southeastern United States. Larvae are frequently encountered in discarded automobile tires, buckets, and neglected birdbaths. They are also sometimes found in small natural container-type habitats, such as water-holding tree holes and tank bromeliads.

Distribution *Aedes albopictus* is not native to North America but was accidentally introduced in the mid-1980s through the port at Houston, Texas, on a ship carrying used automobile tires from Hong Kong. Since its introduction, this species has spread throughout much of the eastern half of the United States, south of the Great Lakes.

Aedes atlanticus Dyar and Knab, 1906, and *Aedes tormentor* Dyar and Knab, 1906

Subgenus *Ochlerotatus*

Woodland floodwater mosquitoes

Aedes atlanticus and *Aedes tormentor* are almost indistinguishable as adults and have overlapping ranges and quite similar biologies. They are therefore treated together in this book. Larvae of the 2 species may be distinguished by differences in the arrangement of the comb scales and siphon setae.

Adults Females of *Aedes atlanticus* and *Aedes tormentor* are medium sized, with brown, white, and silvery-white coloration. The mostly brown scutum has a stripe of silvery-white scales down the center. The abdomen has lateral patches of white scales at the base of each segment. The tarsi and wings are dark scaled. Although adults of the 2 species are very similar, they can be distinguished by minute differences in the dorsal stripe of the scutum. The stripe tapers slightly in *Aedes tormentor* but does not in *Aedes atlanticus.* The adults of 2 other mosquito species in our area, *Aedes infirmatus* and *Aedes dupreei,* also have a silvery stripe on the scutum and are easily mistaken for adults of *Aedes atlanticus* and *Aedes tormentor.* Adults of *Aedes atlanticus* and *Aedes tormentor* reach their greatest abundance in mid to late summer. Females feed on a wide variety of hosts, including mammals, birds, reptiles, and amphibians. They can be quite pestiferous where they are abundant. Females normally seek out

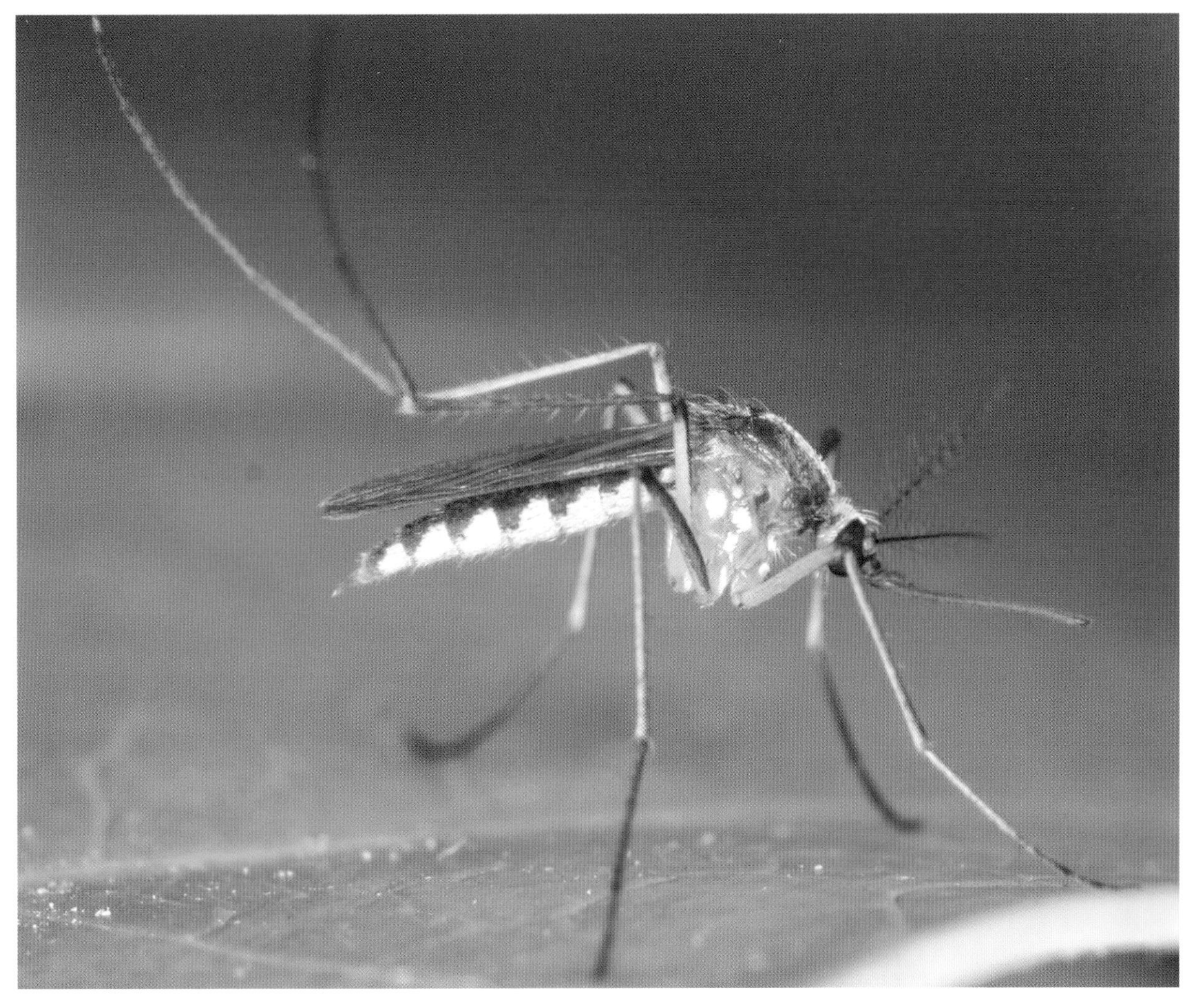

Aedes tormentor female

their hosts in the twilight hours but will bite in broad daylight if their resting sites are disturbed. Although adults of *Aedes atlanticus* and *Aedes tormentor* have been found infected with eastern equine encephalitis, La Crosse encephalitis, and West Nile viruses, they are not considered to be important in the transmission of any human pathogens.

Larvae Larvae of *Aedes atlanticus* and *Aedes tormentor* have a banded appearance due to areas of alternating dark and light pigmentation on the body. They are both medium sized when mature, and both have short, stout siphons. Setae of the head of both species are mostly unbranched, except for the large multibranched setae at the base of each antenna and the

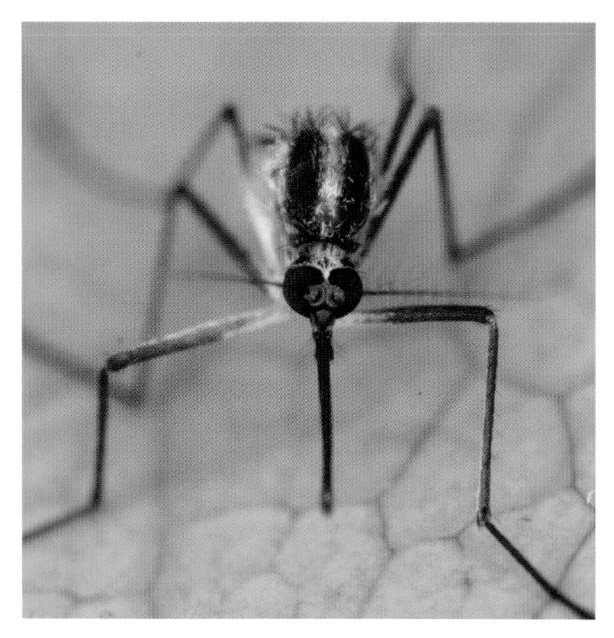

Aedes tormentor female

Aedes atlanticus larva

Aedes tormentor larva

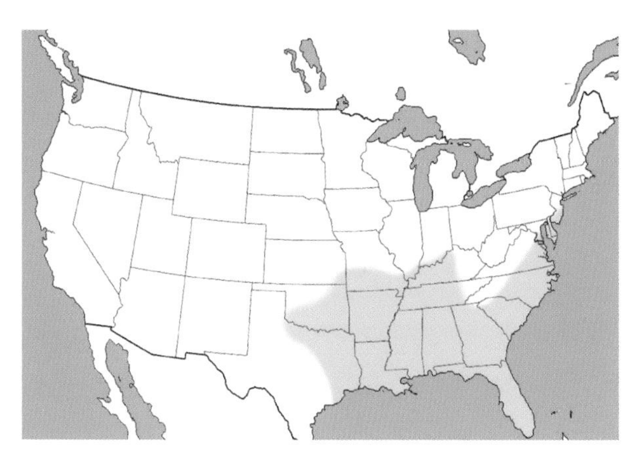

Distribution of *Aedes atlanticus* in the U.S.

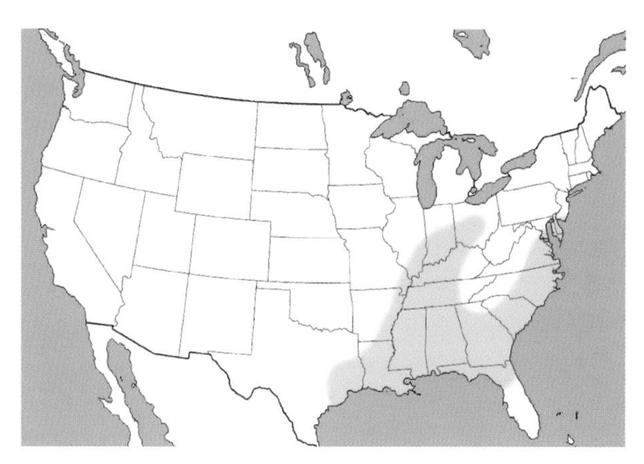

Distribution of *Aedes tormentor* in the U.S.

branched seta arising midway along each antenna. In both species the saddle completely surrounds the anal segment. The pecten of *Aedes atlanticus* larvae reaches to the midpoint of the siphon and does not surpass the siphon seta. The pecten of *Aedes tormentor* larvae reaches well past the siphon midpoint and surpasses the siphon seta. The comb of *Aedes atlanticus* is a single curved row of 4–6 scales. The comb of *Aedes tormentor* is also a single curved row, but of 9–13 smaller scales.

The head, anterior abdominal segments (1–6), and tenth abdominal segment of both species are all dark, while the thorax and posterior abdominal segments (7–8) are pale. Larvae of these 2 species are found in temporary pools in fields and woodlands.

Distribution *Aedes atlanticus* and *Aedes tormentor* have similar distributions, both occurring throughout the southeastern and east-central United States.

Aedes atropalpus (Coquillett, 1902)

Subgenus *Ochlerotatus*

Rock pool mosquito

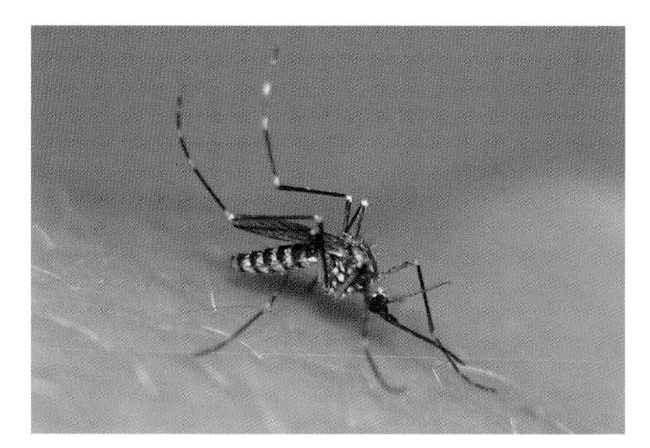

Aedes atropalpus female

Adults Adults of *Aedes atropalpus* are rather small, yet strikingly marked. Their dark bodies are ornamented with patches of golden and white scales. The hind legs have narrow pale bands at the base and the apex of each tarsal segment, a feature that is shared with only 1 other *Aedes* species in our area, *Aedes canadensis*. The dark brown scutum is bordered on each side by a patch of golden setae. The abdomen is dark, with basal patches of pale scales on the side of each segment. Wings have a patch of pale scales at the base of the costal vein. Females are autogenous, laying their first batch of eggs without taking a blood meal. After the first batch is laid, however, females

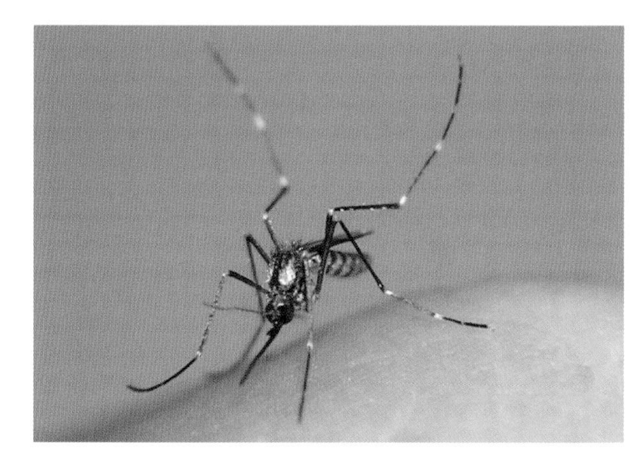

Aedes atropalpus female

Aedes atropalpus larva

Rock pool containing larvae of *Aedes atropalpus*

will take blood meals and can be pests in areas where they are abundant. *Aedes atropalpus* is not known to transmit any human pathogens.

Larvae Larvae of *Aedes atropalpus* are small. Setae of the head and antennae are mostly unbranched, but a few branched setae are present. The siphon is short and broad at its base. The pecten reaches nearly to the tip of the siphon. The teeth of the pecten that are nearer to the tip are spaced farther apart than those near the base of the siphon. The siphon seta arises from within the pecten. The saddle covers only the dorsal half of the anal segment. The comb is a large patch of many (20–60) small scales. As the common name implies, larvae of the rock pool mosquito are found in water-filled cavities of exposed rock outcroppings and stream bedrock. They may also occasionally be found in discarded automobile tires.

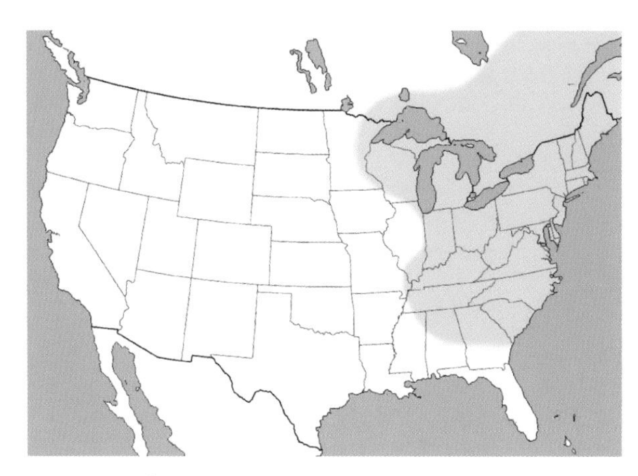

Distribution of *Aedes atropalpus* in the U.S.

Distribution *Aedes atropalpus* is found throughout the eastern half of the United States and southern Canada but is generally restricted to areas bordering waterways with rock pools.

Aedes aurifer (Coquillett, 1903)

Subgenus *Ochlerotatus*

Golden bog mosquito

Adults *Aedes aurifer* is a medium-sized, dark mosquito with patches of golden and white scales on the thorax and abdomen. The palps, proboscis, and tarsi are quite dark and lack pale scales. The dorsal thorax has a median stripe of dark, purplish-black scales, bordered on each side by areas of pale golden scales. Four other species in our area (*Aedes thibaulti, Aedes sticticus, Aedes triseriatus,* and *Aedes hendersoni*) also have a dark median stripe on the thorax bordered by areas of pale scales. These species can be distinguished by subtle differences in the coloration of the dorsal thorax and by patterns of dark and pale scales on the abdomen. Females feed mainly on mammals, including humans. They are most abundant in spring and can be quite bothersome in their boggy habitats during periods of peak abundance. Females do not normally fly far from their breeding habitat. *Aedes aurifer* is not known to be important in the transmission of any human pathogens.

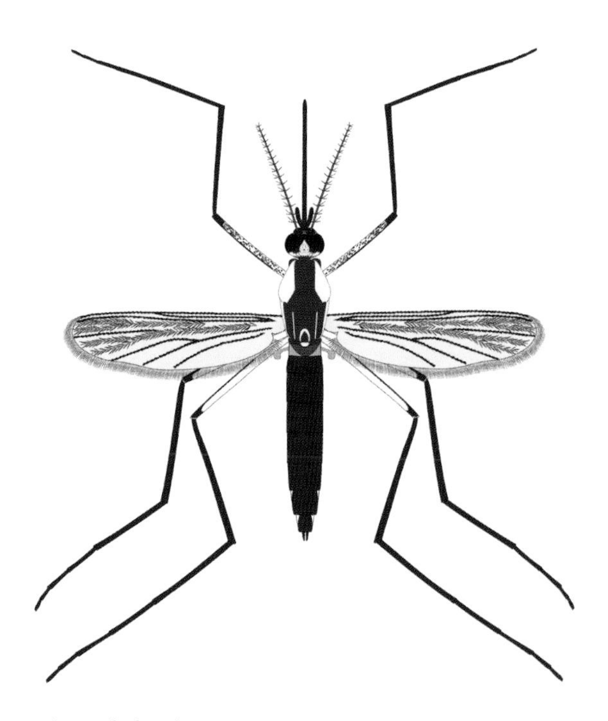

Aedes aurifer female

Larvae Larvae of *Aedes aurifer* are medium sized. The median setae of the head are 2-branched, while those near the base of the antennae and the antennal tuft are many branched. The siphon is rather elongate, with a pecten that reaches to its midpoint and a large multibranched seta. The comb is a patch of 20–30 small scales. The saddle extends downward to near the ventral midline of the anal segment and has many spicules (tiny spines) on its dorsal apical surface. Larvae are found most commonly in bogs and marshes, usually near floating or emergent vegetation over somewhat deep water. They are usually widely dispersed in their larval habitat.

Distribution *Aedes aurifer* is most common in the northeastern United States but is also found as far north as southeastern Canada and as far south as North Carolina.

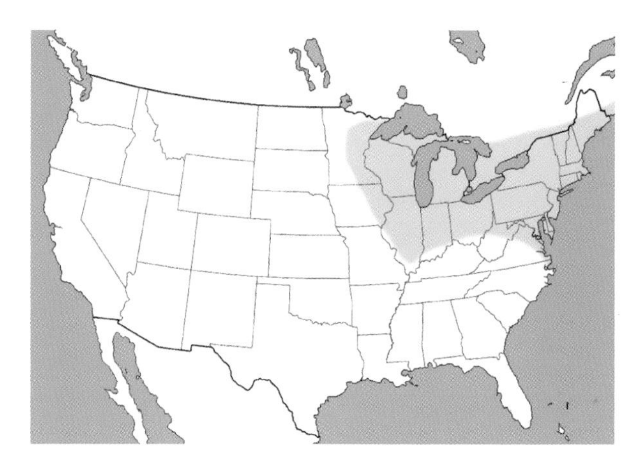

Distribution of *Aedes aurifer* in the U.S.

Aedes canadensis (Theobald, 1901)

Subgenus *Ochlerotatus*

Two subspecies of *Aedes canadensis* are found in our area, *Aedes canadensis canadensis* and *Aedes canadensis mathesoni.* Refer to the identification key (p. 25) for characters to distinguish between the subspecies.

Adults *Aedes canadensis* females are medium sized, with a brown scutum. The dark brown abdomen is conspicuously pointed at its apex. The scutum and abdomen have patches of white scales laterally. The hind legs of adults have pale bands on the basal and apical portions of each tarsal segment. Wings and proboscis have dark scales. Although females will feed from a variety of hosts (including humans), box turtles seem to be preferred hosts. Females can commonly be observed buzzing about a box turtle's head in search of a feeding site. After digesting blood meals, females will lay their eggs in depressions in the ground that will fill with water the following winter. Until that time the eggs will remain dormant in the soil. *Aedes canadensis* has been incriminated as a vector of La Crosse encephalitis virus, eastern equine encephalitis virus, and Jamestown Canyon virus, all of which are human pathogens.

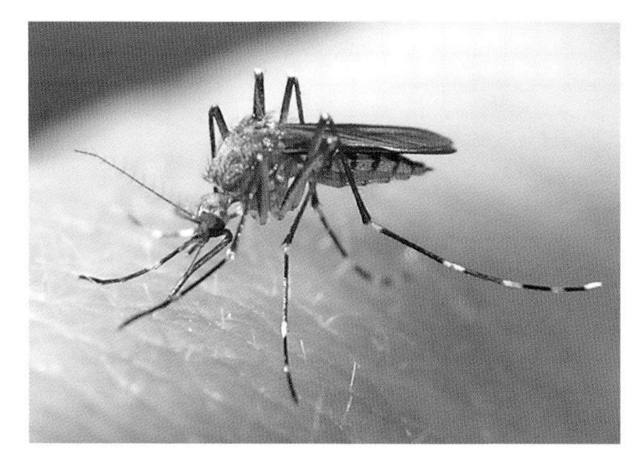

Aedes canadensis female

Aedes canadensis female feeding on a box turtle

Aedes canadensis larva

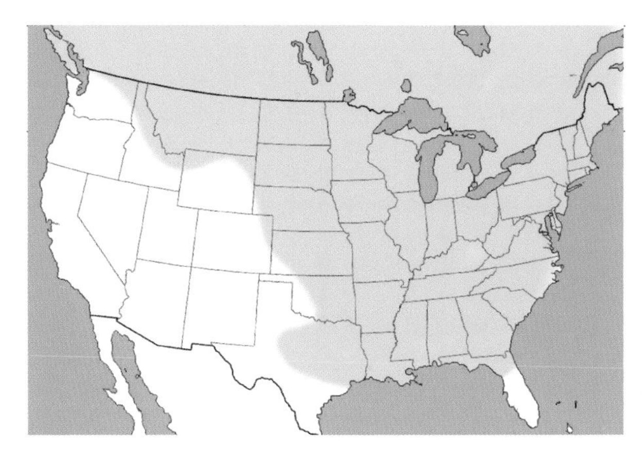

Distribution of *Aedes canadensis canadensis* in the U.S.

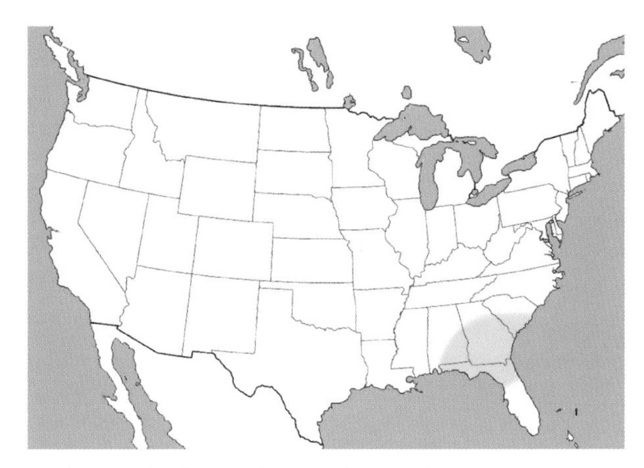

Distribution of *Aedes canadensis mathesoni* in the U.S.

Larvae Larvae of *Aedes canadensis* are medium sized. Most setae of the head and antennae are densely branched. The siphon is slightly elongate, with a pecten that reaches almost to its midpoint. The saddle covers only the dorsal half of the anal segment. The comb is a large patch of 30–60 small scales. Larvae are usually encountered in vernal pools and water-filled puddles created by uprooted trees. They can be tremendously abundant in these ephemeral habitats and are an important food source for young salamanders and predaceous aquatic insects, especially in early spring.

Distribution *Aedes canadensis canadensis* is widely distributed across the eastern half of the United States and southern Canada. *Aedes canadensis mathesoni* is restricted to parts of Alabama, Florida, Georgia, and South Carolina.

Aedes cantator (Coquillett, 1903)

Subgenus *Ochlerotatus*

Brown salt marsh mosquito

Adults *Aedes cantator* females are medium sized, with brown and white markings. The scutum is reddish brown and the proboscis and palps are dark scaled. The abdomen has pale basal bands on each segment, which are narrowest near their midpoint. Wings have only dark scales. The hind tarsi have narrow pale bands on each segment. *Aedes vexans,* a species that is much more common in the southeastern United States, shares many features with *Aedes cantator.* While both species have pale basal markings on the abdomen, *Aedes cantator* has broad basal bands. *Aedes vexans,* on the other hand, has 2 pale basal chevrons on each segment. *Aedes cantator* females are quite pestiferous biters, especially in the springtime. Although they bite mostly at dusk, they will also bite during the day when the vegetation where they rest is disturbed. Females feed mainly on mammals but also on songbirds. *Aedes cantator* is thought to be involved in the transmission of eastern equine encephalitis virus in the northeastern United States.

Aedes cantator female. Courtesy of Bo Zaremba.

Aedes cantator female. Courtesy of Bo Zaremba.

Larvae Larvae of *Aedes cantator* are medium sized. Setae of the head are mostly branched, often densely so. The siphon is slightly elongate. The pecten reaches to the midpoint of the siphon and terminates just before the multibranched siphonal seta. The saddle covers the dorsal two-thirds of the anal segment. The comb is a large patch of many small scales. Larvae of *Aedes cantator* are associated with coastal marshes. They are most commonly found in lower-salinity pools bordering upland areas of salt marshes but may also be found in pools of brackish water within the marsh.

Distribution *Aedes cantator* is primarily a northern Atlantic Coast species. It has also been reported from Kentucky, North Carolina, and Virginia.

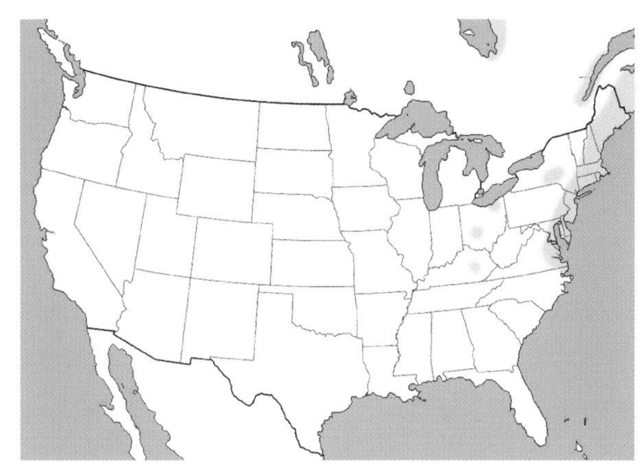

Distribution of *Aedes cantator* in the U.S.

Aedes cinereus Meigen, 1818

Subgenus *Aedes*

Minute floodwater mosquito
or small woodland mosquito

Adults Adults of *Aedes cinereus* are quite small and generally lack distinguishing features. The body is mostly brown, with pale areas on the thorax and abdomen. The lateral patches of pale scales on the abdomen often coalesce to form a continuous broad pale lateral stripe. The underside of the abdomen is entirely pale. The tarsi are solid brown and lack pale bands. *Aedes cinereus* reaches peak abundance in mid to late summer. The palps are very small in both males and females. Females feed mostly on mammals, including deer, raccoons, rabbits, opossums, foxes, and a number of rodents. This mosquito will also bite humans and can be a pest where it is abundant. *Aedes cinereus* is not known to be important in the transmission of any human pathogens.

Larvae Larvae of *Aedes cinereus* are usually small. Setae of the head and antennae are many branched. The siphon is rather elongate and has a small seta just below the dorsal midline, a feature that distinguishes larvae of this species from all other *Aedes* species in our area. However, this seta is exceptionally small and

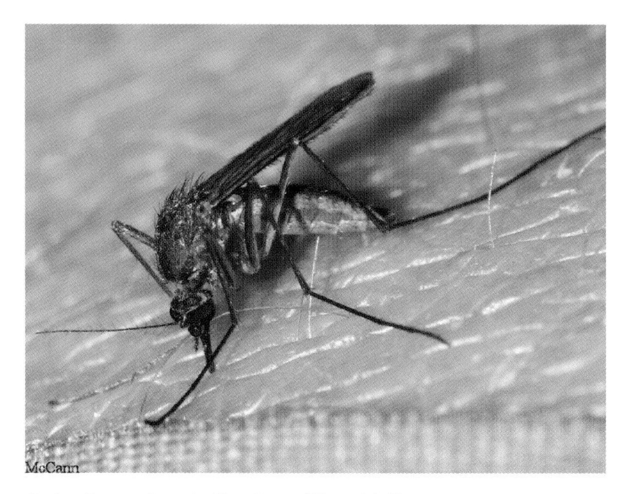

Aedes cinereus female. Courtesy of Sean McCann.

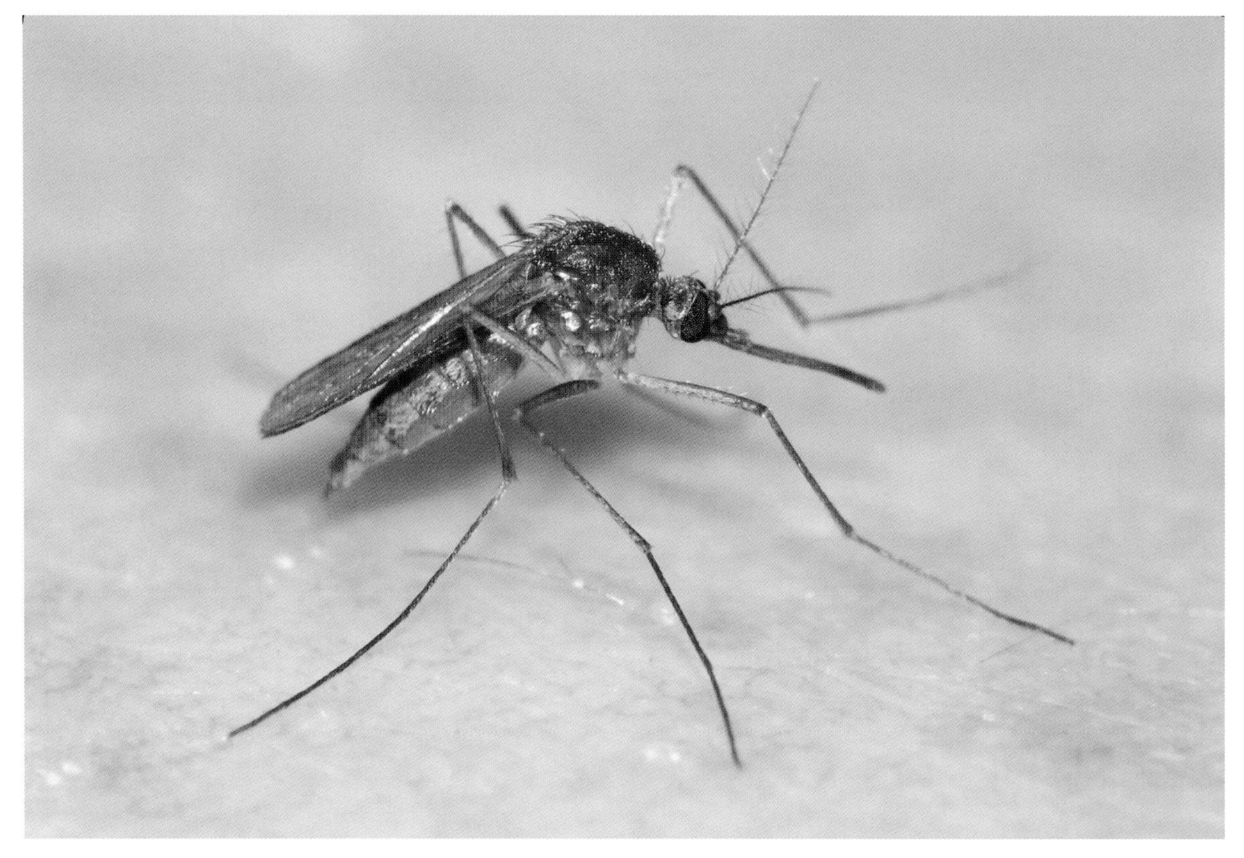

Aedes cinereus female. Courtesy of Sean McCann.

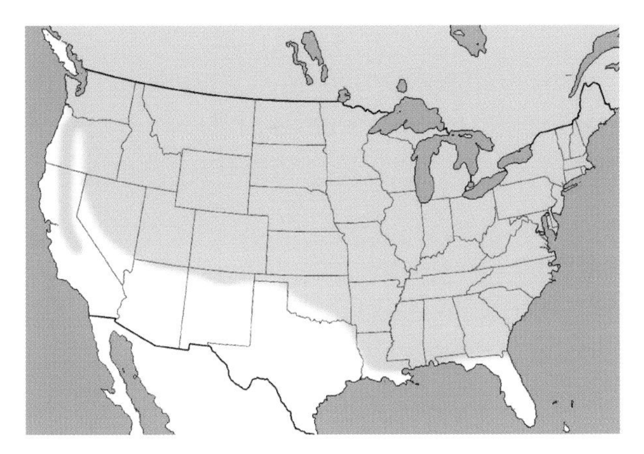

Distribution of *Aedes cinereus* in the U.S.

cannot be seen without high magnification. The pecten extends beyond the midpoint of the siphon, and the 2 apical teeth of the pecten are spaced far apart from the other teeth and from one another. The saddle covers the dorsal two-thirds of the anal segment. The comb is an irregular double row of small scales (about 10–15). Larvae are most commonly encountered in rainwater or snowmelt pools, in both shaded woodlands and sunny locations.

Distribution *Aedes cinereus* is widely distributed throughout much of the United States and Canada. This mosquito is much more common in the northern states than it is in the south.

Aedes dupreei (Coquillett, 1904)

Subgenus *Ochlerotatus*

Dupree's floodwater mosquito

Adults Adults of *Aedes dupreei* are small, with brown and silvery-white coloration. The scutum has a stripe of silver scales down the center, with brown scales laterally. The abdomen is mostly dark, with lateral patches of silvery-white scales at the base of each segment. Tarsi, wings, and proboscis are dark. Adults of 3 other species in our area (*Aedes atlanticus, Aedes tormentor,* and *Aedes infirmatus*) resemble adults of *Aedes dupreei.* Adults of *Aedes dupreei* usually reach their greatest abundance in mid to late summer (August and September). Little is known about the feeding habits of the females, although they are thought to feed mostly on songbirds. *Aedes dupreei* is not known to be important in the transmission of any human pathogens.

Larvae Larvae of *Aedes dupreei* are small and easily recognized by their extremely long and dark anal papillae. The siphon is somewhat short and narrow, with

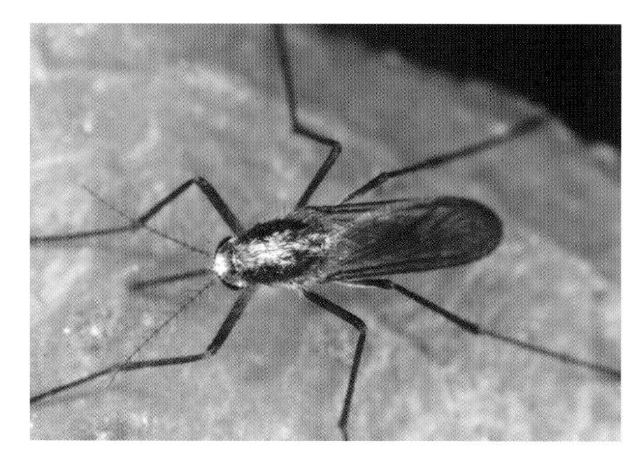

Aedes dupreei female

Aedes dupreei female

Aedes dupreei larvae

the pecten not quite reaching to its midpoint. Setae of the head and antennae are small, with 1–4 branches. The comb is a single curved row of 7–10 scales. The saddle completely surrounds the anal segment. *Aedes dupreei* larvae are most commonly found in summer rainwater pools.

Distribution *Aedes dupreei* is found throughout the eastern half of the United States, south of the Great Lakes.

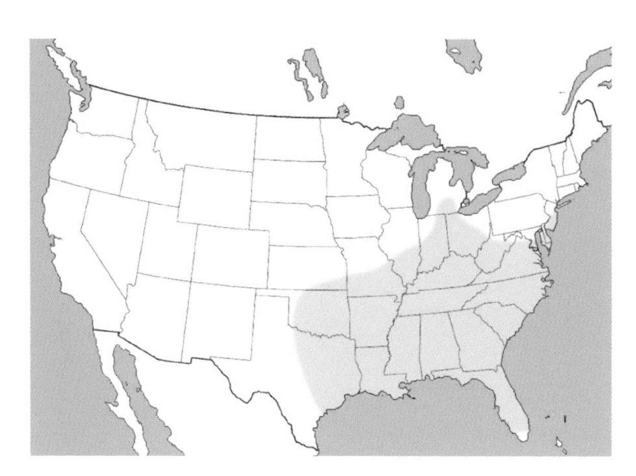

Distribution of *Aedes dupreei* in the U.S.

Aedes fulvus pallens Ross, 1943

Subgenus *Ochlerotatus*

Eastern yellow and black mosquito

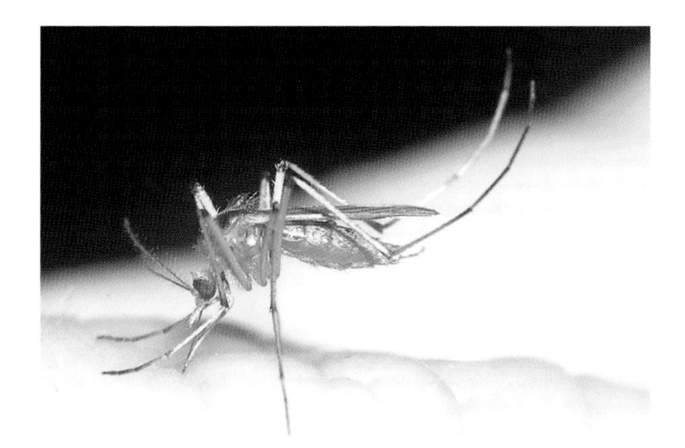

Aedes fulvus pallens female

Adults *Aedes fulvus pallens* is one of the most striking mosquitoes of the southeastern United States and is unlikely to be confused with any other mosquito in our area. Adults are large and bright yellow and black. The legs, proboscis, and palps are yellow with black tips. Two large dark spots covered with black scales adorn the otherwise yellow scutum. Wings are dark scaled. Females will readily bite humans; however, they are rarely encountered in large numbers. They feed mostly on mammals in the wild, including rabbits, deer, armadillos, and rodents. *Aedes fulvus pallens* is not known to transmit any human pathogens.

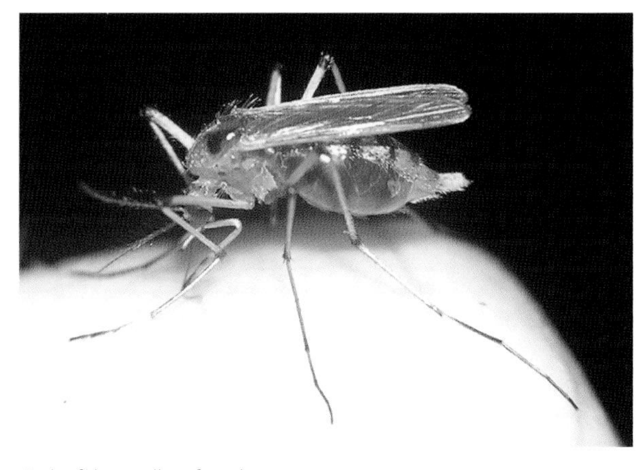

Aedes fulvus pallens female

Larvae Larvae of *Aedes fulvus pallens* are quite large when mature. Setae of the head are variable in size and branching. The siphon is short and stout. The large, branched seta of the siphon arises within the pecten, which has 1–2 teeth separated from the other teeth, and reaches nearly to the tip of the siphon. The saddle surrounds the anal segment. The comb is a large patch of 25 or more scales. Larvae of this species are typically rare; they are found in temporary rainwater depressions along with larvae of *Aedes atlanticus* and *Aedes tormentor.*

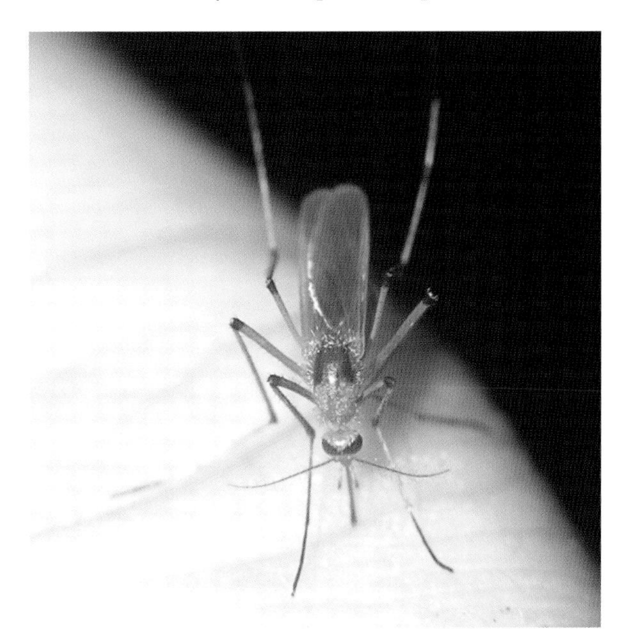

Aedes fulvus pallens female

Distribution *Aedes fulvus pallens* is relatively rare but may be found throughout the southeastern United States, excluding the Appalachian Mountains.

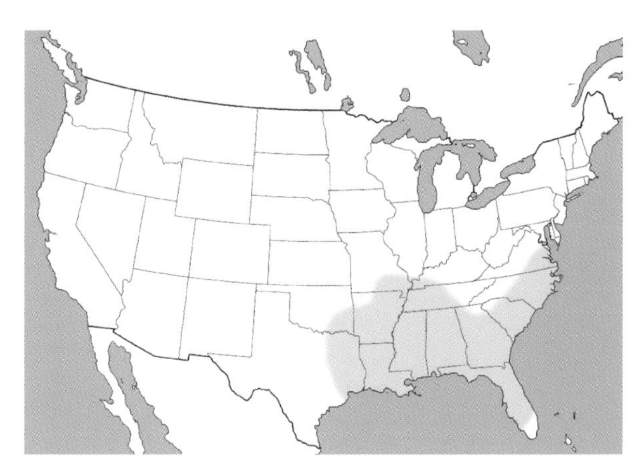

Distribution of *Aedes fulvus pallens* in the U.S.

Aedes grossbecki Dyar and Knab, 1906

Subgenus *Ochlerotatus*

Grossbeck's speckled mosquito

Adults Adults of *Aedes grossbecki* are medium sized to large. The exoskeleton is dark brown and is clad in a mixture of brown and white scales. The legs have broad bands of white scales at the base of each tarsal segment. One of the most distinguishing features is the broad triangular brown and white scales found on the wings. The brown proboscis usually has a few white scales near the base. The scutum has a stripe of dark scales down the dorsal midline that is bordered by patches of white scales laterally. The abdomen is dark brown, with white bands at the base of each segment. Little is known about the feeding habits of *Aedes grossbecki*. Females are thought to feed mostly on mammals and are known to bite humans. Adults are normally encountered only in the late winter and

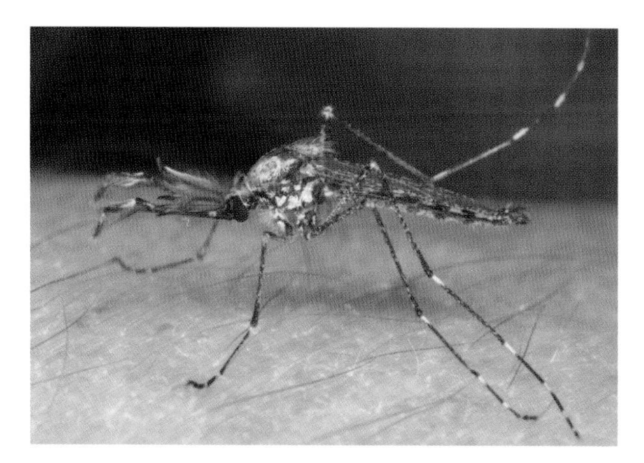

Aedes grossbecki male

Aedes grossbecki female

Aedes grossbecki larva

early spring, prior to the yearly surveillance activities of mosquito control programs. *Aedes grossbecki* is not known to transmit any human pathogens.

Larvae Larvae of *Aedes grossbecki* are large and often dark. Setae of the head are mostly branched. The siphon is somewhat elongate, with a pecten extending about a third of its length. The saddle covers slightly more than half of the dorsum of the anal segment and has numerous small teeth along the apical edge, giving it a rough appearance when viewed from the side. The comb is a patch of 25–30 scales. Larvae are most often encountered in springtime floodwater pools in lowland hardwood forests. Fully grown larvae of *Aedes grossbecki* are larger than those of other species that inhabit the same habitat (*Aedes canadensis* and *Aedes sticticus*).

Distribution *Aedes grossbecki* is found throughout the eastern half of the United States, south of the Great

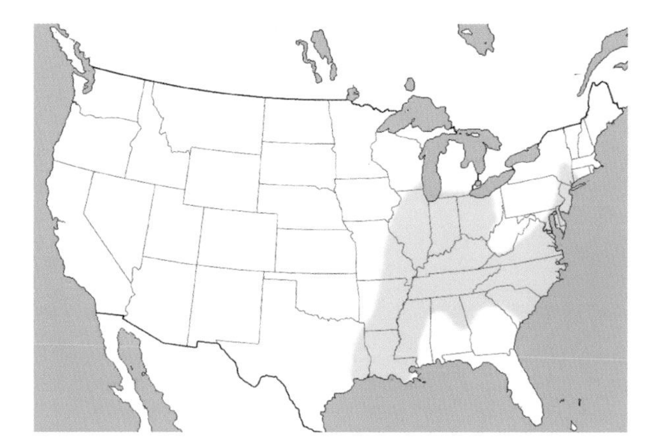

Distribution of *Aedes grossbecki* in the U.S.

Lakes, excluding much of Alabama, Georgia, and Florida. The author has collected females and larvae of *Aedes grossbecki* in Tuskegee National Forest in east-central Alabama, outside the normally reported range for this species.

Aedes infirmatus Dyar and Knab, 1906

Subgenus *Ochlerotatus*

Silverback mosquito

Adults Adults of *Aedes infirmatus* are medium sized, with brown and silvery-white coloration. They have a broad stripe of silver-white (or bright yellowish) scales down the center of the otherwise brown scutum. The abdomen has lateral patches of silvery-white scales at the base of each segment. Tarsi and wings are dark. Adults of 3 other species in our area (*Aedes dupreei, Aedes atlanticus,* and *Aedes tormentor*) are similar to those of *Aedes infirmatus.* All 4 species have a pale median stripe on the scutum, but only females of *Aedes infirmatus* have a stripe that is much broader than the lateral brown areas and does not reach the posterior border of the scutum. Adults of *Aedes dupreei, Aedes atlanticus,* and *Aedes tormentor* have a median stripe that is more or less equal in width to the darker lateral areas of the scutum and reaches the posterior border of the scutum. Adults of *Aedes infirmatus* may be encountered from spring through summer. Females feed on the blood of mammals, including rabbits, armadillos, deer, and humans. They are persistent biters, seeking out their hosts during daylight hours, especially in wooded areas. *Aedes infirmatus* is not known to transmit any human pathogens, although it has been found infected with eastern equine encephalitis virus and trivittatus virus.

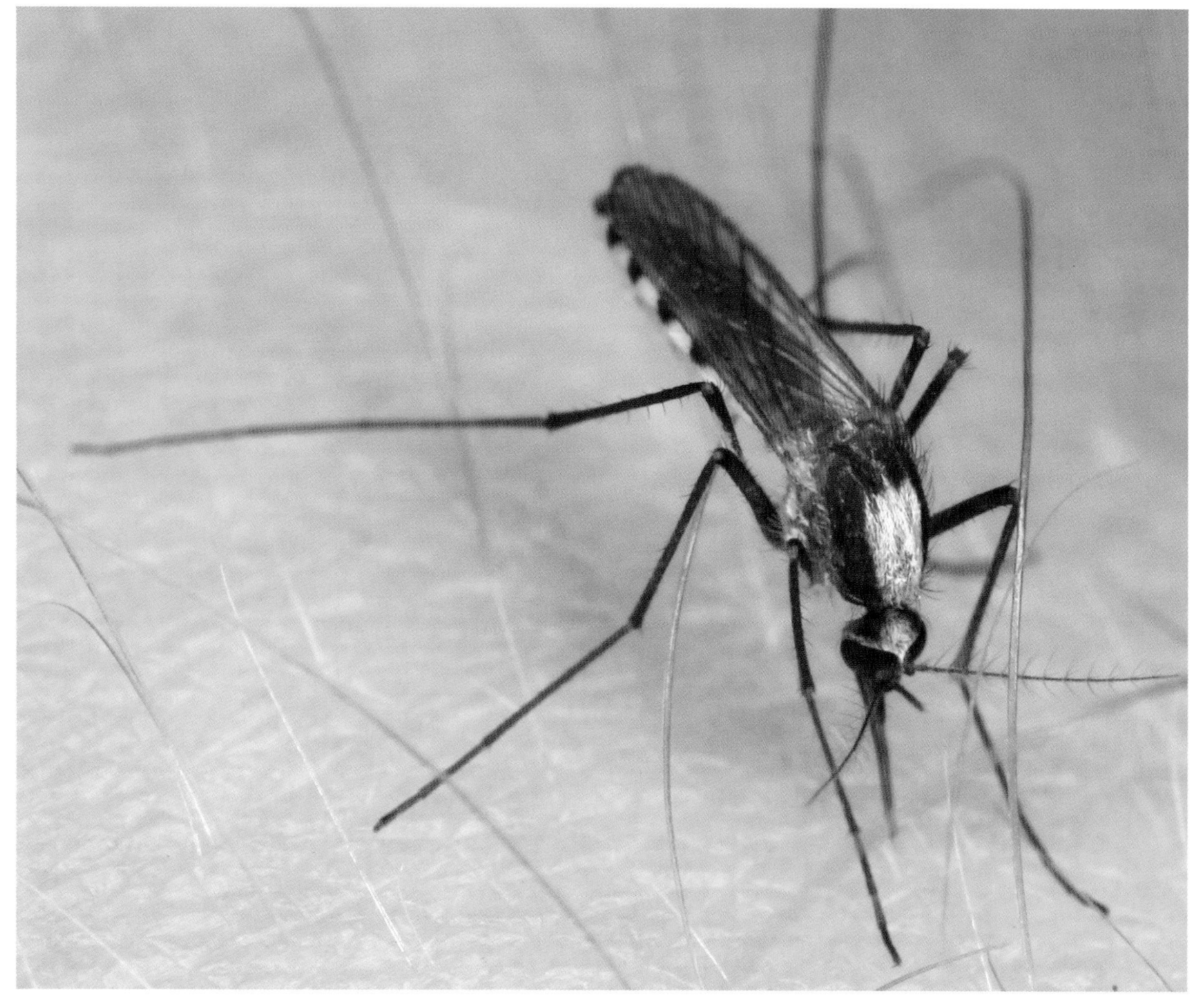

Aedes infirmatus female

Aedes infirmatus female

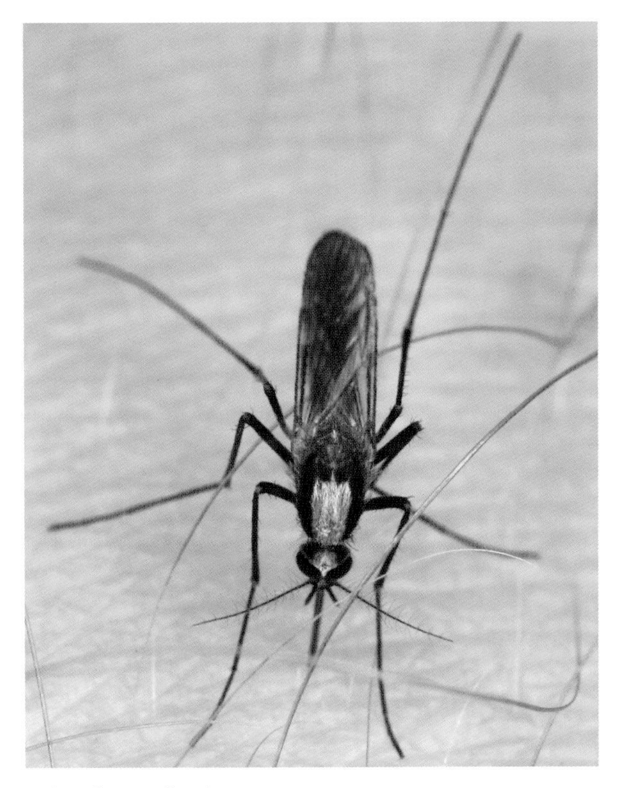

Aedes infirmatus female

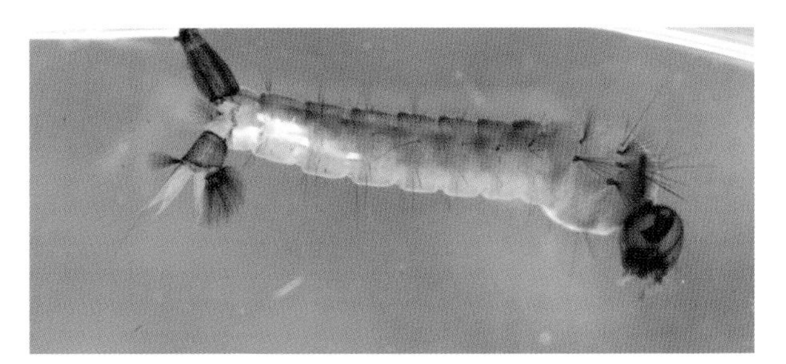

Aedes infirmatus larva

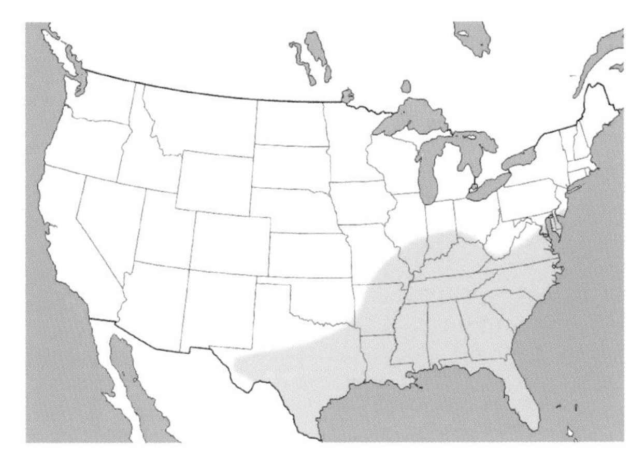

Distribution of *Aedes infirmatus* in the U.S.

Larvae Larvae of *Aedes infirmatus* are medium sized. Setae of the head are mostly unbranched, except for a large multibranched seta at the base of each antenna, and a branched seta arising midway along each antenna. The siphon is short, fairly stout, and slightly swollen at its base. The pecten reaches to the midpoint of the siphon. The saddle completely encircles the anal segment. The comb is a patch or irregular double row of about 20 scales. Larvae occur primarily in ephemeral rainwater pools.

Distribution *Aedes infirmatus* is found in the southern and eastern United States, from western Texas to Florida and north to New Jersey.

Aedes japonicus (Theobald, 1901)

Subgenus *Finlaya*

Asian bush mosquito

Aedes japonicus female

Adults *Aedes japonicus* is a medium-sized mosquito with black, white, and golden markings. The scutum is black, with golden stripes arranged in a pattern reminiscent of the lyre-shaped markings on the scutum of *Aedes aegypti.* The lateral portions of the thorax and abdomen have patches of silvery-white scales. The hind legs have broad pale bands on the first 3 tarsal segments, but the last 2 tarsal segments are entirely dark scaled. Females are persistent biters and attack when the vegetation where they rest is disturbed. They attack most frequently during the early evening hours but will bite at virtually any time of day. Females feed mainly on the blood of mammals, including humans. In its native range, *Aedes japonicus* transmits the virus that causes Japanese encephalitis. However, since this mosquito has been introduced to the United States only relatively recently, it is not yet known what role it will play in the transmission of mosquito-borne pathogens in the United States.

Aedes japonicus female

Aedes japonicus larva

Larvae Larvae of *Aedes japonicus* are medium sized. The large setae of the head are branched and arranged in a straight row along the front of the head. The siphon is slightly elongate and thin, with a pecten that reaches nearly to its tip. The teeth of the pecten that are nearer to the tip are spaced farther apart than those near the base of the siphon. The siphon seta arises from within the pecten. The saddle of the anal segment is highly spiculate, giving it a rough appearance. The saddle does not completely surround the anal segment. Larvae are most commonly encountered in discarded automobile tires and rock pools but may also be collected from a host of other man-made and natural containers.

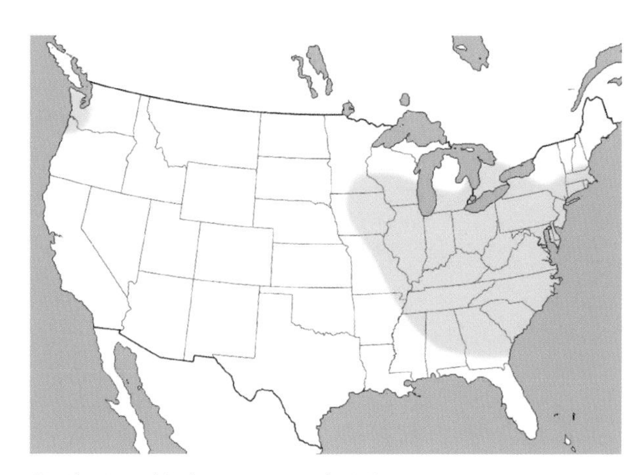

Distribution of *Aedes japonicus* in the U.S.

Distribution *Aedes japonicus* is native to eastern Asia but was accidentally introduced to the United States in 1998. It was first detected in New York, Connecticut, and New Jersey but has since spread to inhabit much of the eastern United States.

Aedes mitchellae (Dyar, 1905)

Subgenus *Ochlerotatus*

Mitchell's mosquito

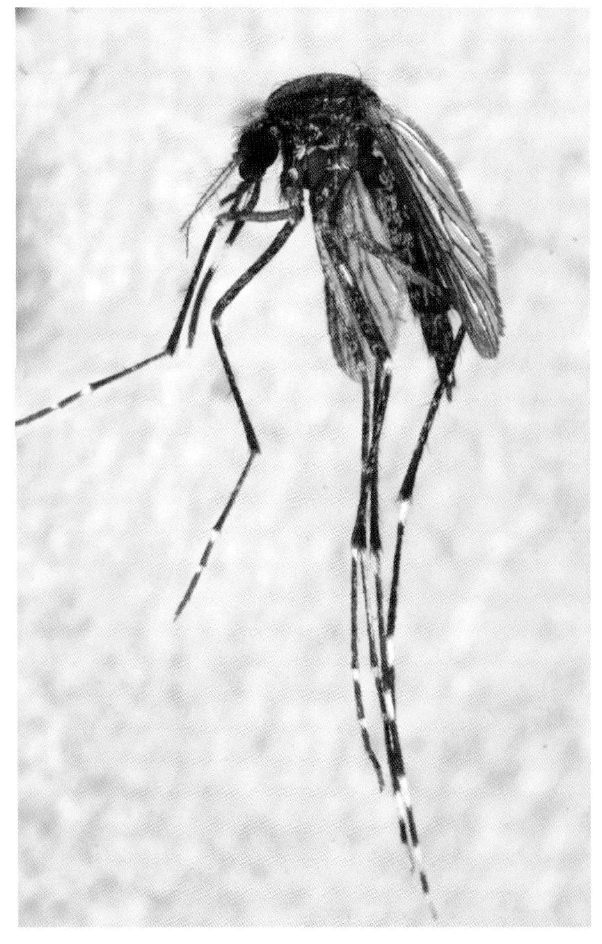
Aedes mitchellae female

Adults Adults of *Aedes mitchellae* are medium sized. The scutum is clad in rich golden-brown scales dorsally and bright white scales laterally. The abdomen is dark and has bands of white scales at the base of each segment and a pale mid-dorsal stripe that passes through the bands. The proboscis is mostly dark scaled but has a pale-scaled ring at its center. The legs are also dark with white markings. Tarsi of the hind legs have broad white rings at the base of each segment. Adults of *Aedes mitchellae* resemble adults of *Aedes sollicitans,* which is very common in coastal areas. Adults of these 2 species are distinguished by differences in their wings: adults of *Aedes mitchellae* have only dark wing scales, while adults of *Aedes sollicitans* have dark and pale wing scales intermixed. Although the females are

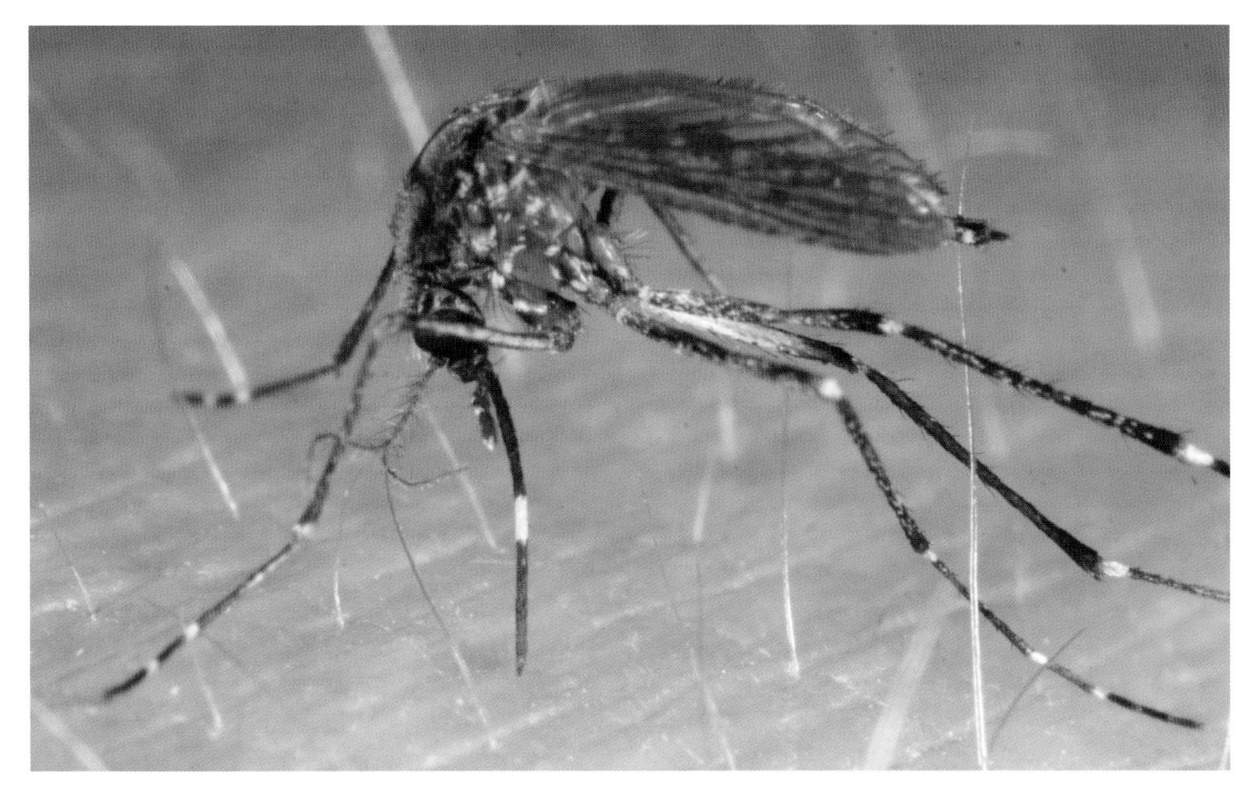
Aedes mitchellae female

rarely abundant, they are strong fliers and persistent biters. *Aedes mitchellae* is suspected to be of minor importance in the transmission of eastern equine encephalitis virus.

Larvae Larvae of *Aedes mitchellae* are medium sized. Setae of the head are mostly unbranched, except for a large multibranched seta at the base of each antenna, and a branched seta arising midway along each antenna. The siphon is somewhat conical, being much broader at the base than at the apex. The pecten reaches to the midpoint of the siphon. The saddle completely encircles the anal segment. The comb is a patch or irregular double row of 15–20 scales. Larvae are found in temporary sunlit rainwater pools that persist for several weeks, especially those that contain grass or other vegetation.

Distribution *Aedes mitchellae* is mainly found in the southeastern United States, but disjunct populations are found in a number of northeastern and midwestern states.

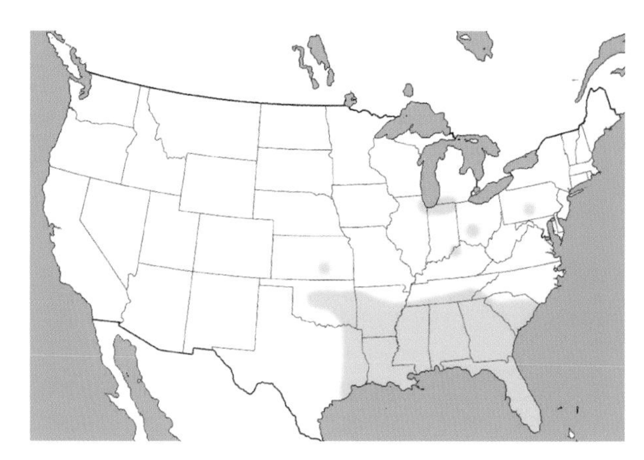

Distribution of *Aedes mitchellae* in the U.S.

Aedes sollicitans (Walker, 1856)

Subgenus *Ochlerotatus*

Golden salt marsh mosquito
or eastern salt marsh mosquito

Aedes sollicitans female

Adults *Aedes sollicitans* is medium sized to large, with black, white, and gold coloration. The scutum is clad in golden-brown scales, while the sides of the thorax are covered in narrow bright white scales. The abdomen is dark and has bands of white scales at the base of each segment and a yellowish-white mid-dorsal stripe that passes through the bands. The dark-scaled proboscis has a bright white ring at its center. The hind legs have broad white rings at the base of each segment, and the first tarsal segment has a white ring at its midpoint. Females of *Aedes sollicitans* resemble those of another species in our area, *Aedes mitchellae*. The wings of *Aedes sollicitans* have both dark and light scales,

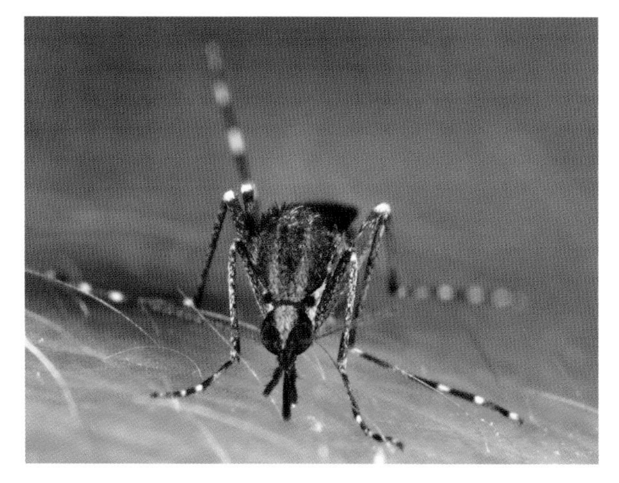

Aedes sollicitans female

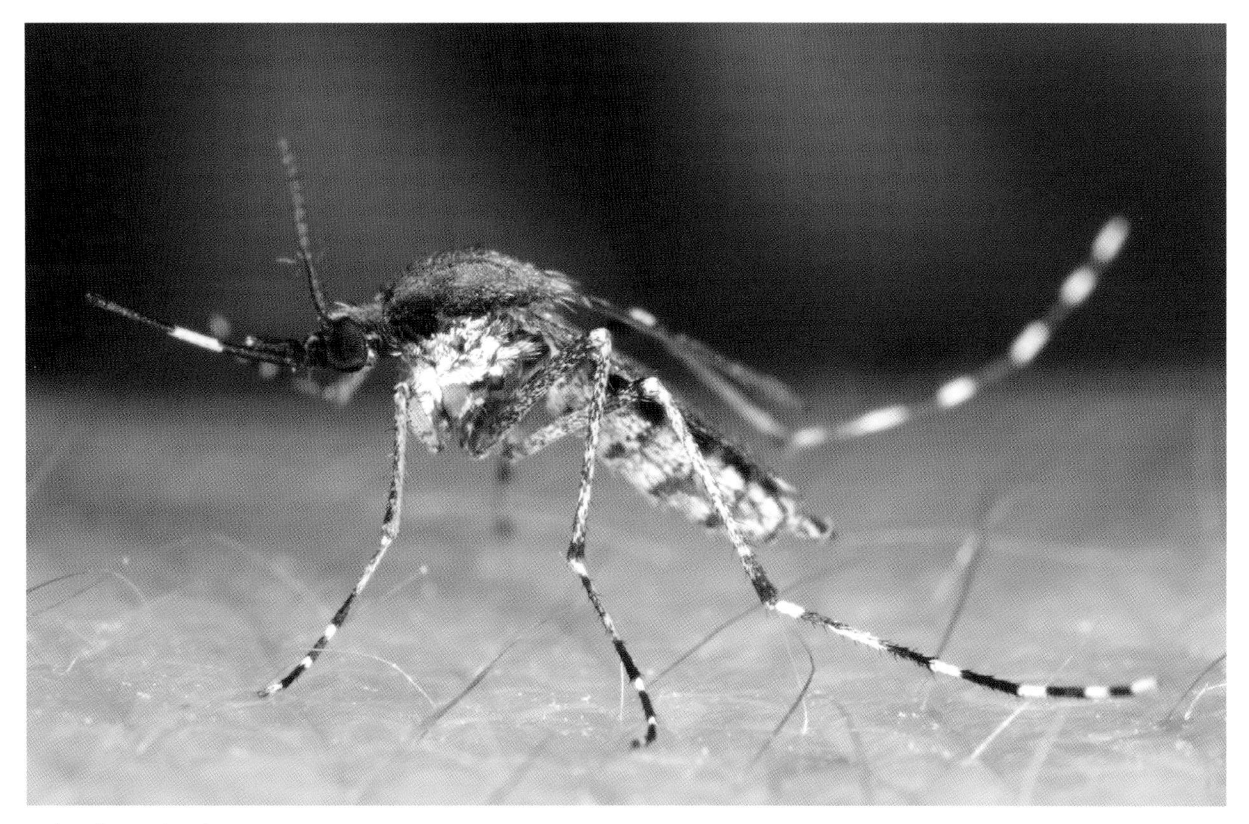

Aedes sollicitans female

Aedes sollicitans larva

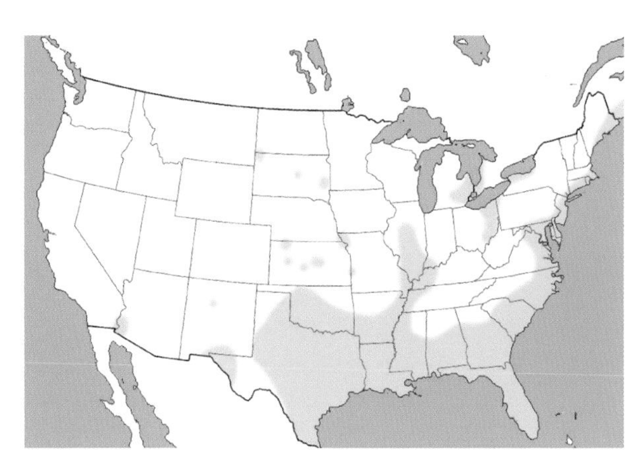

Distribution of *Aedes sollicitans* in the U.S.

while those of *Aedes mitchellae* have only dark scales. Females of *Aedes sollicitans* feed mostly on mammals and take blood meals from rabbits, deer, humans, dogs, and a variety of livestock. They can be incredibly numerous in coastal areas and are very pestiferous biters in these situations. Females are strong fliers and may fly against the wind to reach a host. Adults rest in grassy vegetation. *Aedes sollicitans* transmits the virus that causes eastern equine encephalitis and the parasite that causes heartworm disease in dogs in the United States.

Larvae Larvae of *Aedes sollicitans* are medium sized. Setae of the head are mostly unbranched, except for a large multibranched seta at the base of each antenna, and a branched seta arising midway along each antenna. The siphon is short, with its pecten reaching to (or just beyond) its midpoint. The saddle surrounds the anal segment and the anal papillae are quite small. The comb is an irregular patch of about 10–20 scales. Larvae are found in coastal pools produced by high tides and/or heavy rainfall. These pools often dry up quickly, so larvae must develop rapidly. Under optimal conditions, larvae may complete their development in as little as 4 days.

Distribution *Aedes sollicitans* is most commonly encountered in coastal areas of the eastern United States but may also be encountered in inland areas of Texas and Oklahoma, as well as a few other states.

Aedes sticticus (Meigen, 1838)

Subgenus *Ochlerotatus*

Northern floodwater mosquito

Adults *Aedes sticticus* is a medium-sized mosquito, with brown, cream, and white coloration. The scutum has a dark median stripe down its center bordered with patches of lighter-colored scales. The abdomen is dark brown, with basal bands of white scales. The tarsi and wings are dark scaled. Females of 4 other mosquitoes in our area (*Aedes aurifer, Aedes thibaulti, Aedes triseriatus,* and *Aedes hendersoni*) somewhat resemble *Aedes sticticus* females. Each of these similar species also has a dark median stripe on the scutum bordered by pale areas and tarsi without pale bands. However, only *Aedes sticticus* females have complete pale bands on all abdominal segments. The other species have lateral patches of white scales or pale bands on only a few segments. *Aedes sticticus* females feed on a wide variety of animals, including mammals, birds, and reptiles. Hosts include deer, foxes, rabbits, squirrels, raccoons, and humans, among many others. In the southeastern United States, adults are most abundant in the spring. *Aedes sticticus* is not known to transmit any human pathogens in the southeastern United States.

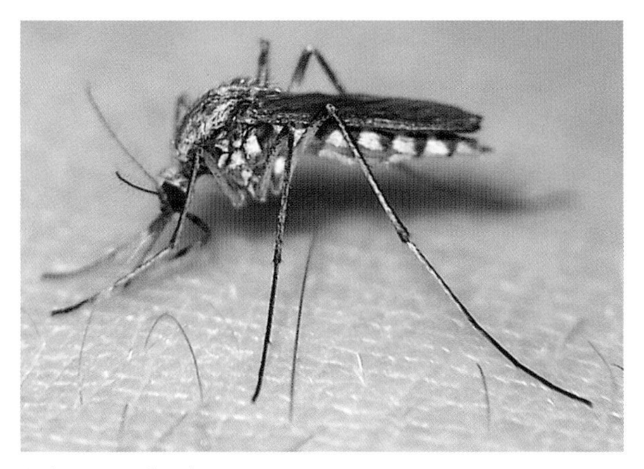

Aedes sticticus female

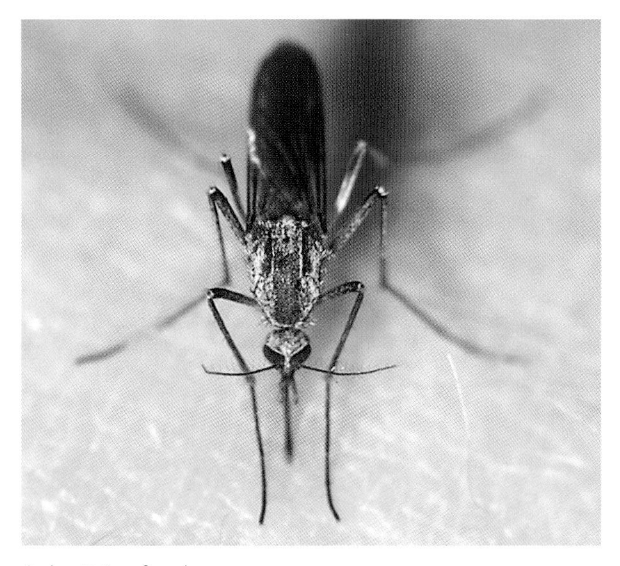

Aedes sticticus female

Larvae Larvae of *Aedes sticticus* are medium sized. Most setae of the head are highly branched. The siphon is somewhat narrow, with the pecten reaching to its midpoint. The comb is an irregular patch of 20–25 scales. The saddle nearly surrounds the anal segment. Considered a riverine floodwater species throughout much of its range, *Aedes sticticus* is found in a variety of habitats in the southeastern United States, including vernal and floodwater pools.

Distribution *Aedes sticticus* is found throughout the United States and southern Canada.

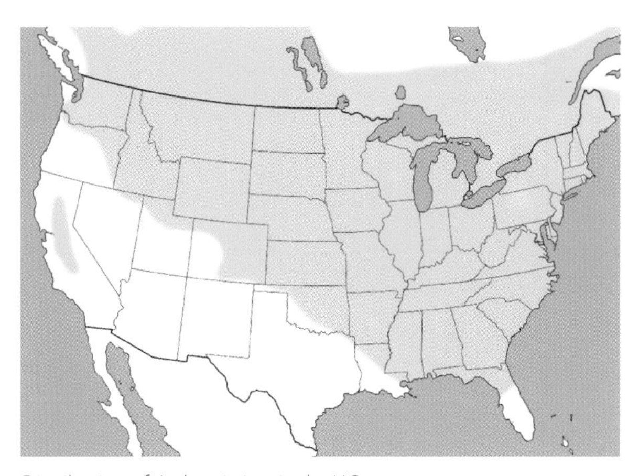

Distribution of *Aedes sticticus* in the U.S.

Aedes taeniorhynchus (Wiedemann, 1821)

Subgenus *Ochlerotatus*

Black salt marsh mosquito

Adults Adults of *Aedes taeniorhynchus* are small to medium sized and have a dark body that is ornamented with patches of white scales. The scutum is covered in brownish scales. The abdomen is dark, with narrow basal bands of white scales on each segment and patches of white scales laterally. The legs of adults are also dark with white markings. The tarsi of the hind legs have white rings at the base of each segment. The proboscis is dark and is encircled by a broad pale ring in the center. However, approximately 1.5% to 2% of adult females lack the pale band on the proboscis. The palps are short and dark and have white scales at their tips. The wings are dark. Females feed mostly on mammals and can be huge pests in coastal areas. They will bite during the day or night. Females are strong fliers and may disperse many miles in search of a blood meal. Adults rest in vegetation. Females transmit the parasites that cause dog heartworm disease in the United States.

Larvae Larvae of *Aedes taeniorhynchus* are small to medium sized. Setae of the head are mostly unbranched, except for a large multibranched seta at the base of each antenna, and a smaller branched

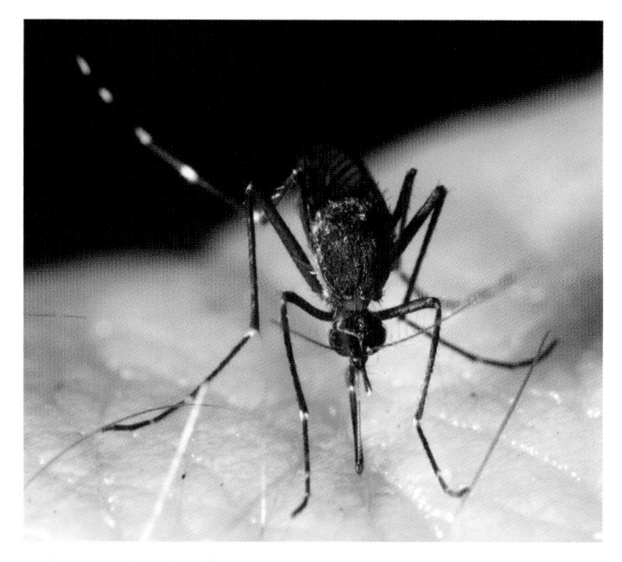

Aedes taeniorhynchus female

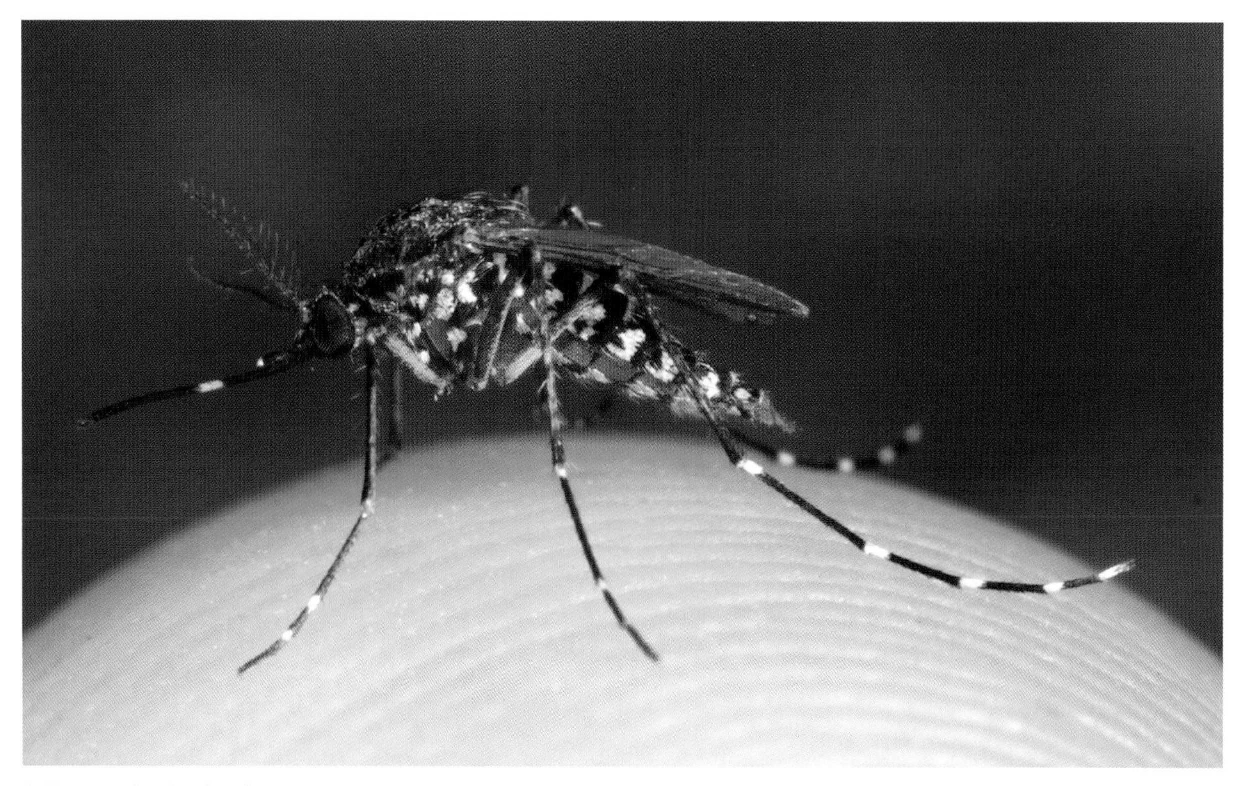

Aedes taeniorhynchus female

Aedes taeniorhynchus larva

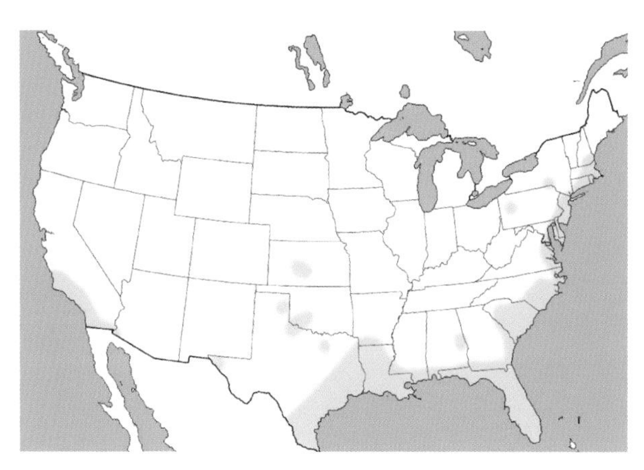

Distribution of *Aedes taeniorhynchus* in the U.S.

seta arising midway along each antenna. The thorax is covered in tiny spines, giving it a rough appearance. The siphon is quite short, with a pecten extending about half its length. The saddle encircles the anal segment. Anal papillae are quite small, usually much shorter than the saddle. Larvae occur mostly in brackish coastal marshes but are occasionally encountered in inland areas. They often aggregate in large masses of thousands of larvae, called "balls."

Distribution *Aedes taeniorhynchus* is most common along coastal areas of the Gulf of Mexico and Atlantic Ocean. Adults are occasionally encountered as far inland as Oklahoma and Arkansas.

Aedes thibaulti (Dyar and Knab, 1910)

Subgenus *Ochlerotatus*

Thibault's mosquito

Adults *Aedes thibaulti* is a medium-sized mosquito with black, brown, and golden coloration. The palps, proboscis, and tarsi are black. The scutum has a median stripe of dark brown scales, bordered on each side by areas of golden scales. Adults of 4 other species in our area (*Aedes aurifer, Aedes sticticus, Aedes triseriatus,* and *Aedes hendersoni*) are similar to those of *Aedes thibaulti.* These species can be distinguished by differences in the coloration of the scutum and patterns of dark and pale scales on the ventral abdomen. Females of *Aedes thibaulti* feed mainly on mammals, especially raccoons, but will readily bite humans that enter their habitat. They are most abundant in early spring but may also be encountered in early summer. Females do

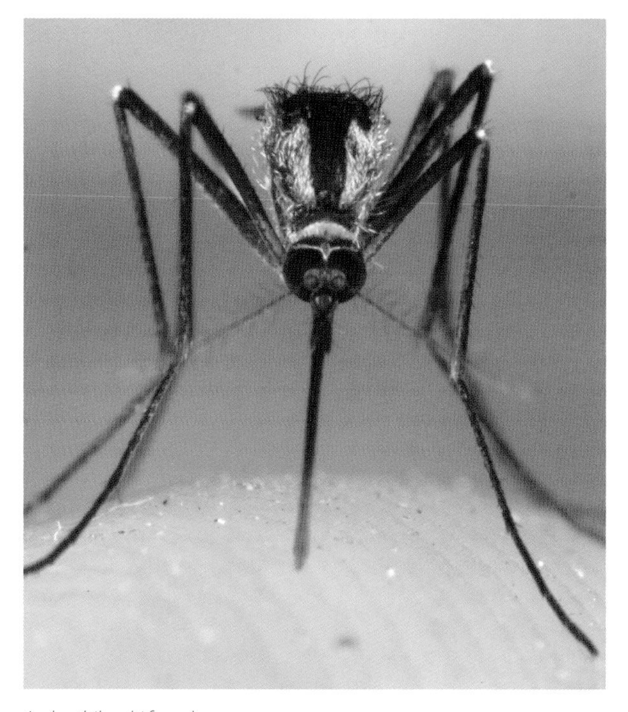

Aedes thibaulti female

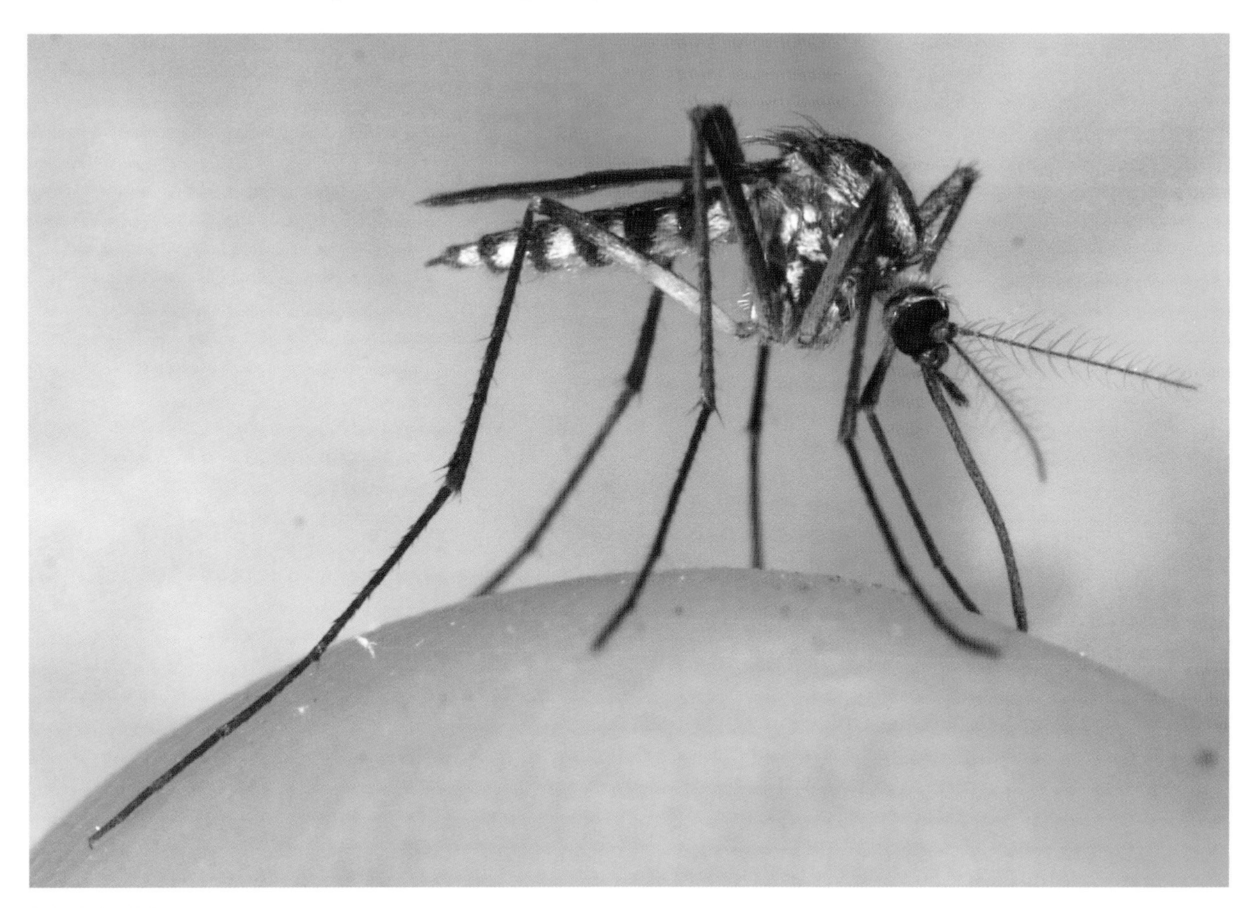

Aedes thibaulti female

Aedes thibaulti larva

not fly far from their breeding habitat. *Aedes thibaulti* is not known to transmit any human pathogens.

Larvae Larvae of *Aedes thibaulti* are medium sized. The setae of the head are mostly large and many branched. The abdominal segments are grayish brown, with a pale anterior band. The alternating dark and light bands give larvae a ringed appearance. The siphon is rather elongate, compared to the other *Aedes* species in our area. The pecten reaches to the midpoint of the siphon. The comb is a patch of 25–35 scales. The saddle covers the dorsal two-thirds of the anal segment and has many spicules (tiny spines) on its dorsal surface that resemble small teeth. Larvae are somewhat subterranean, being found almost exclusively in dark, crypt-like, water-filled holes in the ground, particularly under tree roots.

Distribution *Aedes thibaulti* is found throughout the northern and eastern United States and southern Canada.

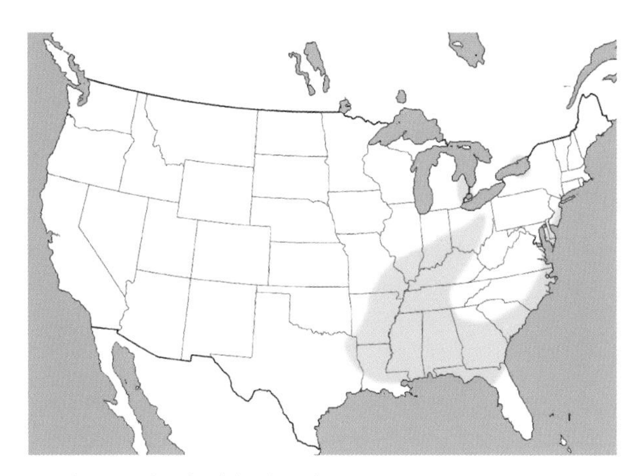

Distribution of *Aedes thibaulti* in the U.S.

Aedes triseriatus (Say, 1823) and *Aedes hendersoni* (Cockerell, 1918)

Subgenus *Protomacleaya*

Eastern treehole mosquitoes

Aedes triseriatus and *Aedes hendersoni* are difficult to distinguish as adults and have overlapping ranges and quite similar biologies. They are therefore treated together in this book. Larvae of the 2 species may be distinguished by differences in the sizes of their anal papillae.

Adults *Aedes triseriatus* and *Aedes hendersoni* are medium-sized mosquitoes with black and silvery-white coloration. The scutum has a jet-black median stripe bordered with patches of silvery-white scales.

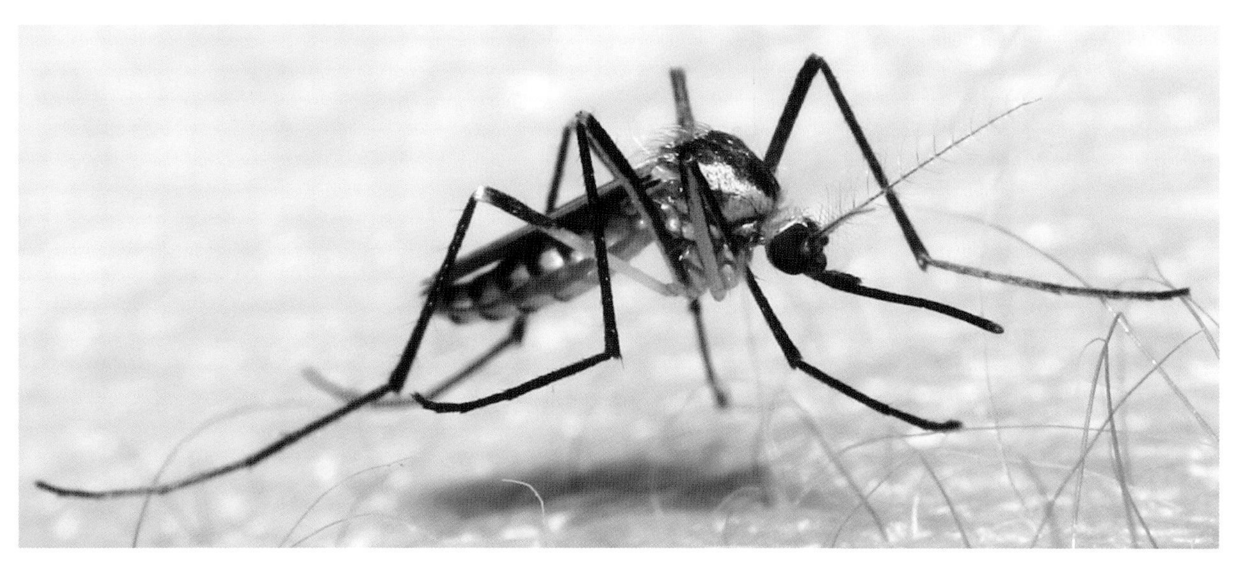

Aedes triseriatus female

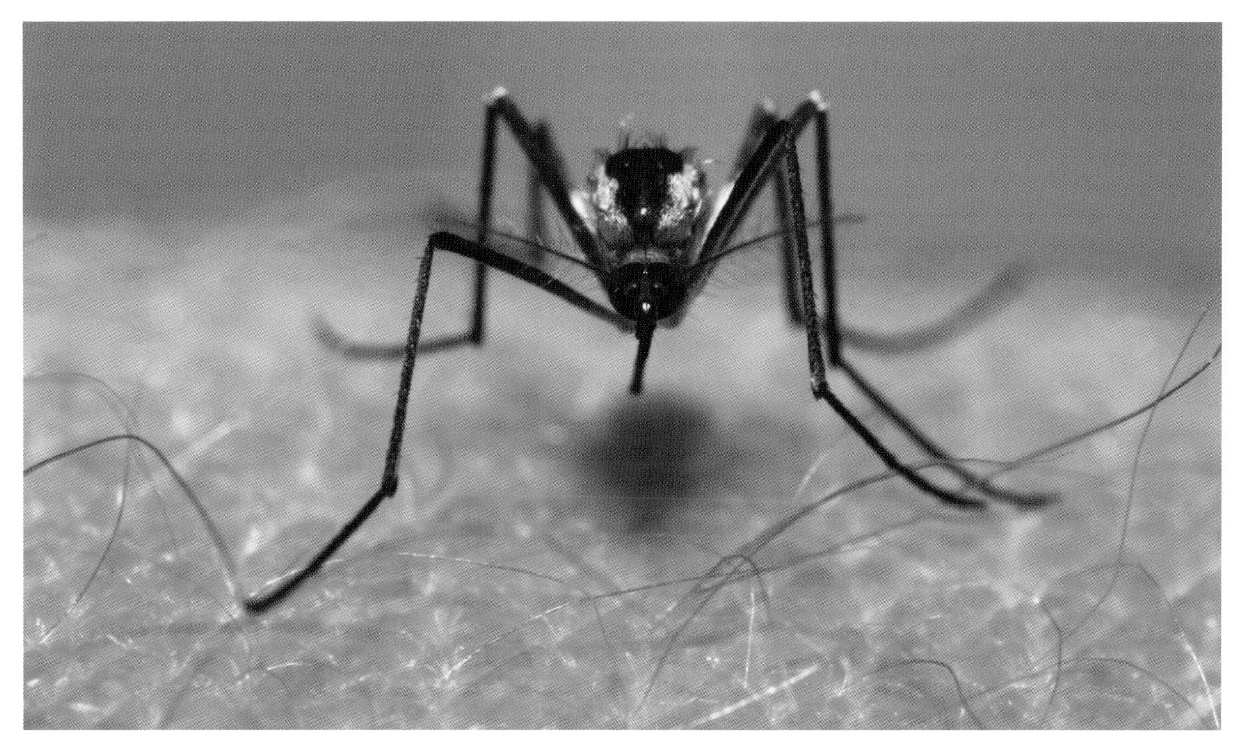

Aedes triseriatus female

Aedes triseriatus larva

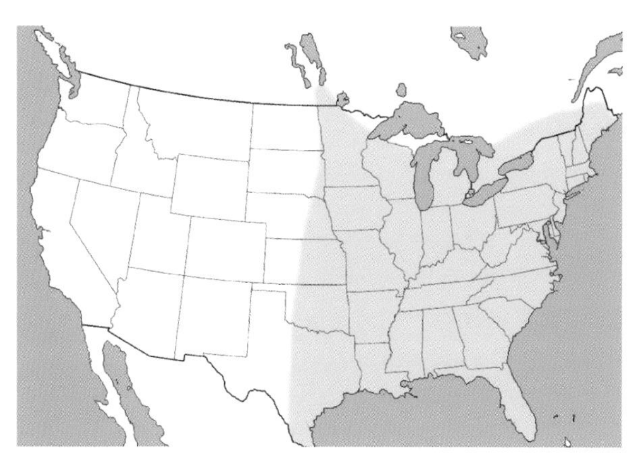

Distribution of *Aedes triseriatus* in the U.S.

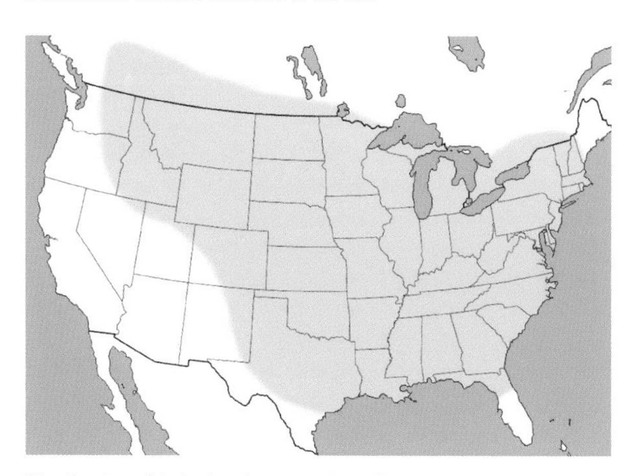

Distribution of *Aedes hendersoni* in the U.S.

The abdomen is also black and, like the scutum, is ornamented with patches of silvery white on the lateral margin of each segment. The tarsi, proboscis, palps, and wings are dark scaled. Adults of 3 other mosquitoes in our area (*Aedes aurifer, Aedes sticticus,* and *Aedes thibaulti*) resemble the adults of *Aedes triseriatus* and *Aedes hendersoni* in some respects. Females of *Aedes triseriatus* and *Aedes hendersoni* feed on a variety of small mammals and a few species of reptiles, particularly turtles. They do not, however, usually feed on amphibians or birds. Females of both species will readily bite humans, and *Aedes triseriatus* is very important in the transmission of La Crosse encephalitis virus in the United States.

Larvae Larvae of *Aedes triseriatus* and *Aedes hendersoni* are pale to grayish, with a dark siphon and head capsule. Setae of the heads of both species are generally small but otherwise quite variable. The seta of the antenna is unbranched. The siphon is moderately long and stout in both species, with pectens that reach to the midpoint. The saddle covers the dorsal half of the anal segment in both species. Larvae of *Aedes triseriatus* and *Aedes hendersoni* are distinguished by differences in the length and shape of the anal papillae. The papillae of *Aedes hendersoni* are bulbous and the dorsal and ventral papillae are of equal length. The papillae of *Aedes triseriatus* are fingerlike and the dorsal papillae are longer than the ventral papillae. Larvae of *Aedes triseriatus* and *Aedes hendersoni* are most commonly encountered in water-filled cavities of deciduous trees (tree holes) but are also occasionally found in water-filled man-made containers such as discarded automobile tires.

Distribution *Aedes triseriatus* is widely distributed across the eastern half of the United States as well as parts of southern Canada. *Aedes hendersoni* is distributed throughout most of the United States, excluding the Pacific and southwestern states.

Aedes trivittatus (Coquillett, 1902)

Subgenus *Ochlerotatus*

Striped floodwater mosquito *or* plains floodwater mosquito

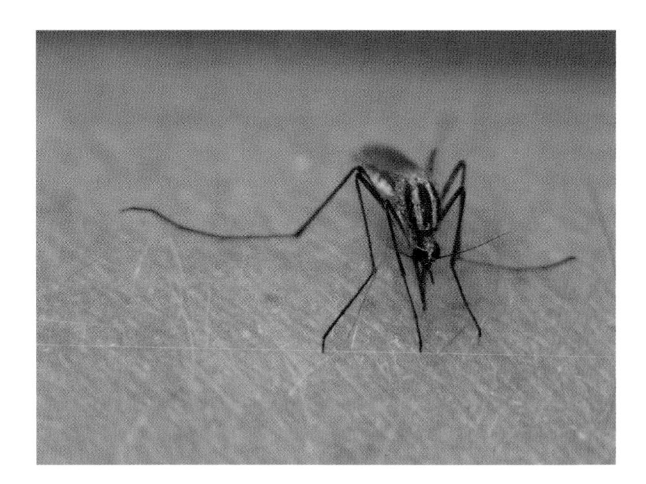
Aedes trivittatus female

Adults *Aedes trivittatus* is a medium-sized mosquito with unique markings on the scutum. These markings consist of a dark median stripe flanked by 2 pale stripes, which are flanked by 2 more dark stripes. The result is a pattern of alternating dark and light stripes on the scutum. The tarsi, palps, proboscis, and wings are dark scaled. Females feed mainly on small to medium-sized mammals, including raccoons, squirrels, opossums, and foxes. Females will readily bite humans, and the bite can be quite painful. They are most abundant in summer in open, sparsely wooded areas, especially following heavy rains. *Aedes trivittatus* is not known to transmit human pathogens but does transmit heartworm to dogs.

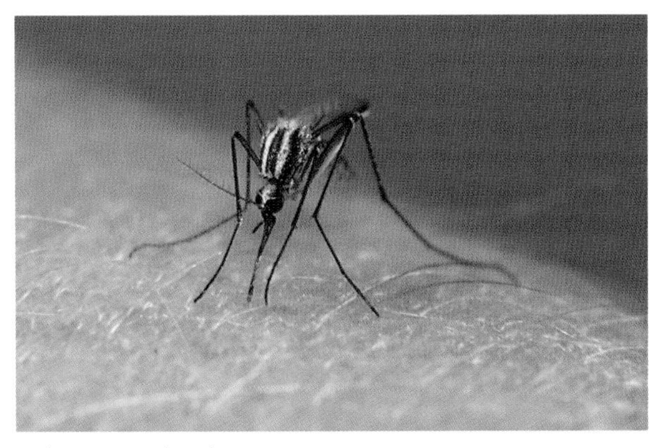
Aedes trivittatus female

Larvae Larvae of *Aedes trivittatus* are medium sized. Setae of the head are mostly unbranched, except for a large multibranched seta at the base of each antenna, and a smaller branched seta arising midway along each antenna. The siphon is short, stout, and slightly tapered apically. The pecten extends past the midpoint of the siphon. The saddle encircles the anal segment. The comb consists of a patch of 18–24 scales. Larvae are found primarily in floodwater pools in summer months.

Distribution *Aedes trivittatus* is found throughout much of the United States but is absent from Florida and several western states. This mosquito is far more common in the northern parts of its range than it is in the south.

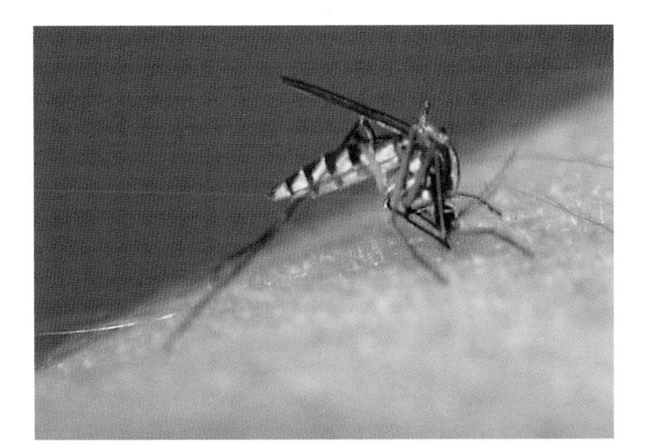
Aedes trivittatus female

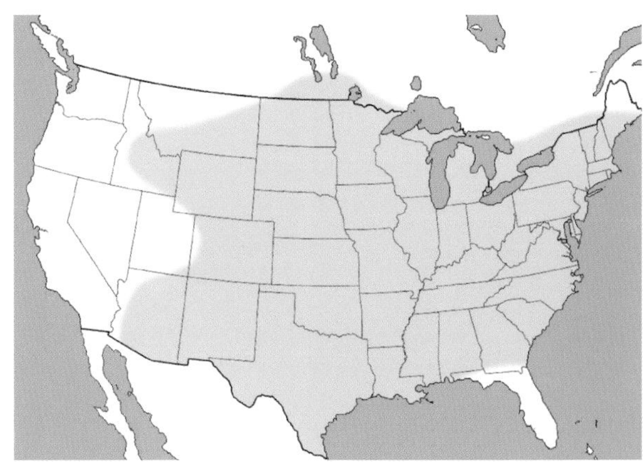
Distribution of *Aedes trivittatus* in the U.S.

Aedes vexans (Meigen, 1830)

Subgenus *Aedimorphus*

Common floodwater mosquito *or* inland floodwater mosquito

Adults *Aedes vexans* females are small to medium sized, with brown and white markings. They can be recognized by the narrow pale bands on their hind legs, plain brown scutum, and 2 white basal chevrons (upside-down triangles) on each abdominal segment. Wings have only dark scales. *Aedes cantator,* a northern species that also occurs in parts of North Carolina, is similar to *Aedes vexans.* The 2 species are distinguished by differences in the coloration of the abdominal tergites. *Aedes cantator* adults have broad pale bands on the basal part of each segment, in contrast to the pale chevrons of *Aedes vexans* adults. *Aedes vexans* females can be very annoying where they are abundant and will readily bite humans. Females feed on a wide variety of hosts, including birds and mammals. They rest in grassy vegetation and seek out their hosts in the twilight hours. Due to its propensity for feeding on humans and its abundance in nature, *Aedes vexans* may be important in the transmission of some human pathogens, such as the viruses that cause West Nile encephalitis and eastern equine encephalitis. *Aedes vexans* females also transmit the filarial nematode that causes heartworm disease in dogs.

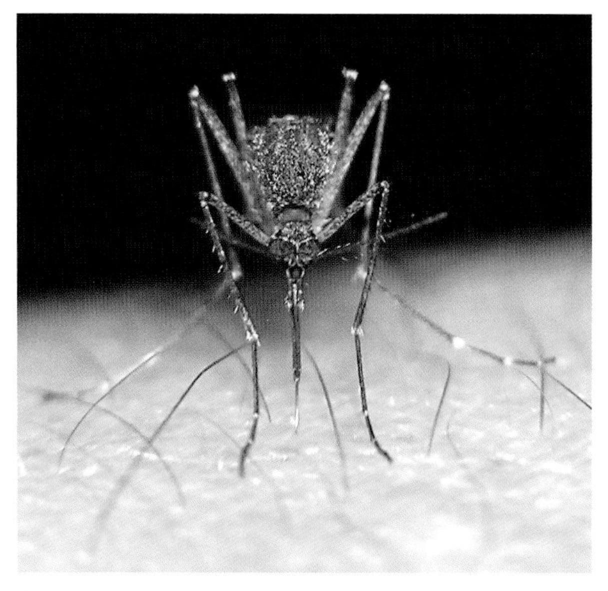

Aedes vexans female

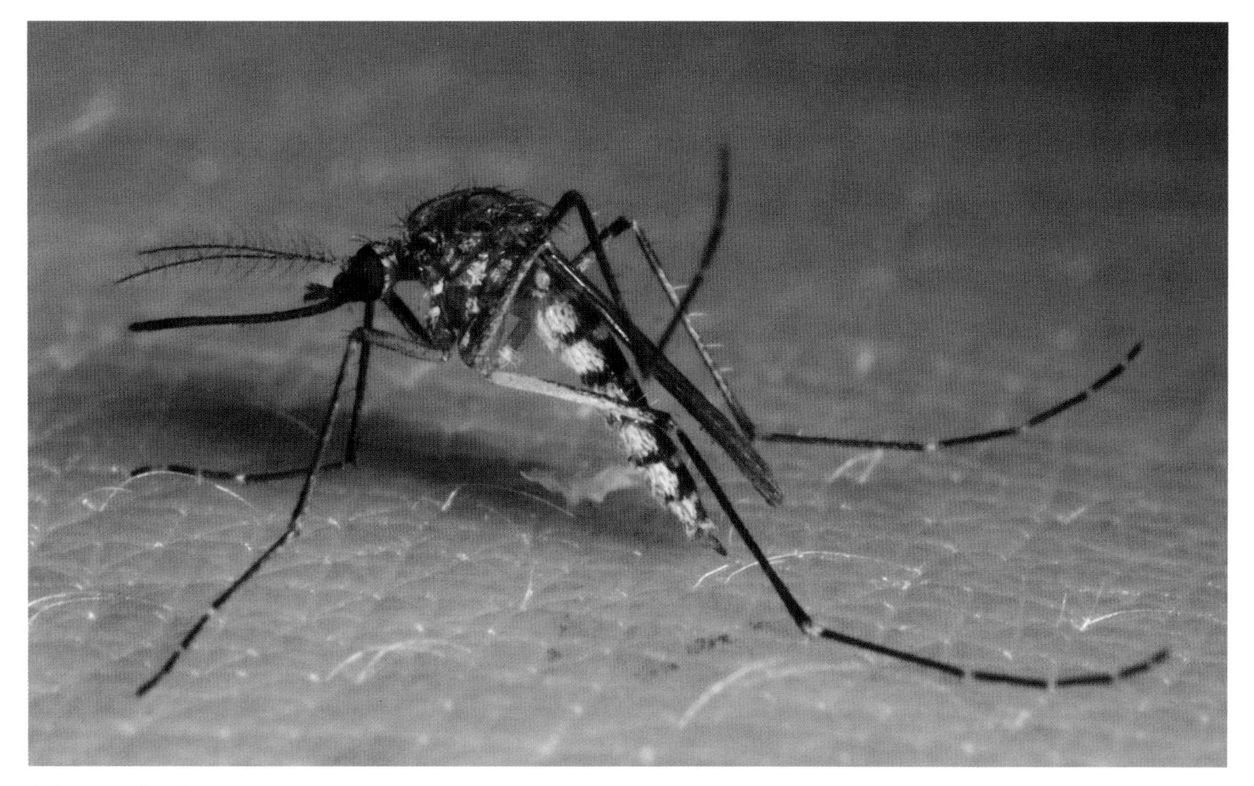

Aedes vexans female

Aedes vexans larva

Larvae Larvae of *Aedes vexans* are small to medium sized. Setae of the head are mostly branched. The siphon is somewhat elongate. The pecten reaches to the midpoint of the siphon. The apical 1–2 teeth of the pecten are spaced apart from the basal teeth and from one another. The saddle nearly encircles the anal segment. The comb is a single or irregular double row of 10–12 scales. Larvae occur in a variety of permanent and temporary bodies of water but are most common in ephemeral floodwater pools and ditches. The temporary nature of their larval habitat makes it necessary for the larvae to develop quickly. In some cases they can complete their development (from egg to adult) in as little as 1 week. Larvae often exhibit an interesting behavior in which they congregate together in a tight group that resembles a grayish ball suspended in the water. This behavior, termed "balling" because of the spherical shape of the mass of larvae, is thought to be a defense against predators.

Distribution *Aedes vexans* is found throughout the United States and much of southern Canada and is very common throughout its range. Prior to the arrival of the Asian tiger mosquito, *Aedes vexans* was considered the number one nuisance mosquito across the United States.

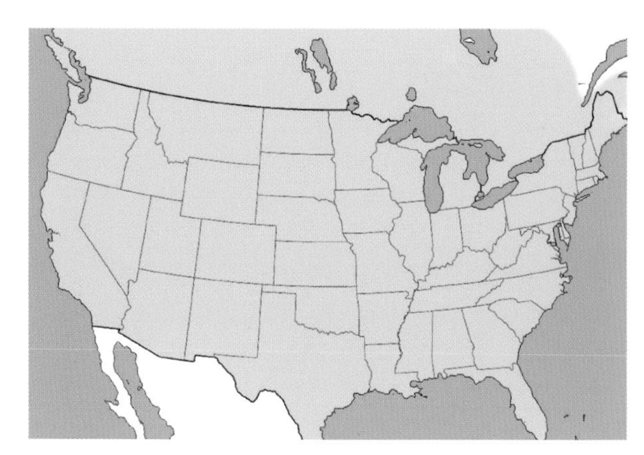

Distribution of *Aedes vexans* in the U.S.

Anopheles

Anopheles mosquitoes are easily distinguished from all others in our area by their long palps (as long as the proboscis) and distinctive posture when feeding. Nearly 500 species of *Anopheles* are recognized worldwide, at least 14 of which are found in the southeastern United States. Several *Anopheles* species are very difficult to separate from one another (even by specialists); therefore, it is common to treat these similar species as groups. The two recognized examples in our area are the *Anopheles quadrimaculatus* group and the *Anopheles crucians* complex. *Anopheles* females feed mostly on mammals, although some species will occasionally bite birds or even reptiles. Most *Anopheles* females are active at night or at dusk. During the day, adults of most species rest in cool, damp places. Most species are medium sized or large, but one small species, *Anopheles barberi,* occurs in our area. *Anopheles* larvae are very different from those of other mosquito species, as they have palmate setae on the abdomen and lack the respiratory siphon that is characteristic of other mosquito larvae. *Anopheles* larvae occur in a wide variety of temporary and permanent aquatic habitats, including marshes, lakes, ponds, tree holes, man-made containers, and floodwater pools. *Anopheles* females lay floating eggs on the surface of the water. Some species of *Anopheles* transmit the parasite that causes malaria in humans.

Anopheles atropos Dyar and Knab, 1906

Subgenus *Anopheles*

Salt-marsh *Anopheles*

Adults Adults of *Anopheles atropos* are medium sized, with dark brown coloration and long, thin legs. Their wings are dark scaled and have 4 rather indistinct dark spots. *Anopheles atropos* adults resemble those of *Anopheles walkeri* and members of the *Anopheles quadrimaculatus* group. These species are distinguished by differences in the palps, halteres, and patterns of the wing scales. Females feed on the blood of large mammals, including humans. *Anopheles atropos* is capable of transmitting the parasite that causes malaria but is not considered to have played a very important role in the transmission of malaria in the United States.

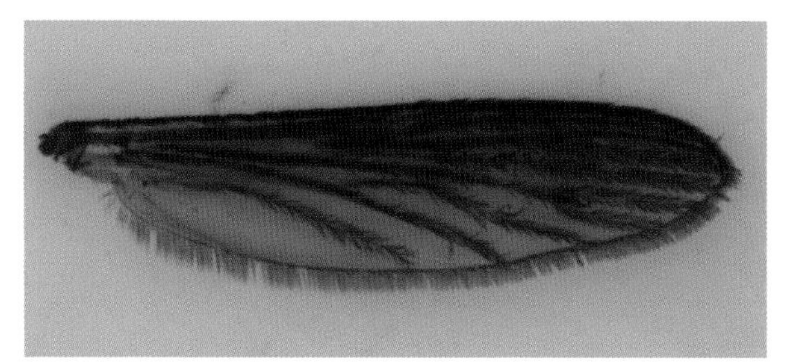

Wing of *Anopheles atropos*

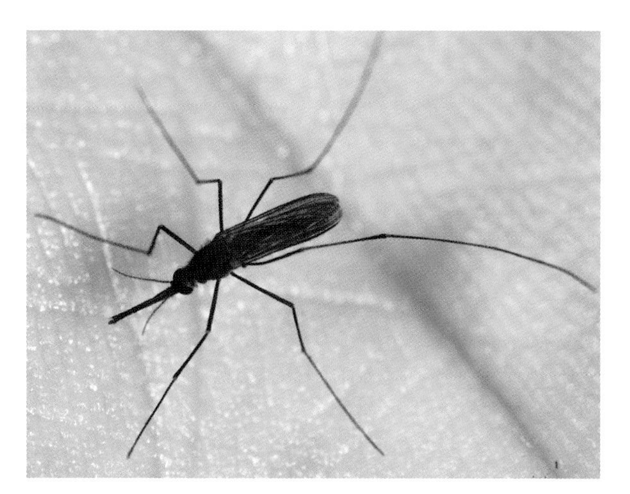

Anopheles atropos female

Larvae Larvae of *Anopheles atropos* are medium sized. Outer frontal setae of the head have 10 or fewer branches. Larvae are found in semipermanent saltwater pools and marshes in coastal areas.

Distribution *Anopheles atropos* is a salt-marsh mosquito, found in coastal areas along the Gulf of Mexico and Atlantic Ocean.

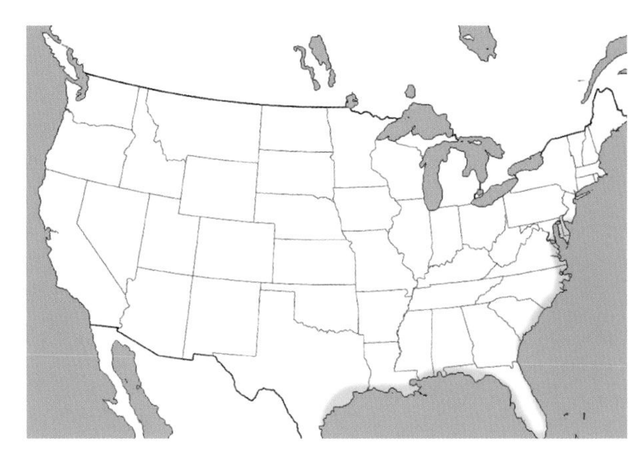

Distribution of *Anopheles atropos* in the U.S.

Anopheles barberi Coquillett, 1903

Subgenus *Anopheles*

Barber's treehole *Anopheles*

Adults Adults of *Anopheles barberi* are quite small, grayish-brown mosquitoes that lack the patterns and spots on their wings that are characteristic of the other *Anopheles* species in our area. This combination of features makes them easy to distinguish from all other species of *Anopheles* in the Southeast. Females feed on the blood of mammals and readily bite humans that venture into their habitat. However, this mosquito is not commonly encountered in numbers great enough to cause annoyance. *Anopheles barberi* is not recognized as a vector of any human pathogens.

Larvae *Anopheles barberi* larvae are quite small, usually dark, and have unbranched setae on the head. They are unique among the species of *Anopheles* in our area in that they are found almost exclusively in water-filled cavities of deciduous trees. *Anopheles barberi* normally survives the winter in the larval stage. During the winter, second-instar larvae may freeze along with the water in their tree hole habitat. When the water thaws they will thaw along with it and resume their development.

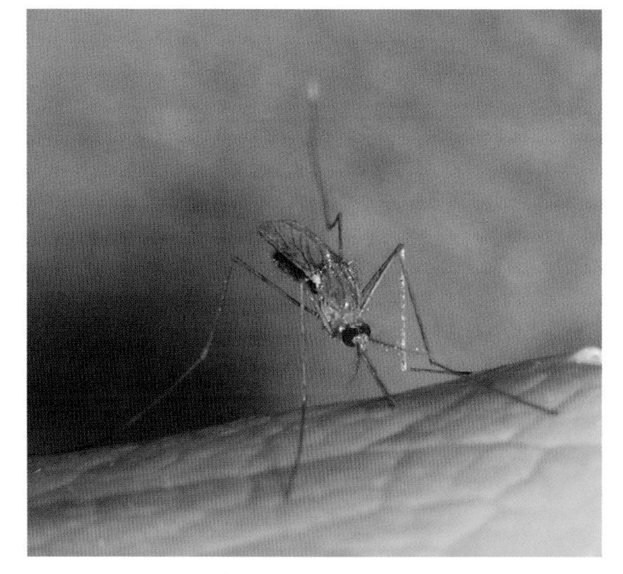

Anopheles barberi female

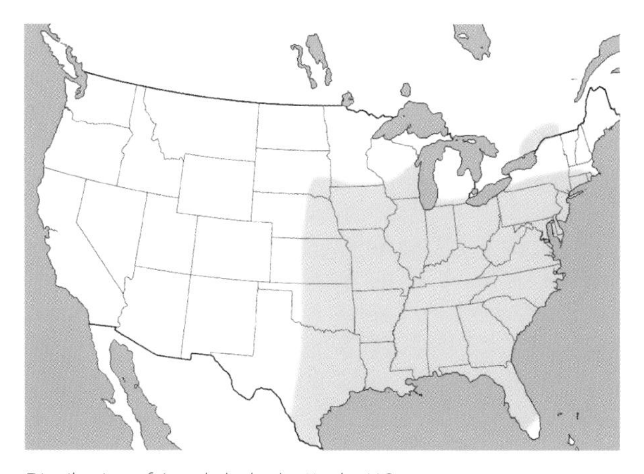

Distribution of *Anopheles barberi* in the U.S.

Distribution *Anopheles barberi* is found throughout the eastern half of the United States but is relatively rare in most areas.

Anopheles barberi larva

Anopheles crucians complex

Subgenus *Anopheles*

Anopheles bradleyi King, 1939; *Anopheles crucians* Wiedemann, 1828; and *Anopheles georgianus* King, 1939, and others

The *Anopheles crucians* complex is a group of (at least) 7 mosquito species that are nearly indistinguishable as adults and have overlapping ranges and similar biologies. Some of the members of the complex can be identified as larvae, while most cannot be differentiated morphologically. At least 4 species have not been formally described. For these reasons, members of this complex are treated together in this book.

Adults Mosquitoes of the *Anopheles crucians* complex are medium sized to large, with distinctive patterns of dark and light scales on the wings and palps. Their wings have a spotted appearance due to alternating patches of dark and light scales along the veins. The wing tips have patches of white scales. Dark palps are ringed with pale scales at the apex of segments 2–5. Females feed on the blood of a variety of mammals, large and small, including rabbits, raccoons, deer, livestock, and humans. Members of the *Anopheles crucians* complex were important vectors of malaria in the Southeast prior to the eradication of the disease in the United States in the first half of the twentieth century.

Larvae Larvae of *Anopheles bradleyi* are found in brackish pools in coastal areas. Larvae of *Anopheles crucians* are found in a variety of aquatic habitats including woodland pools, ponds, swamps, and marshes. Larvae of *Anopheles georgianus* are found in

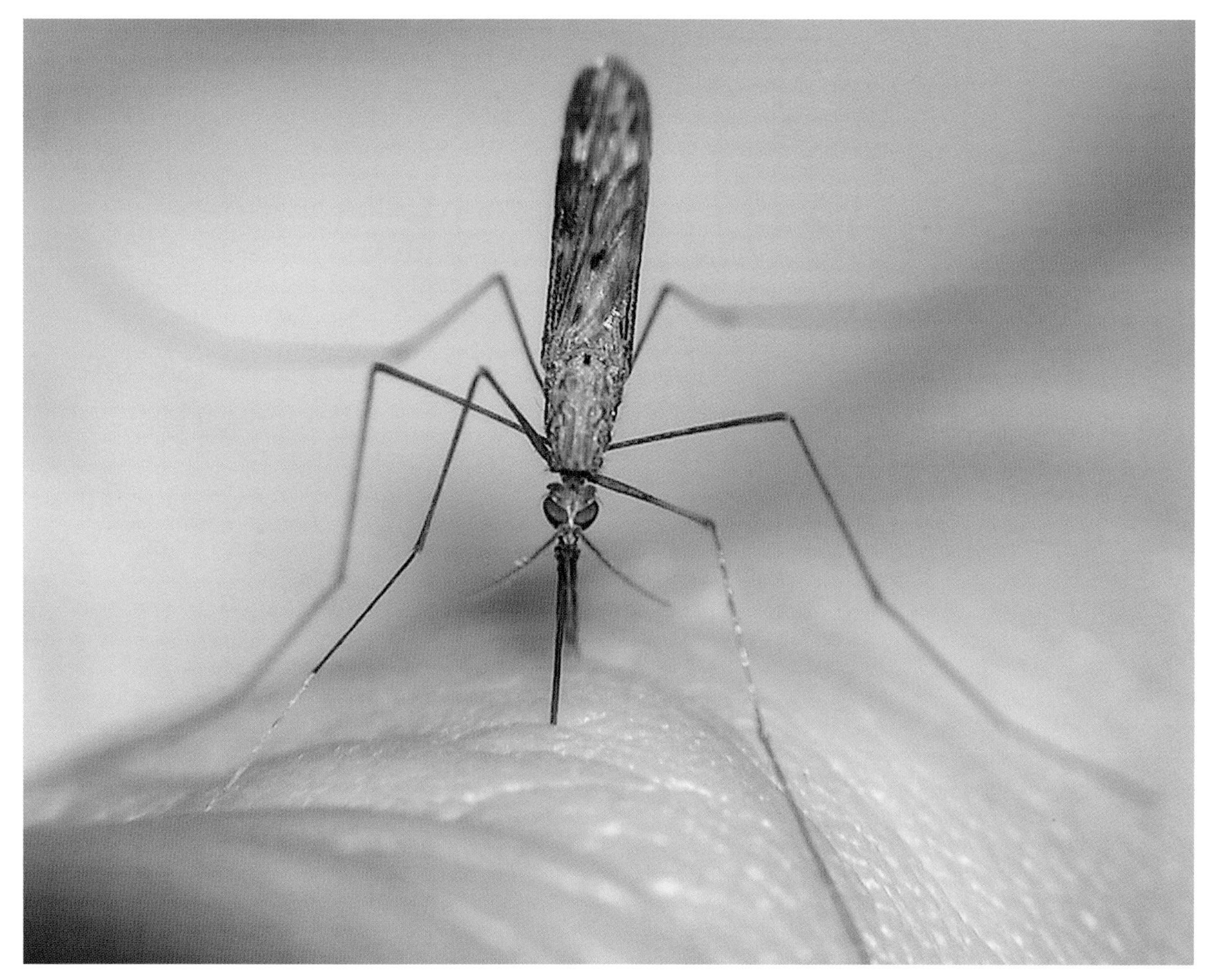

Anopheles crucians female

Anopheles crucians female

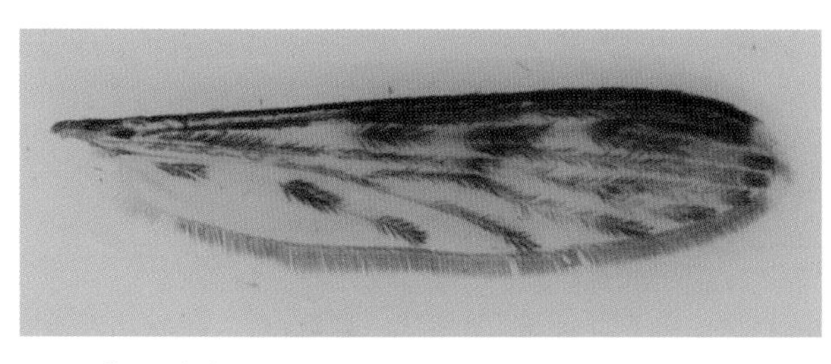

Wing of *Anopheles crucians*

puddles and seepage wetlands. Refer to the key (p. 59) for characteristics used to identify members of the *Anopheles crucians* complex.

Distribution *Anopheles bradleyi* is a coastal species, occurring along the Atlantic and Gulf Coasts. *Anopheles crucians* is common throughout the southeastern United States. *Anopheles georgianus* occurs from Louisiana through North Carolina.

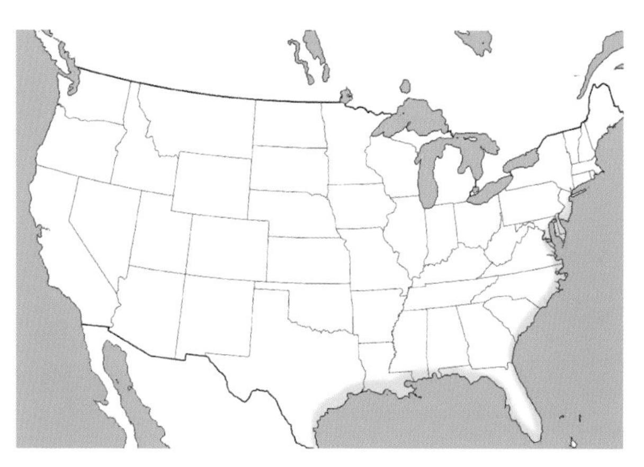

Distribution of *Anopheles bradleyi* in the U.S.

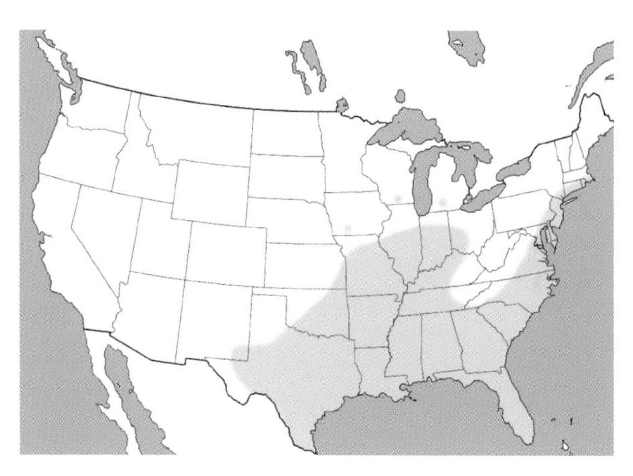

Distribution of *Anopheles crucians* in the U.S.

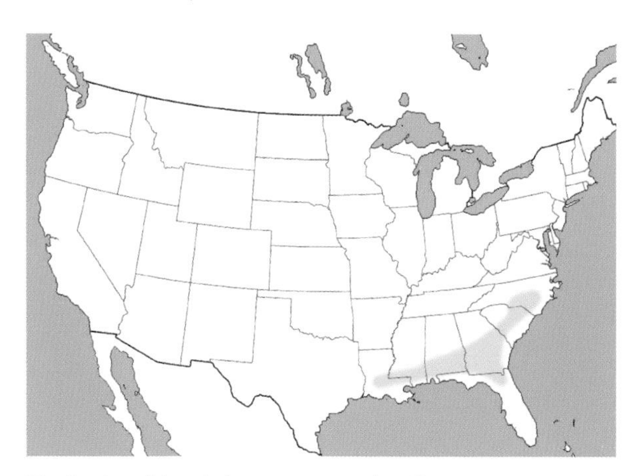

Distribution of *Anopheles georgianus* in the U.S.

Anopheles perplexens Ludlow, 1907

Subgenus *Anopheles*

Adults Adults of *Anopheles perplexens* are medium sized, with highly patterned wings. Alternating patches of dark and light scales on the wings give them a spotted appearance. Two small pale spots on the leading edge of the wing can be used to identify adults. One other species of *Anopheles* in our area, *Anopheles punctipennis,* has a similar arrangement of pale spots on the wings. The relative sizes of the pale spots have been used to tell these 2 mosquitoes apart. However, recent studies have shown that the sizes of the pale spots can be so variable that this may not be a reliable way of differentiating between these 2 species. Distinct differences in their eggs may be the only reliable way of telling these 2 mosquitoes apart. The legs and palps of *Anopheles perplexens* adults are dark scaled. Females feed on the blood of mammals, including humans.

Anopheles perplexens female

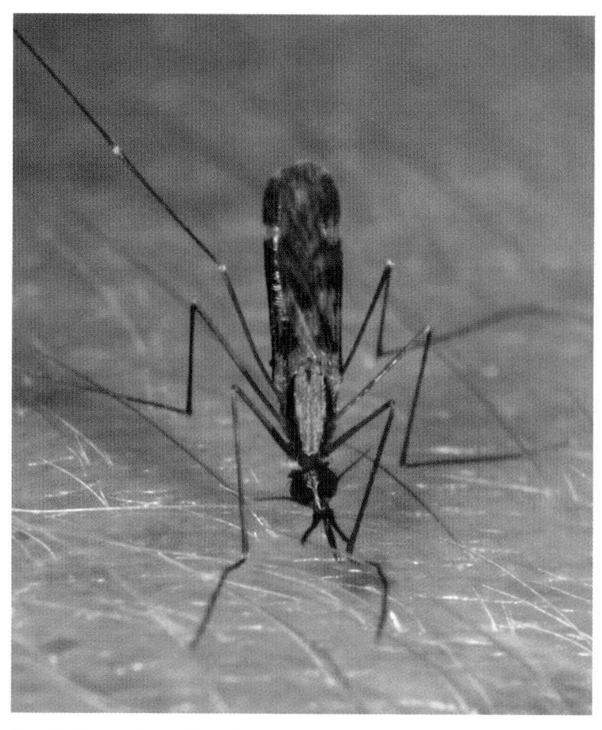

Anopheles perplexens female

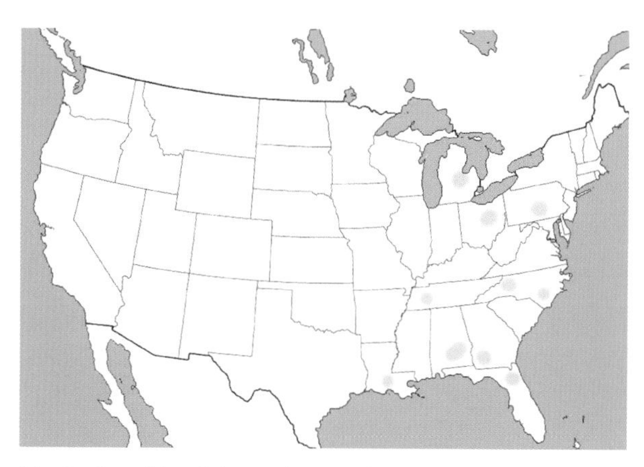

Distribution of *Anopheles perplexens* in the U.S.

Anopheles perplexens is not known to transmit any human pathogens in the United States.

Larvae *Anopheles perplexens* larvae are found predominantly in limestone springs, spring runs, and sinkholes but may also be encountered in intermittently flooded swamps. Refer to the key (p. 59) for characteristics used to identify the larvae of this species.

Distribution *Anopheles perplexens* is an uncommon mosquito with a rather spotty distribution. It occurs in isolated sites throughout the eastern United States.

Anopheles pseudopunctipennis Theobald, 1901

Subgenus *Anopheles*

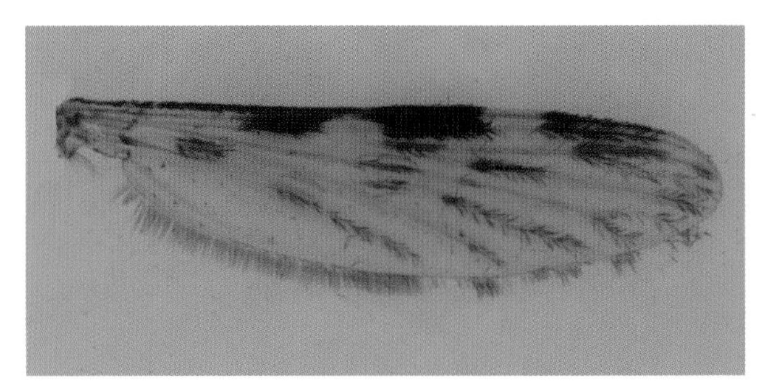
Wing of *Anopheles pseudopunctipennis*

Adults Adults of *Anopheles pseudopunctipennis* are medium sized, with dark and pale scales on the wings and palps. Two other *Anopheles* species in our area, *Anopheles punctipennis* and *Anopheles perplexens*, have wings with very similar patterns of light and dark scales. *Anopheles pseudopunctipennis* adults may be distinguished from those of these 2 similar species by examining the palps. The palps of *Anopheles pseudopunctipennis* adults have pale bands at the apex of segments 2–5, while the palps of *Anopheles punctipennis* and *Anopheles perplexens* are entirely dark scaled. *Anopheles pseudopunctipennis* females feed on the blood of mammals and are known to bite humans. *Anopheles pseudopunctipennis* is considered an important vector of malaria in Latin America but does not transmit any human pathogens in the United States.

Anopheles pseudopunctipennis female

Larvae Larvae of *Anopheles pseudopunctipennis* are medium sized. The setae along the front edge of head (frontal setae) are unbranched. Larvae are most commonly encountered in pools in sluggish streams, especially during the dry season. They are often associated with mats of *Spirogyra* algae and floating debris.

Distribution In the southeastern United States, *Anopheles pseudopunctipennis* is found in Louisiana, western and southern Mississippi, and southwestern Tennessee but is reported to be very rare in those states.

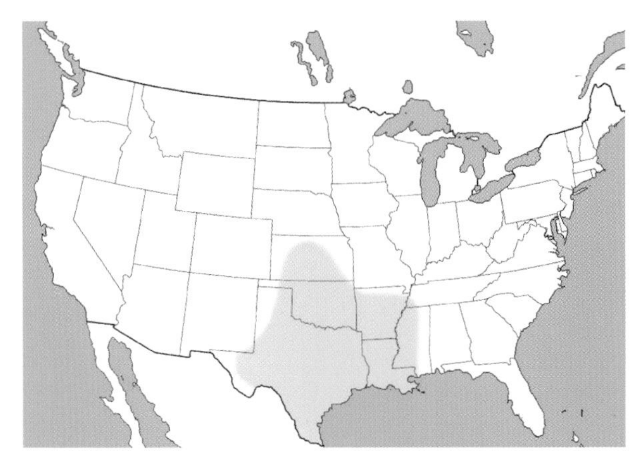
Distribution of *Anopheles pseudopunctipennis* in the U.S.

Anopheles punctipennis (Say, 1823)

Subgenus *Anopheles*

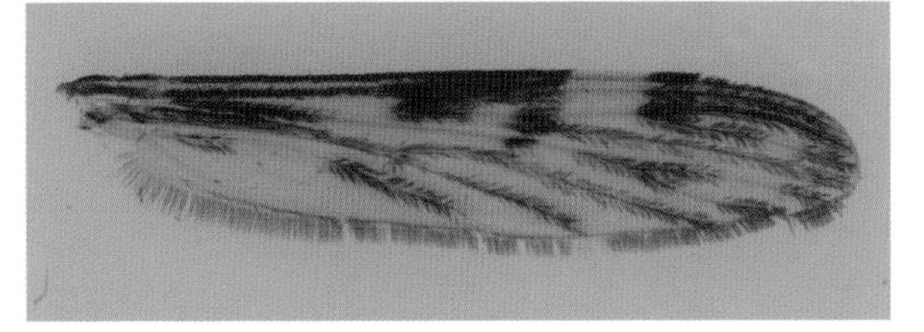

Wing of *Anopheles punctipennis*

Adults Adults of *Anopheles punctipennis* are medium sized to large, with 2 pale spots on the leading edge of their highly patterned wings. Adults of 1 other species in our area, *Anopheles perplexens,* are very similar to adults of *Anopheles punctipennis.* The relative sizes of the pale spots have been used to tell these 2 mosquitoes apart. However, recent studies have shown that the sizes of the pale spots can be so variable that this may not be a reliable way of differentiating between these 2 species. For now, distinct differences in their eggs may be the only reliable way of telling these 2 mosquitoes apart. The scutum of *Anopheles punctipennis* adults has a gray median stripe bordered by brown. The tarsi and palps are entirely dark scaled. Females feed on the blood of mammals, including deer, rabbits, livestock, and humans. *Anopheles punctipennis* is a very common pest species in neighborhoods, parks, and woodlands. Females usually bite just after sunset. Prior to the eradication of malaria in the United States, *Anoph-*

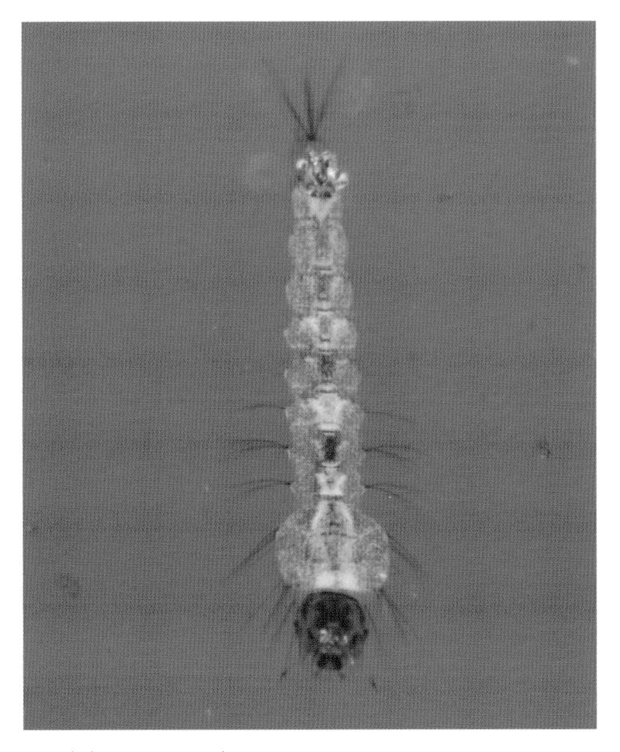

Anopheles punctipennis larva

Anopheles punctipennis female

eles punctipennis was important in malaria transmission in the southeastern states.

Larvae *Anopheles punctipennis* breeds in a wide variety of permanent and semipermanent bodies of water, including sluggish streams, puddles, ponds, and stream overflow pools. Larvae can also be found in water-filled man-made containers, especially rain-filled barrels and watering troughs. Refer to the key (p. 59) for characteristics used to identify *Anopheles punctipennis.*

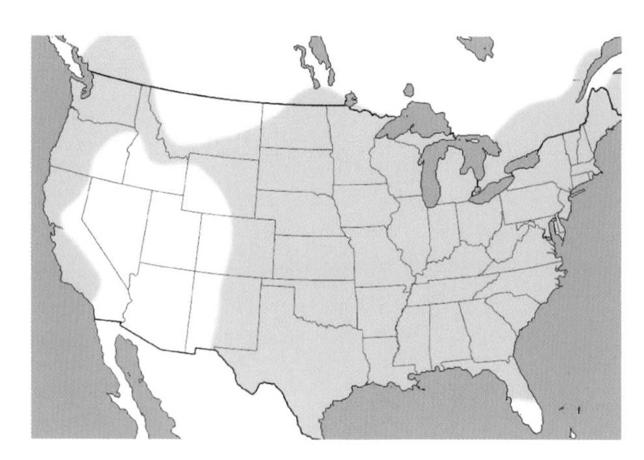

Distribution of *Anopheles punctipennis* in the U.S.

Distribution *Anopheles punctipennis* is a common and widespread mosquito, found throughout much of the United States and southern Canada.

Anopheles quadrimaculatus group

Subgenus *Anopheles*

Anopheles diluvialis Reinert, 1997; *Anopheles inundatus* Reinert, 1997; *Anopheles maverlius* Reinert, 1997; *Anopheles quadrimaculatus* Say, 1824; and *Anopheles smaragdinus* Reinert, 1997

Once considered a single species, the mosquito known as *Anopheles quadrimaculatus* is now considered to be at least 5 species that closely resemble one another. These 5 species, *Anopheles quadrimaculatus, Anopheles smaragdinus, Anopheles diluvialis, Anopheles inundatus,* and *Anopheles maverlius,* have overlapping ranges and similar biologies, and they can only be distinguished by slight differences in the number and coloration of bodily scales and setae. For these reasons, these 5 species are treated as the *Anopheles quadrimaculatus* group in this book.

Adults Members of the *Anopheles quadrimaculatus* group are dark colored and medium sized to large. Their wings and palps have only dark scales. Four distinct dark spots can be seen on the wings. The spots are composed of patches of dark scales located at the junctures of wing veins. Females feed mainly on the blood of large mammals, such as deer, livestock, and humans. Adults can often be observed resting in great numbers in hollow trees near the swamps and marshes where they breed. They can also be found resting under bridges over streams and other wet areas. Members of the *Anopheles quadrimaculatus* group were very important in the transmission of malaria in the

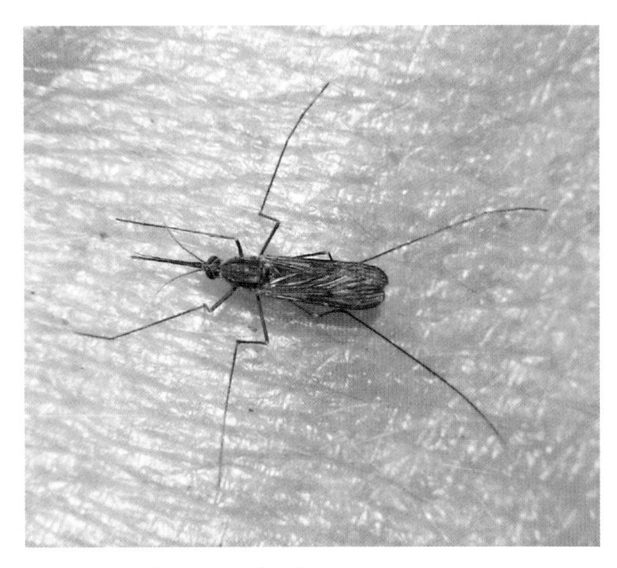

Anopheles quadrimaculatus female

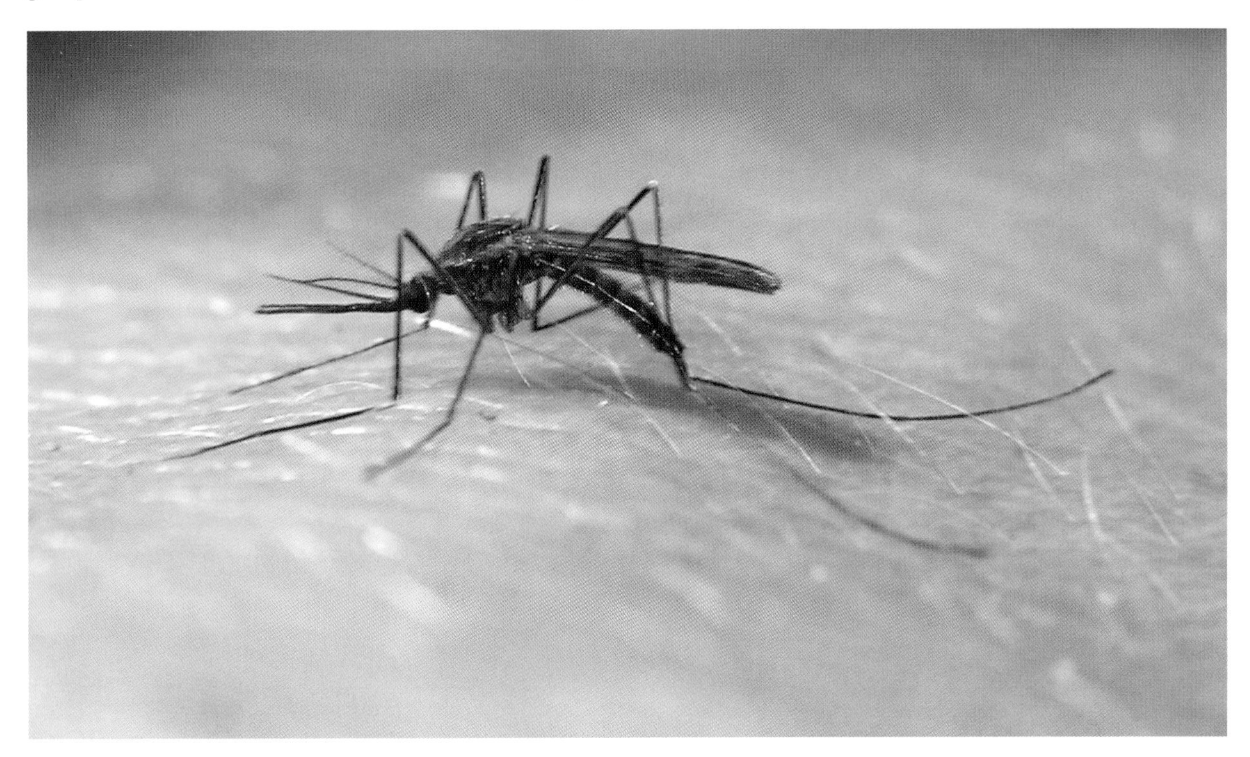

Anopheles quadrimaculatus female

Wing of *Anopheles quadrimaculatus*

southeastern states prior to eradication of the disease in the United States. Females can be pestiferous biters and will enter houses in search of a blood meal. They will often overwinter inside human dwellings and can sometimes be found flying around indoors on warm winter days.

Larvae Larvae of the *Anopheles quadrimaculatus* group are found in permanent bodies of water that support emergent aquatic vegetation. Refer to the key (p. 59) for characteristics used to distinguish members of the *Anopheles quadrimaculatus* group from other species.

Distribution Members of the *Anopheles quadrimaculatus* group are found throughout much of the eastern United States as well as portions of southern Canada and Mexico.

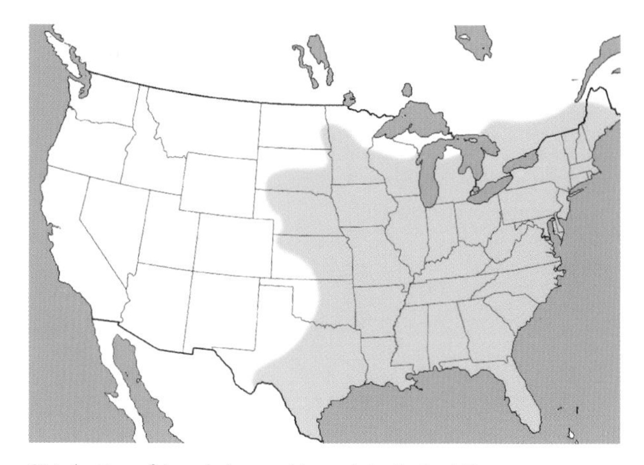

Distribution of *Anopheles quadrimaculatus* in the U.S.

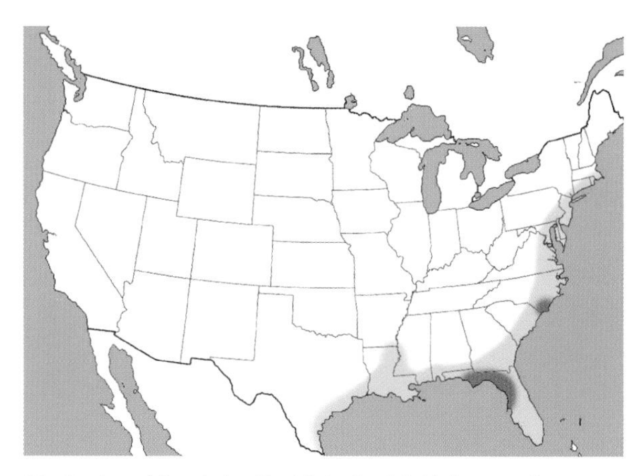

Distribution of *Anopheles diluvialis* in the U.S. (dark gray indicates confirmed range, light gray probable range)

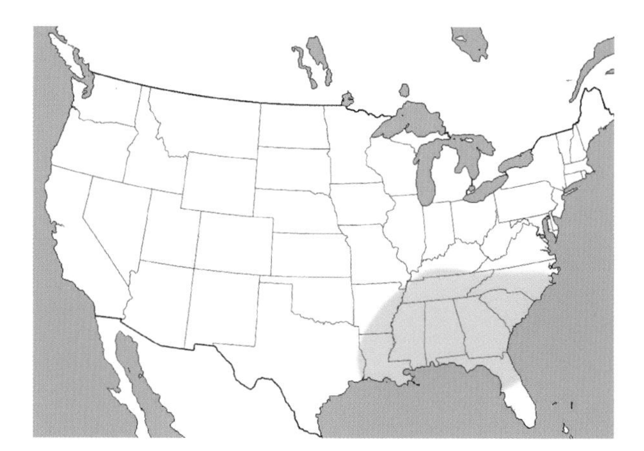

Distribution of *Anopheles smaragdinus* in the U.S.

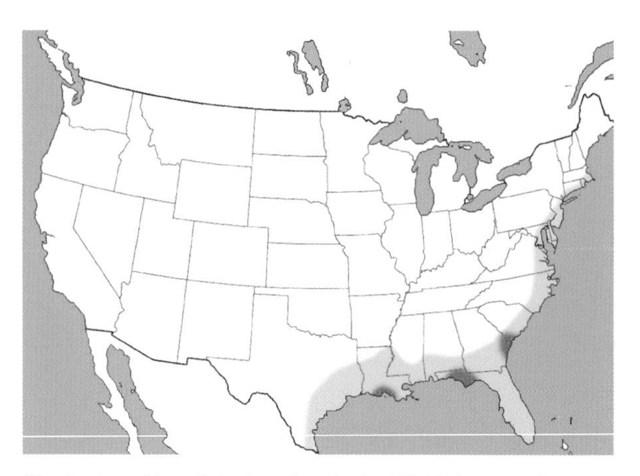

Distribution of *Anopheles inundatus* in the U.S. (dark gray indicates confirmed range, light gray probable range)

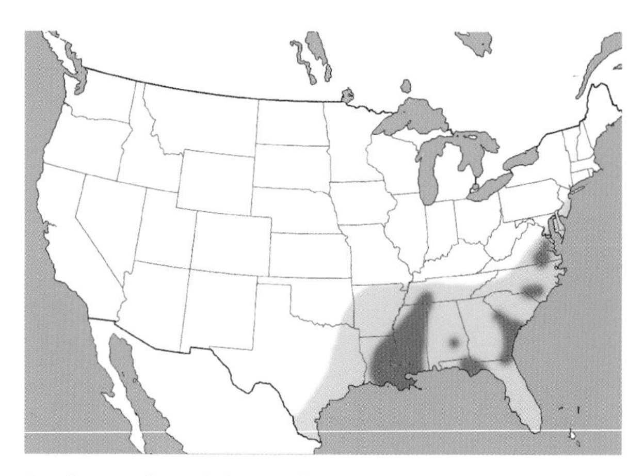

Distribution of *Anopheles maverlius* in the U.S. (dark gray indicates confirmed range, light gray probable range)

Anopheles walkeri Theobald, 1901

Subgenus *Anopheles*

Walker's *Anopheles*

Adults Adults of *Anopheles walkeri* are medium-sized, dark brown mosquitoes. The wings have only dark scales, with 4 indistinct dark spots. Adults of *Anopheles walkeri* resemble adults of *Anopheles atropos* and members of the *Anopheles quadrimaculatus* group. Pale bands on the proboscis of *Anopheles walkeri* adults distinguish them from members of the *Anopheles quadrimaculatus* group. To differentiate *Anopheles walkeri* adults from those of *Anopheles atropos,* one must examine the halteres, the knob-like vestiges of the hind wings. Adults of *Anopheles walkeri* have pale-scaled halteres, while adults of *Anopheles atropos* have dark-scaled halteres. Females of *Anopheles walkeri* feed on the blood of large mammals (including humans). Adults can be found resting in dense vegetation near their larval habitat. *Anopheles walkeri* is not important in the transmission of human pathogens.

Larvae Larvae of *Anopheles walkeri* are medium sized. The inner frontal setae of the head have sparse aciculae (small branch-like spines) near their tips. Larvae are usually encountered in permanent bodies of water that have aquatic grasses, especially those growing under the shade of buttonbush *(Cephalanthus).*

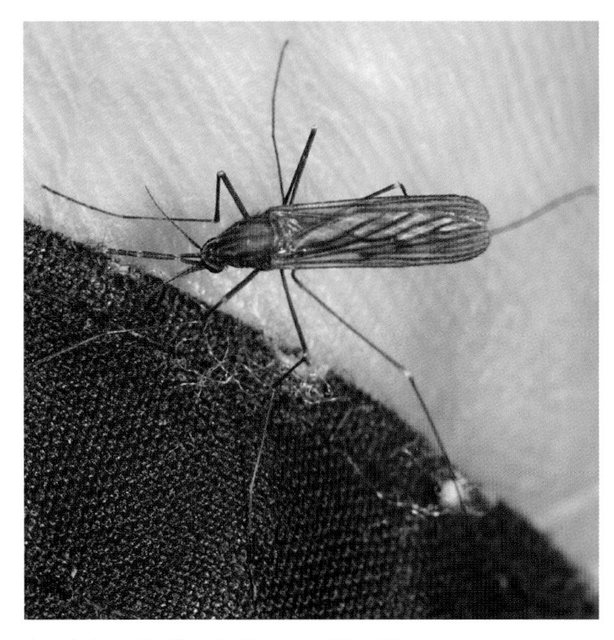

Anopheles walkeri female. Courtesy of Tom Murray.

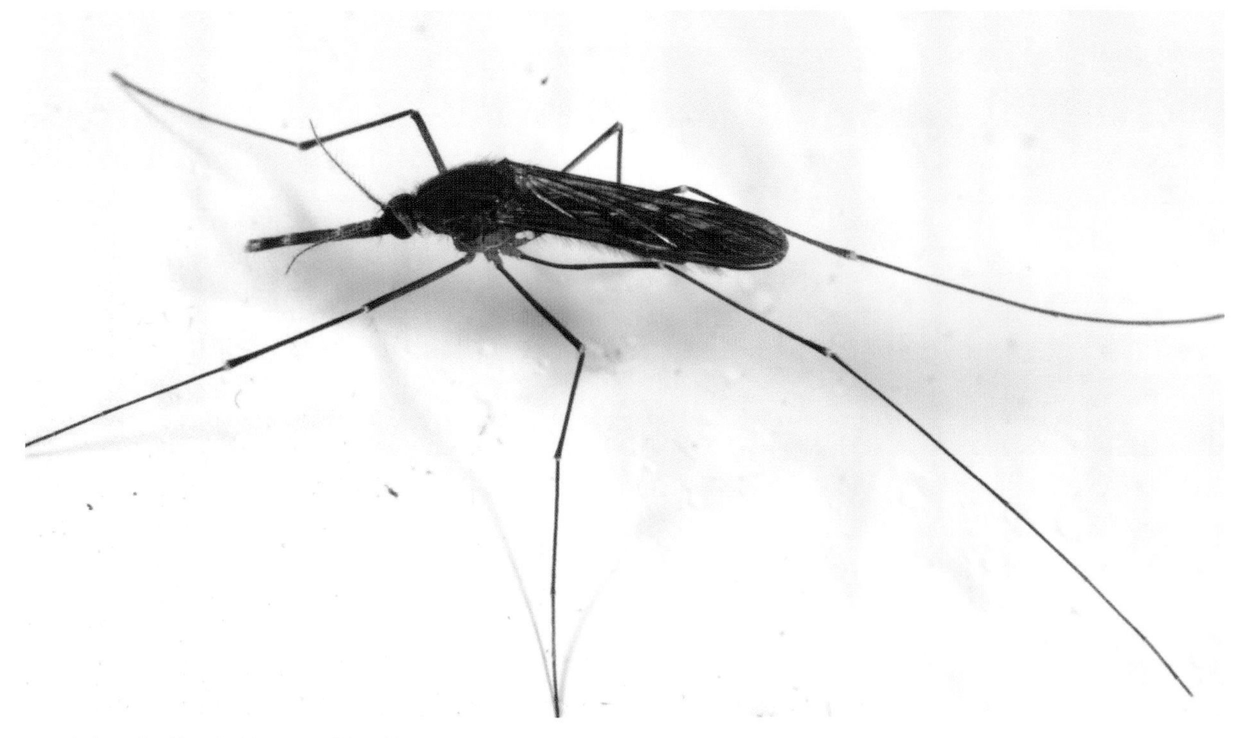

Anopheles walkeri female. Courtesy of Tom Murray.

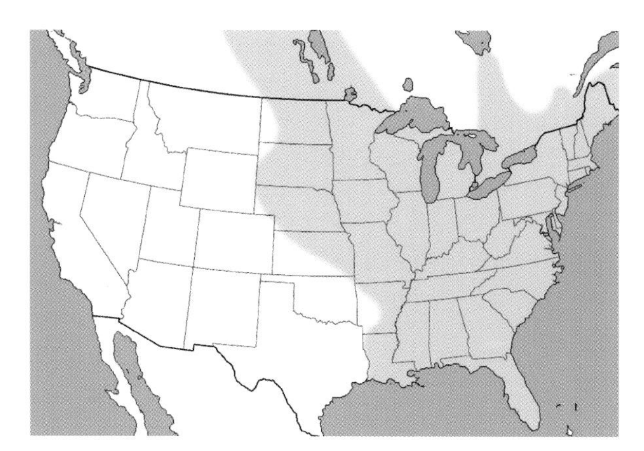

Distribution of *Anopheles walkeri* in the U.S.

Distribution *Anopheles walkeri* is found throughout much of the eastern United States and Canada.

Coquillettidia

Nearly 60 species of *Coquillettidia* occur worldwide, but the genus is most diverse in the tropics. Only 1 species is found in our area. Adults of most species are medium sized to large, and quite variable in appearance. Scales of the wings may be broad or narrow, and either all dark or a mixture of dark and pale. Most species feed on birds and/or mammals. Several species attack humans, and some are important in the transmission of human pathogens, such as eastern equine encephalitis virus. Both daytime and nighttime biters are known. Larvae of *Coquillettidia* are similar to those of *Mansonia* in having a siphon modified for piercing plant tissue to reach air stores within. The modified siphon is short and tapers to a sharp point. *Coquillettidia* larvae are found in ponds, lakes, and marshes with rooted aquatic plants.

Coquillettidia perturbans (Walker, 1856)

Subgenus *Coquillettidia*

Cattail mosquito

Adults *Coquillettidia perturbans* adults are medium sized, stout bodied, and speckled. The body is clad in a mixture of dark and light broad scales. The scutum is brown and lacks any recognizable pattern. The legs are dark brown, with broad white bands at the base of each tarsal segment. The first tarsal segment of the hind leg has an additional pale band near its middle. The tibia of the hind leg has a broad preapical band of pale scales. The proboscis has a similar white band at its center. The wings have broad scales of brown and white scattered throughout. Females can be extremely perturbing in situations where they are abundant and are persistent and painful biters. They feed on the blood of mammals and birds. Adults may be encountered during spring and summer, reaching their greatest abundance from June through August. *Coquillettidia perturbans* is considered a "bridge vector" of eastern equine encephalitis virus in the United States, transmitting the virus from birds to humans and horses.

Larvae Larvae of *Coquillettidia perturbans* are medium sized. Setae of the head are large and branched. The antennae are quite long and each bears

Coquillettidia perturbans pupae, attached to roots of cattail (*Typha*)

Coquillettidia perturbans female

Coquillettidia perturbans larvae

a very large, profusely branched seta. The siphon resembles a short, pointed saw and is specialized for piercing tissue of aquatic plants to reach the stores of air within. The saddle surrounds the anal segment. The comb is a single row of 8–15 small scales. Larvae are found only in permanent and semipermanent bodies of water with rooted aquatic vegetation, particularly cattails *(Typha)*.

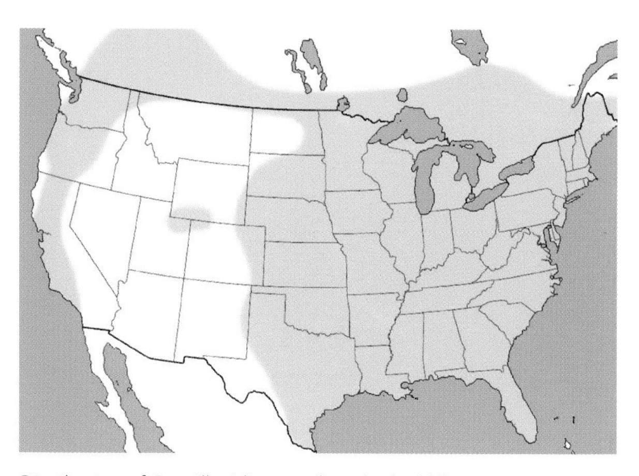

Distribution of *Coquillettidia perturbans* in the U.S.

Distribution *Coquillettidia perturbans* is widely distributed across the eastern half of the United States, parts of southern Canada, and several western states.

Culex

Adult *Culex* mosquitoes of the southeastern United States are generally brownish, with small patches of white scales. More than 1,200 species of *Culex* are known worldwide, 11 of which are found in our area. As a group, *Culex* mosquitoes feed on a wide variety of animals. Some species feed almost exclusively on birds, while others specialize on reptiles and amphibians. Others feed on birds and mammals. Some species feed on nearly any animal they come in contact with. Adults range from very small to medium sized. Most *Culex* larvae have a long respiratory siphon. They occur in a wide variety of temporary and permanent aquatic habitats, including marshes, lakes, ponds, tree holes, man-made containers, and floodwater pools. *Culex* females lay their eggs in floating clusters, called "rafts," on the surface of the water. Many species of *Culex* are important in the transmission of viruses that cause disease in humans.

Culex coronator Dyar and Knab, 1906

Subgenus *Culex*

Adults *Culex coronator* adults are medium sized and brown and white. This mosquito is one of just 2 species of *Culex* in our area that have pale bands on the hind legs, the other being *Culex tarsalis.* The 2 species are distinguished by the presence or absence of a white band on the proboscis. *Culex tarsalis* adults have a distinct white ring on the proboscis, while *Culex coronator* adults do not, although they may have a pale patch on the underside of the proboscis. Females of *Culex coronator* feed primarily upon mammals, such as deer and domestic animals, but also occasionally feed on birds. Although *Culex coronator* is not considered to be a species of major health importance, several pathogens have been isolated from field-collected females. Venezuelan equine encephalitis virus has been isolated from Central American females, and St. Louis encephalitis virus has been detected in females from the Caribbean.

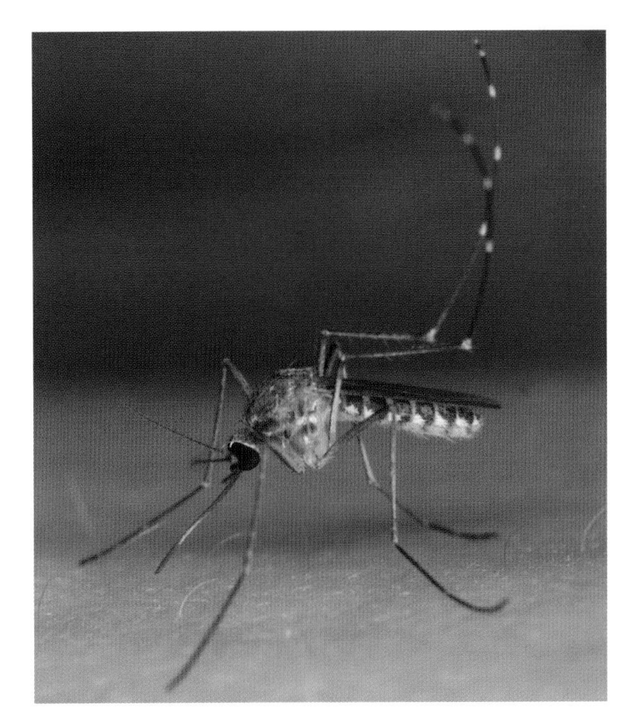

Culex coronator female

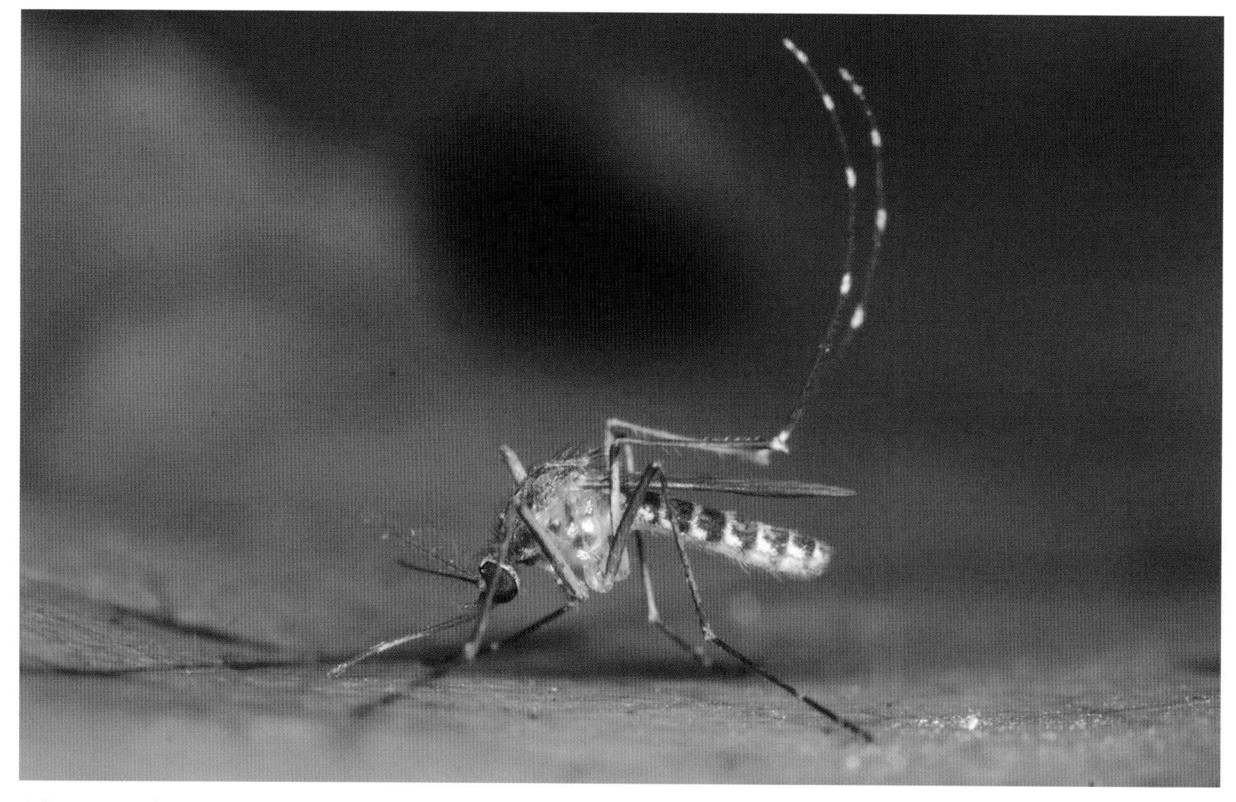

Culex coronator female

Culex coronator larva

Larvae Larvae of *Culex coronator* are small to medium sized. Setae of the head are multibranched. The siphon is quite long and thin and bears several small spines just before its apex. Four pairs of branched setae arise along the length of the siphon. Larvae are found in a variety of habitats, including rain-filled depressions and water-filled man-made containers, especially those in direct sunlight.

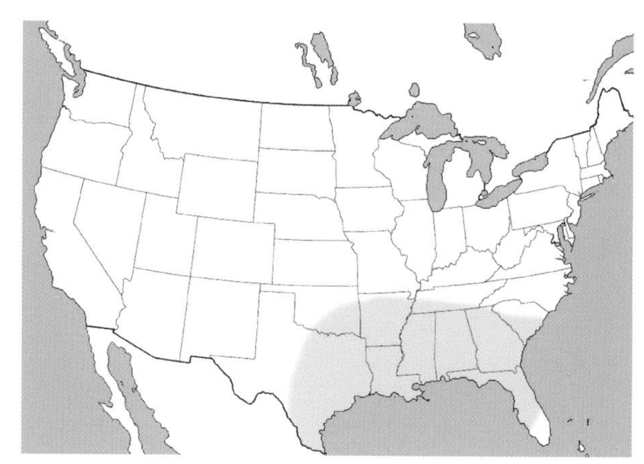

Distribution of *Culex coronator* in the U.S.

Distribution *Culex coronator* is a recent arrival to the southeastern United States. Prior to the year 2000, this mosquito was found only in Texas, New Mexico, and Arizona (also in Central and South America). Since 2001, *Culex coronator* has been found in Louisiana, Mississippi, Alabama, Florida, Georgia, North Carolina, and South Carolina.

Culex erraticus (Dyar and Knab, 1906)

Subgenus *Melanoconion*

Adults Adults of *Culex erraticus* are small, dark brown, and ornamented with patches of white scales on the scutum and abdomen. The abdomen is often dark scaled dorsally but may have narrow white basal bands. Patches of white scales are found on the lateral and ventral areas of the abdomen. Alternating patches of light and dark scales on the abdomen give it a somewhat striped appearance. Females of *Culex erraticus* are often difficult to distinguish from 2 other *Culex* species in our area, *Culex peccator* and *Culex pilosus.* All 3 species are small and dark, with small patches of white scales. However, only females of *Culex erraticus* have a median patch of white scales on the mesepimeron (a lateral body plate just below the base of the wing). *Culex erraticus* females feed on the blood of a wide variety of animals, including many species of mammals, birds, reptiles, and amphibians. Some of the most common hosts include deer and large wading birds (herons and egrets). *Culex erraticus* may be important in the transmission of eastern equine encephalitis virus in the southeastern United States.

Larvae Larvae of *Culex erraticus* are small. Seta 6C of the head is unbranched or has only 2 branches. The

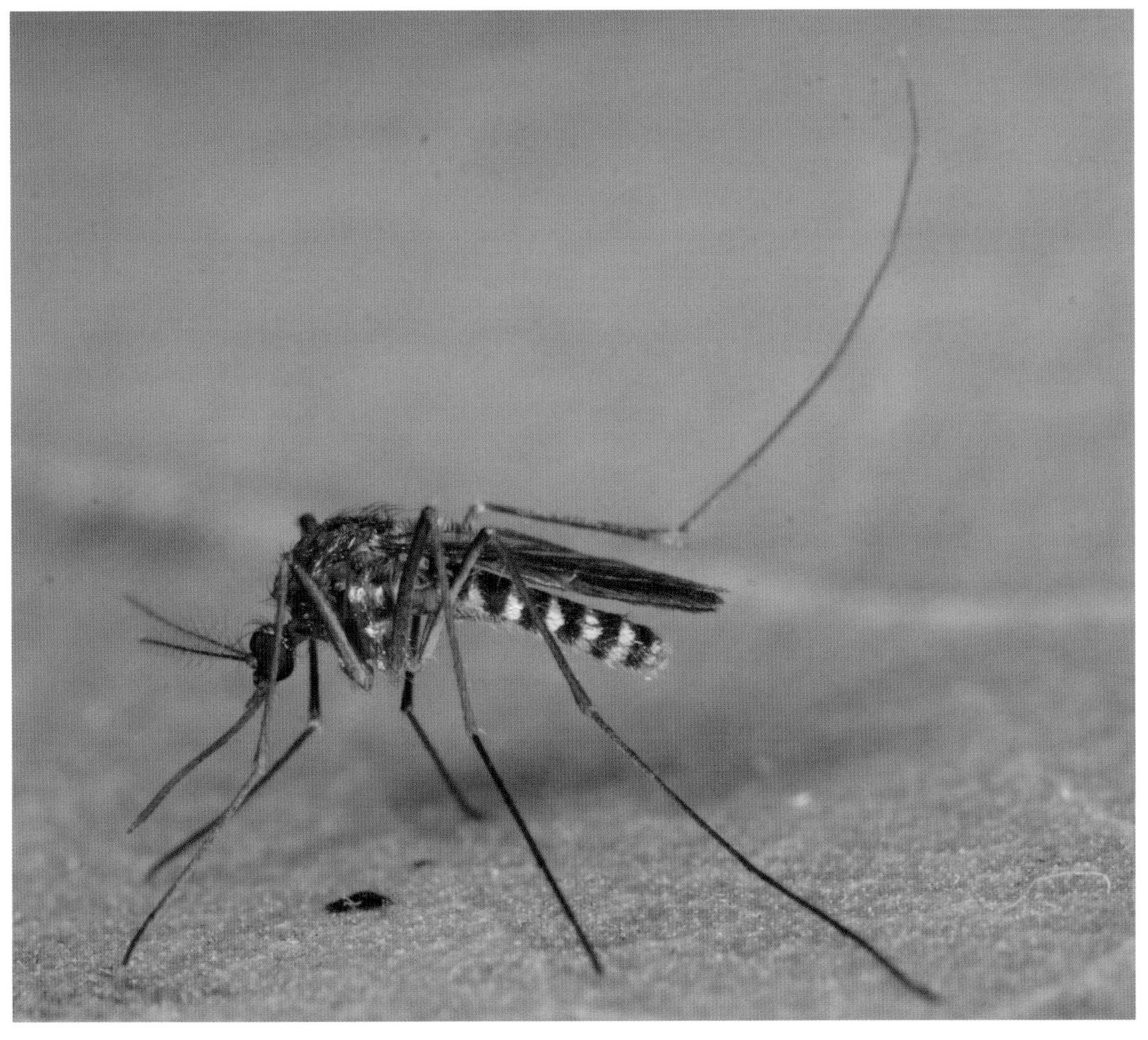

Culex erraticus female

Culex erraticus larva

Culex erraticus larva

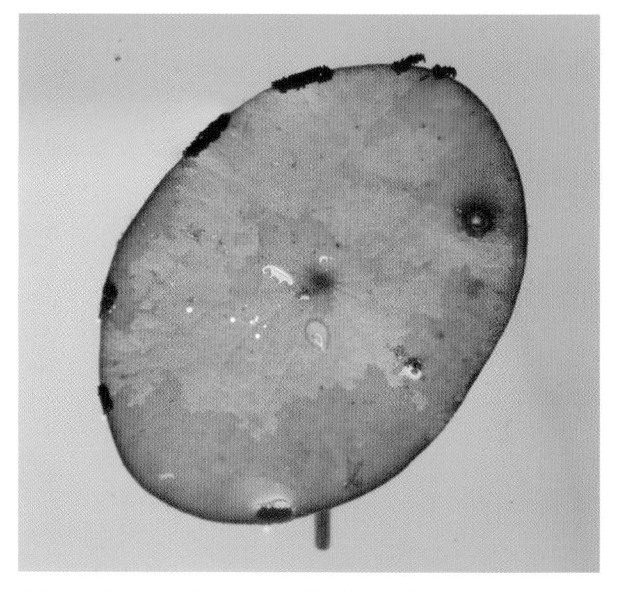

Culex erraticus egg rafts on watershield (*Brasenia*)

antennae are white with dark tips. The color of the thorax and abdomen is quite variable and may have a greenish hue or be cream or straw-colored, or even black. The siphon is long and thin and slightly curved dorsally. Five pairs of densely branched setae are borne in a line along the ventral edge of the siphon, and 1 pair of branched setae occurs dorsal to those. Larvae are found in permanent and semipermanent ponds and marshes, especially those with rooted aquatic plants. Eggs are typically laid in clusters on the edges of leaves of aquatic plants, such as watershield *(Brasenia schreberi)* and duckweed *(Lemna).*

Distribution *Culex erraticus* is found throughout the eastern United States, south of the Great Lakes. This common mosquito also occurs in Central America and northern South America.

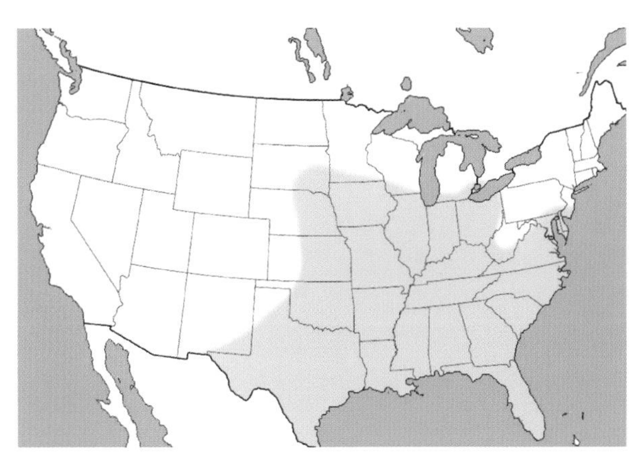

Distribution of *Culex erraticus* in the U.S.

Culex nigripalpus Theobald, 1901

Subgenus *Culex*

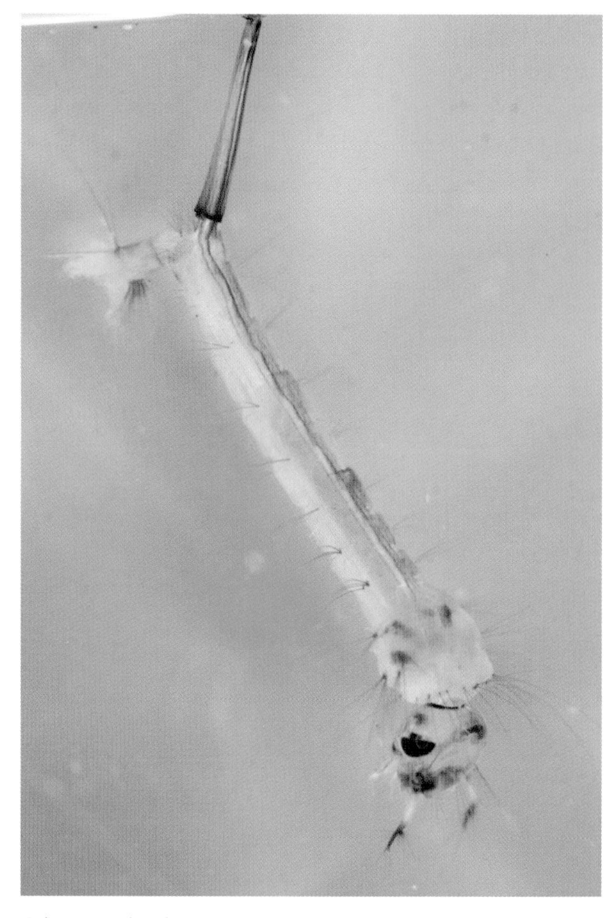
Culex nigripalpus larva

Adults Adults of *Culex nigripalpus* are small to medium sized, with a light brown thorax and dark brown abdomen. The patches of white scales on the thorax are small and rarely consist of more than 6 scales. Abdominal segments have basal patches of white scales laterally and may also have narrow pale basal bands. The palps and proboscis are dark. The tarsi and wings are dark scaled. Females feed mainly on the blood of mammals and birds but will occasionally take blood from amphibians and reptiles. *Culex nigripalpus* is important in the transmission of the viruses that cause St. Louis encephalitis, eastern equine encephalitis, and West Nile encephalitis.

Larvae Larvae of *Culex nigripalpus* are medium sized. Setae of the head are multibranched. The thorax has

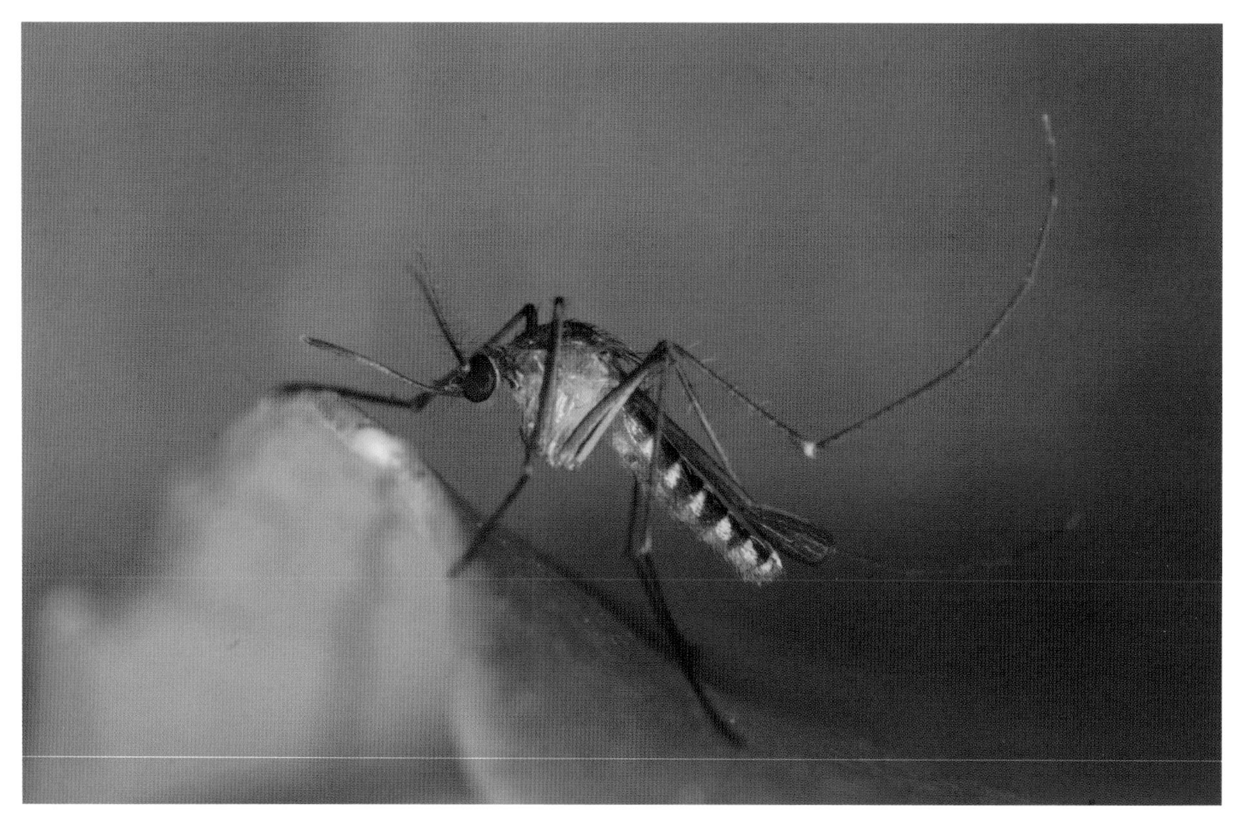
Culex nigripalpus female

many tiny spines, giving it a rough appearance. The siphon is long and thin and bears 4 pairs of branched or unbranched setae along its length. Larvae are most commonly found in ditches, grassy pools, and marshes but are occasionally found in water-filled man-made containers.

Distribution *Culex nigripalpus* is found in the southeastern coastal states and western portions of Kentucky and Tennessee.

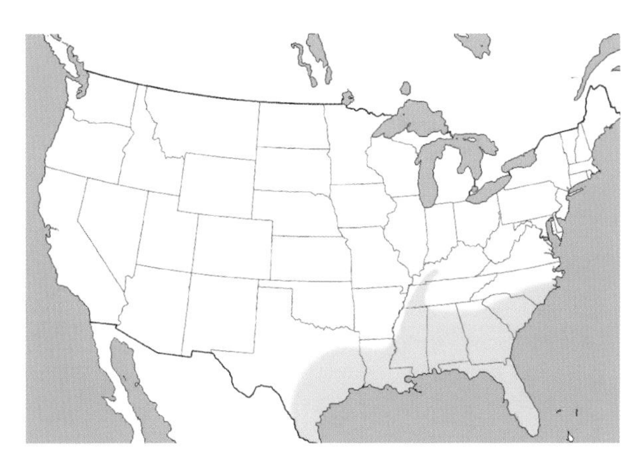

Distribution of *Culex nigripalpus* in the U.S.

Culex peccator Dyar and Knab, 1909

Subgenus *Melanoconion*

Adults *Culex peccator* adults are small and dark brown and have small patches of white scales on the thorax and abdomen. They resemble adults of *Culex erraticus* and *Culex pilosus.* These species are distinguished from one another by differences in the number and arrangement of pale scales on the thorax and abdomen. *Culex peccator* females feed mainly on the blood of amphibians and reptiles but occasionally feed on mammals and birds. Common hosts include bullfrogs, cottonmouth snakes, wading birds, white-tailed deer, and alligators. Eastern equine encephalitis virus has been detected in wild-caught females of *Culex peccator,* suggesting that this mosquito may be involved in the transmission of this virus in the southeastern United States.

Larvae Larvae of *Culex peccator* are quite small. Middle head setae (5C and 6C) are unbranched or have up to 3 branches. The siphon is long, thin, and slightly curved and bears a broad dark band near its center. Five pairs of densely branched setae are borne in a line along the ventral edge of the siphon, and 2 pairs of 2-branched setae occur dorsal to those. Larvae are

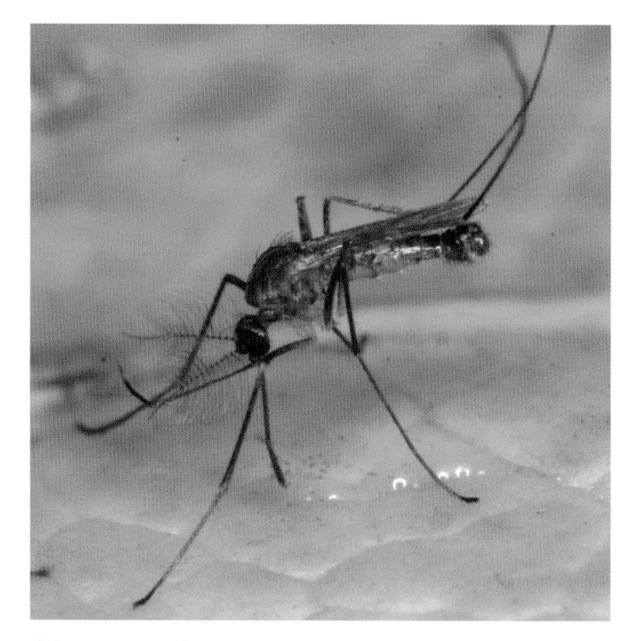

Culex peccator male

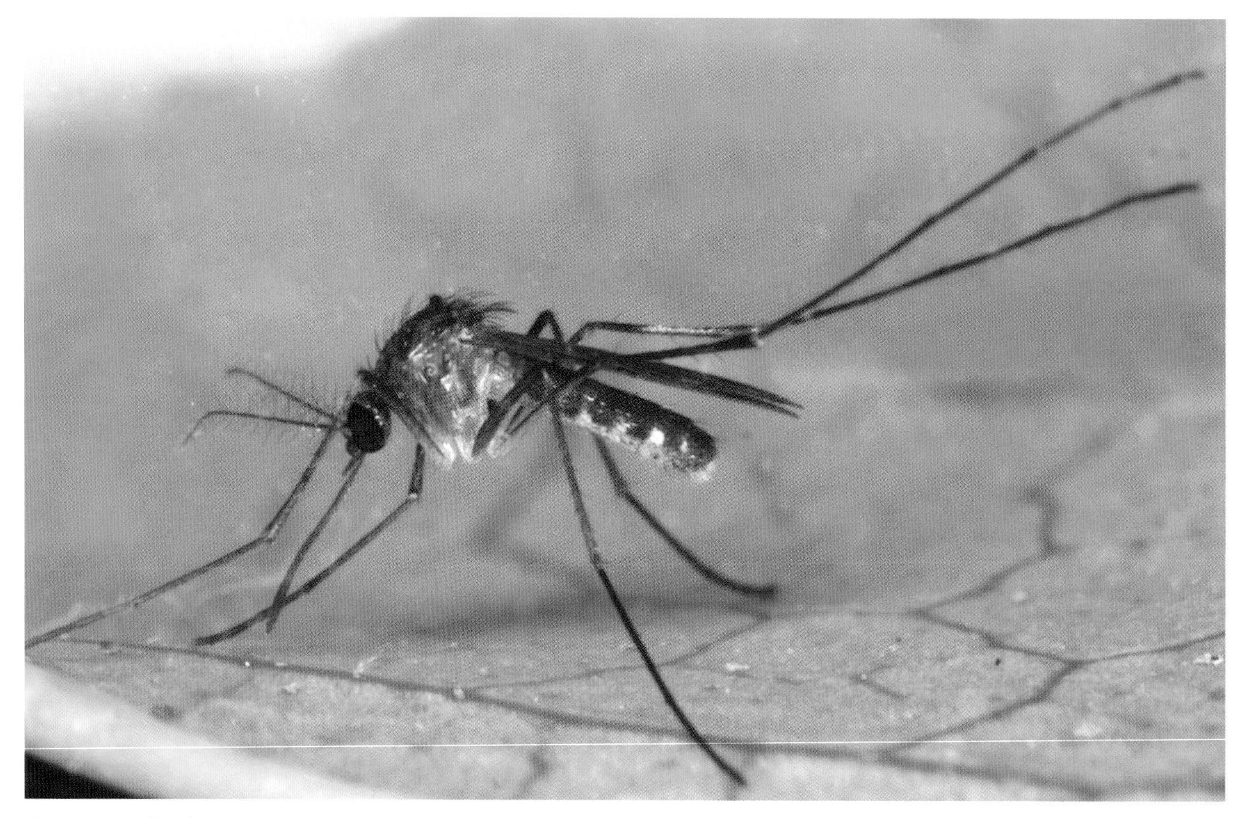

Culex peccator female

Culex peccator larva

most commonly encountered in vegetated wetlands and sluggish streams, especially those dammed by beavers.

Distribution *Culex peccator* occurs throughout the southeastern United States.

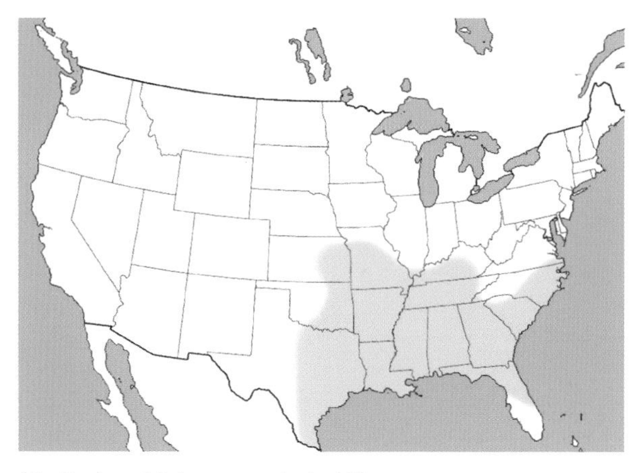

Distribution of *Culex peccator* in the U.S.

Culex pilosus (Dyar and Knab, 1906)

Subgenus *Melanoconion*

Adults Adults of *Culex pilosus* are small, dark brown, and ornamented with patches of white scales on the thorax and abdomen. Females of *Culex pilosus* closely resemble females of 2 other *Culex* species in our area, *Culex erraticus* and *Culex peccator.* These species are distinguished by slight differences in the shading of their exoskeleton (see key to adults) and differences in the size and location of small patches of white scales on the thorax. *Culex pilosus* females feed mainly on the blood of reptiles and mammals, with lizards being the most common hosts. *Culex pilosus* is not known to be important in the transmission of any human pathogens in the United States.

Larvae Larvae of *Culex pilosus* are small. Middle head setae (5C and 6C) are unbranched or have only 2 branches. The siphon is quite short, slightly curved, and bears numerous long, multibranched setae along its entire length. When disturbed, larvae will sub-

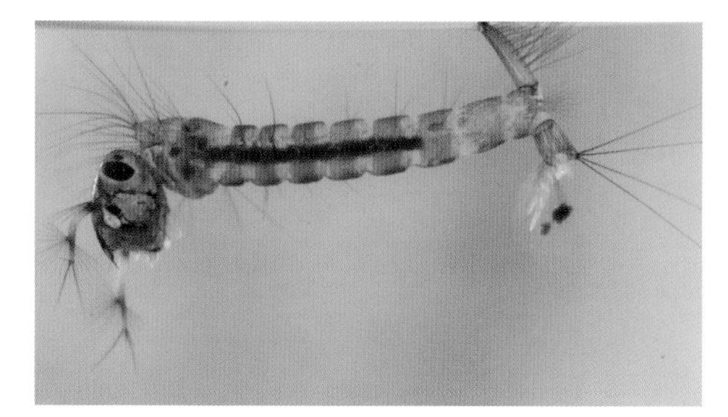

Culex pilosus larva

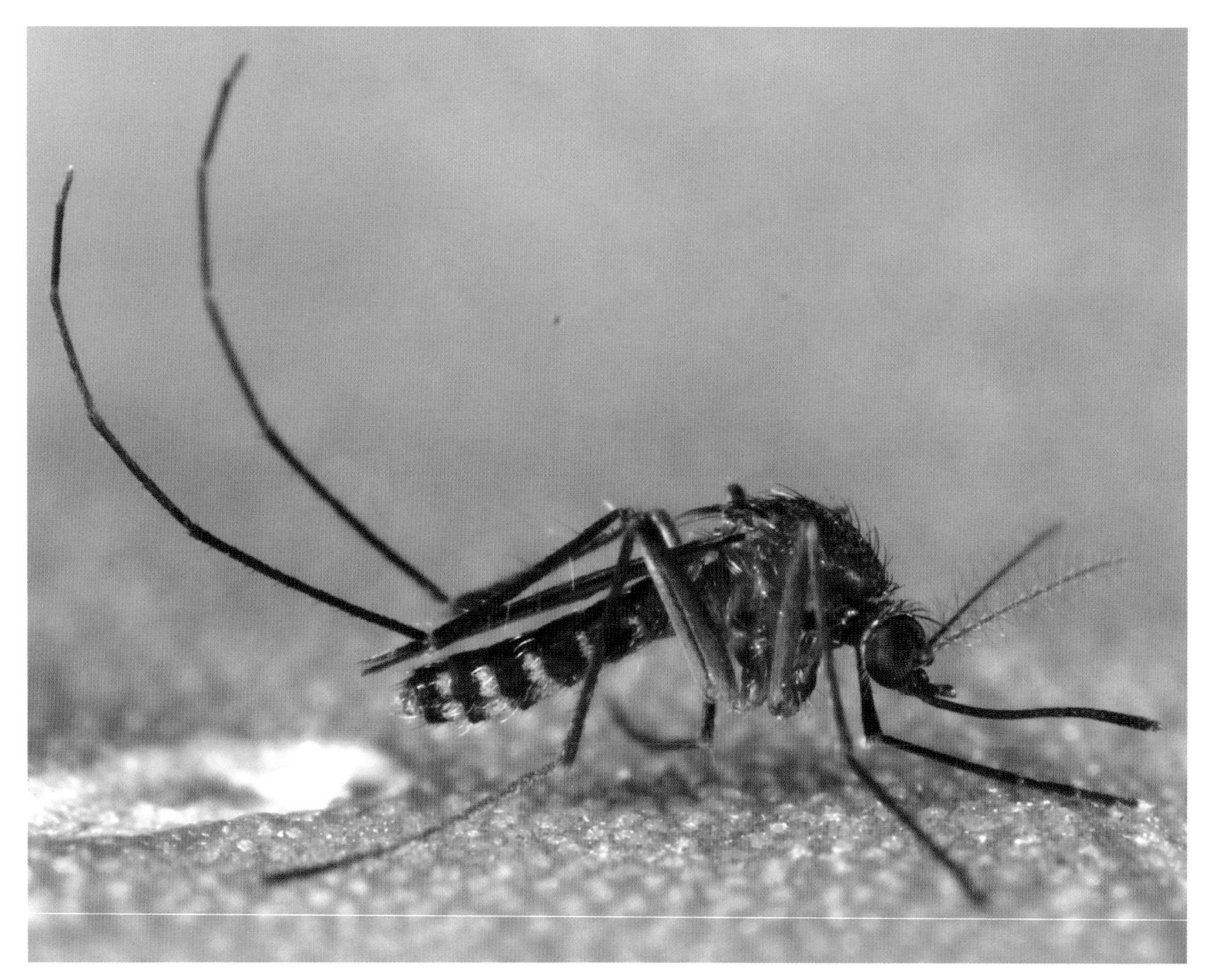

Culex pilosus female

merge, lie on their backs, and wiggle. They are found in temporary pools following heavy summer rains.

Distribution *Culex pilosus* is found in the southeastern coastal United States, and occasionally elsewhere. It is rare throughout most of its range.

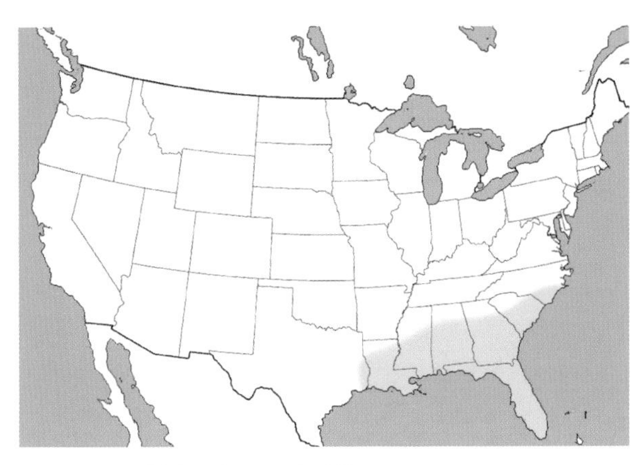

Distribution of *Culex pilosus* in the U.S.

Culex pipiens Linnaeus, 1758, and *Culex quinquefasciatus* Say, 1823

Subgenus *Culex*

Northern house mosquito *and* southern house mosquito

Culex pipiens and *Culex quinquefasciatus* are almost indistinguishable as adults, have overlapping ranges and quite similar biologies, and interbreed in places where they occur together. They are therefore treated together in this book.

Adults Adults of *Culex pipiens* and *Culex quinquefasciatus* are small to medium sized, with a light brown thorax and darker brown abdomen. Abdominal segments have a pale dorsal band along the basal edge, which narrows where it joins the lateral patches of pale scales. The proboscis, palps, tarsi, and wings are dark. Females feed mostly on songbirds but also on humans and other mammals. They will enter homes in search of blood—hence the common names "northern house mosquito" (*Culex pipiens*) and "southern house mosquito" (*Culex quinquefasciatus*). *Culex pipiens* and *Culex quinquefasciatus* are important in the transmission of many bird pathogens that cause diseases in humans. In our area, both of these species transmit West Nile virus, eastern equine encephalitis virus, and St. Louis encephalitis virus, as well as the filarial worm that causes heartworm disease in dogs. In other parts of the world, females of *Culex quinquefasciatus*

Culex quinquefasciatus larva

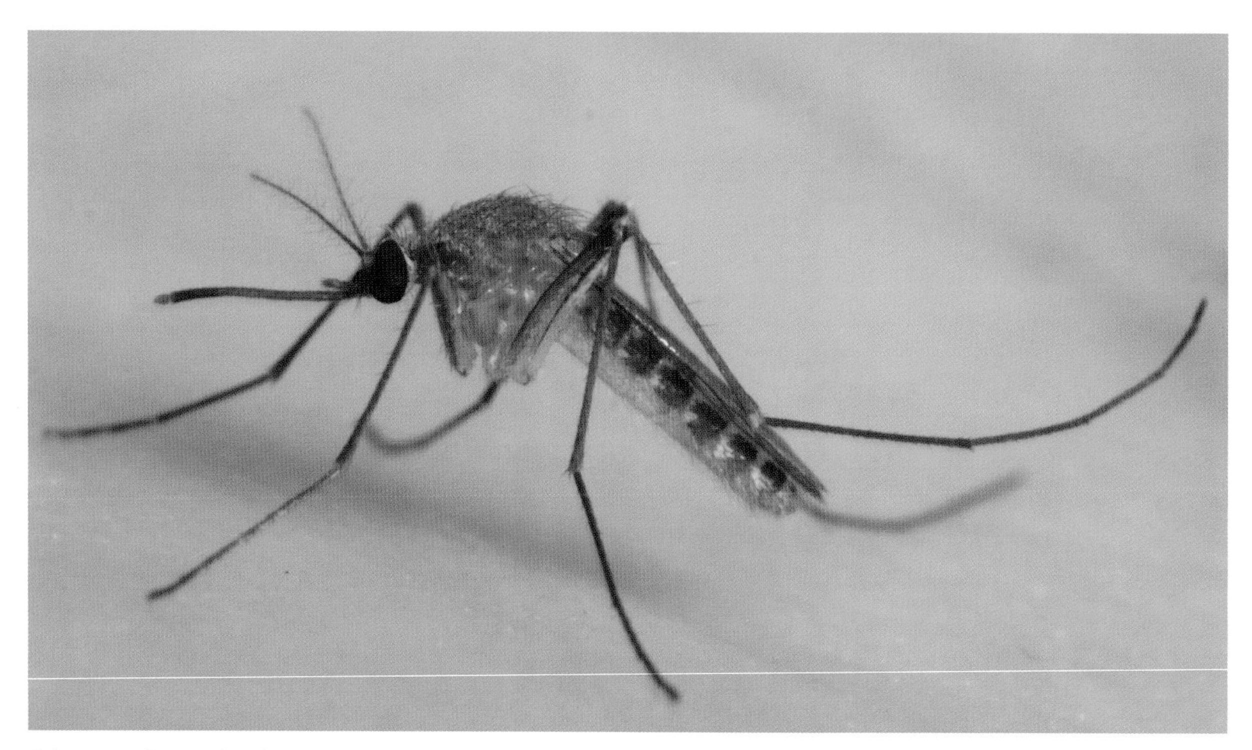

Culex quinquefasciatus female

Culex quinquefasciatus egg rafts

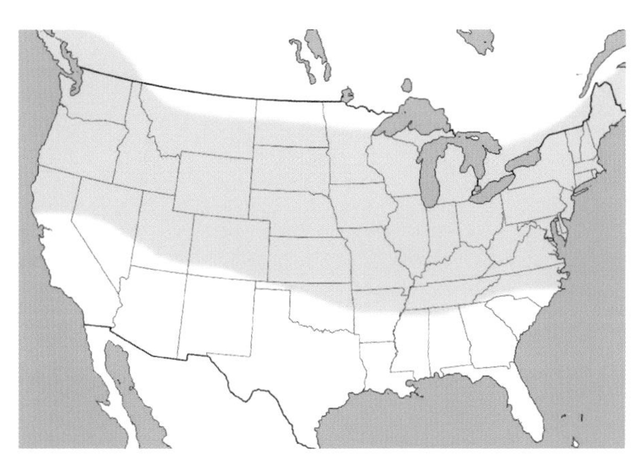

Distribution of *Culex pipiens* in the U.S.

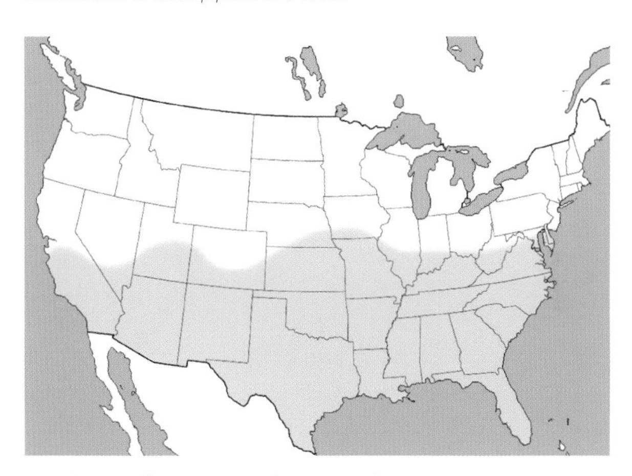

Distribution of *Culex quinquefasciatus* in the U.S.

transmit another filarial worm, *Wuchereria bancrofti*, to humans. Chronic infection with this filarial worm blocks lymph vessels and may lead to enlarged, grotesque limbs, a condition known as elephantiasis.

Larvae Larvae of *Culex pipiens* and *Culex quinquefasciatus* are medium sized. Setae of the head are multibranched. The siphon is shorter than that of most other *Culex* species in our area. It bears 4 to 5 pairs of branched setae, 1 of which is dorsally out of line with the others. Larvae of these 2 species can be found in a variety of habitats, often tolerating high levels of organic pollutants. They may be found in marshes, roadside ditches, discarded automobile tires, cesspools, sewage catch basins, and any number of water-filled man-made containers.

Distribution *Culex pipiens* is found throughout the northern United States and parts of Canada, while *Culex quinquefasciatus* is limited to southern states. Where the two species co-occur, hybrids are more common than either of the individual species.

Culex restuans Theobald, 1901

Subgenus *Culex*

Adults Adults of *Culex restuans* are medium sized, with brown and white coloration. The abdomen has a broad white band at the base of each segment and the scutum has 2 small pale dots (on most specimens). These features can be used to tell adults of *Culex restuans* apart from those of 2 similar species, *Culex pipiens* and *Culex quinquefasciatus*. The tarsi, wings, and palps of *Culex restuans* adults are dark. The probos-

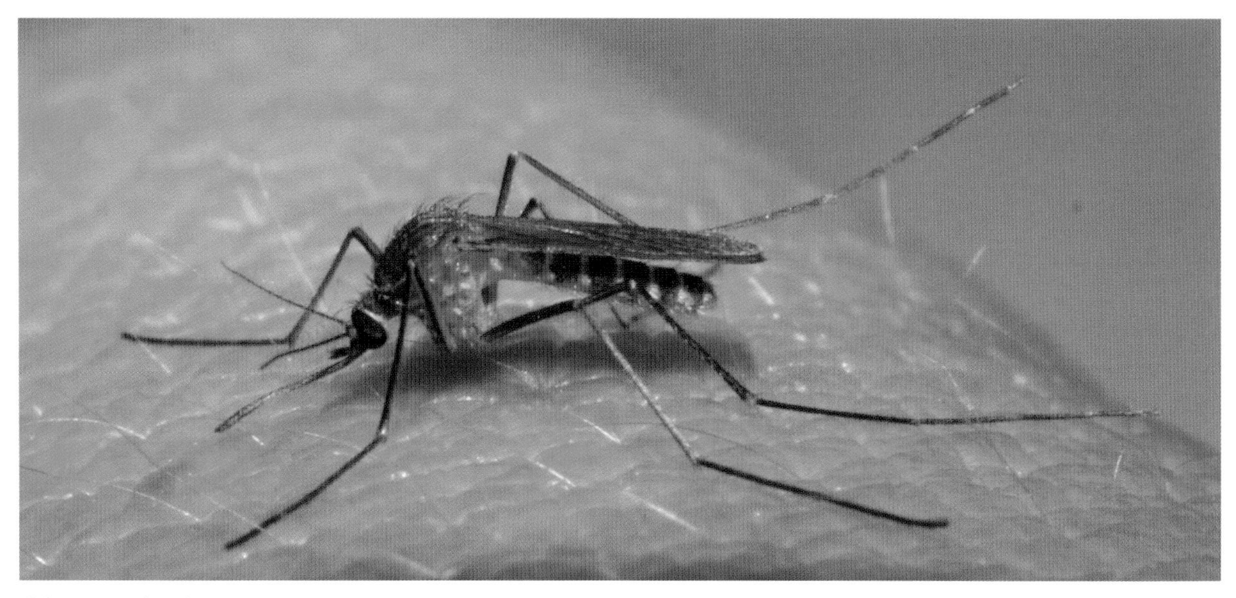

Culex restuans female

Scutum and abdomen of *Culex restuans* female

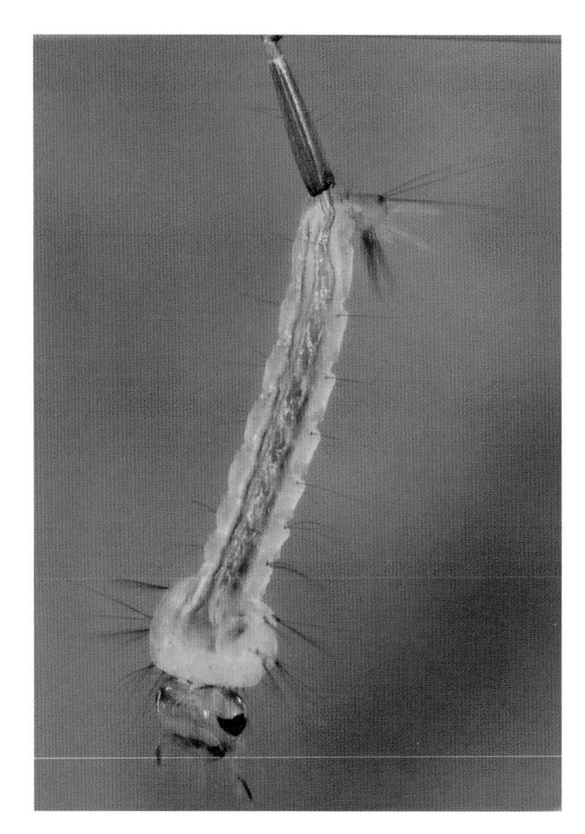

Culex restuans larva

Culex restuans egg raft

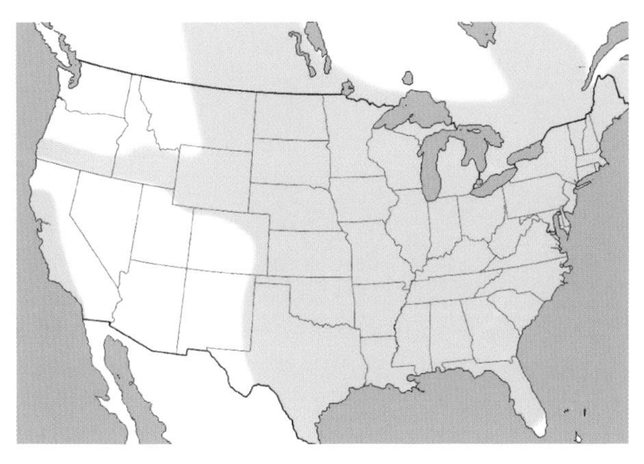

Distribution of *Culex restuans* in the U.S.

cis is dark scaled but may have a patch of pale scales on the underside, near its midpoint. Females feed mainly on songbirds but are reported to occasionally feed on humans. *Culex restuans* transmits the viruses that cause West Nile encephalitis and eastern equine encephalitis in humans.

Larvae Larvae of *Culex restuans* are medium sized. Setae of the head are multibranched. The siphon is moderately long and bears 3 pairs of unbranched setae placed irregularly along its length and 1 pair of smaller branched setae near its apex. Spring is the most common time of year to encounter larvae of *Culex restuans* in the southeastern United States. They are found in semipermanent pools, puddles, and ditches but are also encountered in man-made containers.

Distribution *Culex restuans* is found throughout the eastern half of the United States and southern Canada as well as several western U.S. states.

Culex salinarius Coquillett, 1904

Subgenus *Culex*

Salt-marsh *Culex*

Abdomen of *Culex salinarius* female

Adults Adults of *Culex salinarius* are medium-sized, brown mosquitoes with patches of pale scales on the thorax and abdomen. Abdominal segments have lateral patches of pale bronze-colored scales basally, and some segments have a narrow band of bronze-colored scales. The terminal segments (7 and 8) of the abdomen are covered in pale bronze-colored scales, sometimes described as "dingy yellow." Females of *Culex salinarius* are most likely to be confused with those of *Culex nigripalpus. Culex salinarius* has a prominent patch of pale scales on the mesepimeron, whereas *Culex nigripalpus* does not. *Culex salinarius* reaches its greatest abundance in coastal areas (and is some-

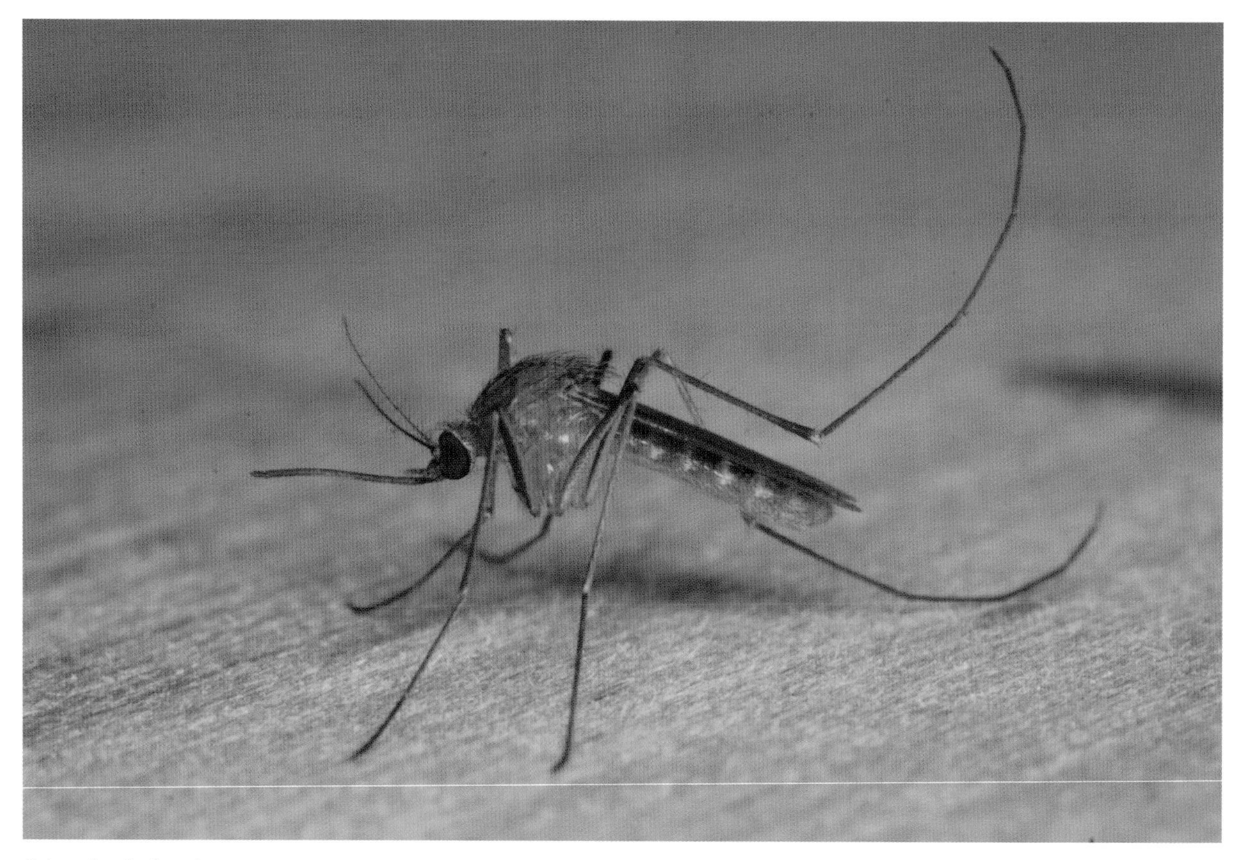

Culex salinarius female

Culex salinarius larva

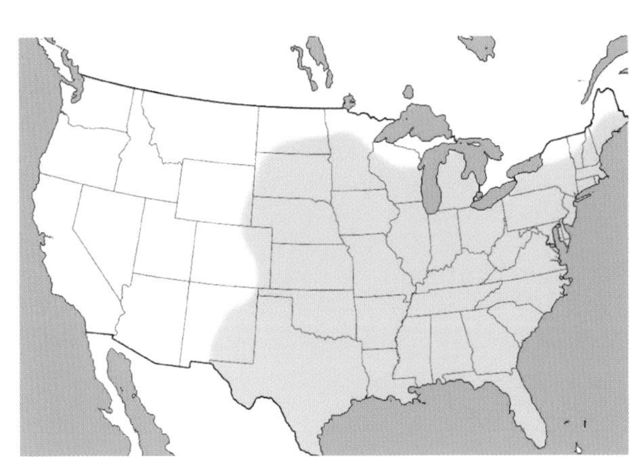

Distribution of *Culex salinarius* in the U.S.

times called the "salt-marsh *Culex*") but also occurs inland. Females feed on the blood of mammals and birds. *Culex salinarius* is considered to be a "bridge vector" of West Nile virus, eastern equine encephalitis virus, and St. Louis encephalitis virus, transmitting the virus from birds to humans and horses.

Larvae Larvae of *Culex salinarius* are medium sized. Setae of the head are long and multibranched. The siphon is long and thin and bears 4 pairs of branched setae. Larvae are most commonly found in densely vegetated areas of freshwater impoundments and coastal marshes.

Distribution *Culex salinarius* occurs throughout the eastern two-thirds of the United States and in a few western locations.

Culex tarsalis Coquillett, 1896

Subgenus *Culex*

Western encephalitis mosquito

Adults *Culex tarsalis* adults are medium sized and, like other *Culex* species in our area, brown with patches of white scales. However, unlike most other *Culex* species in our area, adults of *Culex tarsalis* have white bands on the tarsal segments of their hind legs. Only 1 other *Culex* species in our area, *Culex coronator*, has white bands on its hind legs. The 2 species are distinguished by the presence or absence of a white band on the proboscis; *Culex tarsalis* has a white band, while *Culex coronator* does not. Females of *Culex tarsalis* feed mainly on the blood of birds but will also bite humans and other mammals. *Culex tarsalis* is important in the transmission of a number of viruses in the western United States, such as West Nile virus, western equine encephalitis virus, and California encephalitis virus. Due to the rarity of this mosquito in the east, it is unlikely to be important in the transmission of human pathogens in our area.

Larvae Larvae of *Culex tarsalis* are medium sized. Setae of the head are long and multibranched. The siphon has 5 pairs of multibranched setae arising in a straight line along the ventral portion. Larvae are

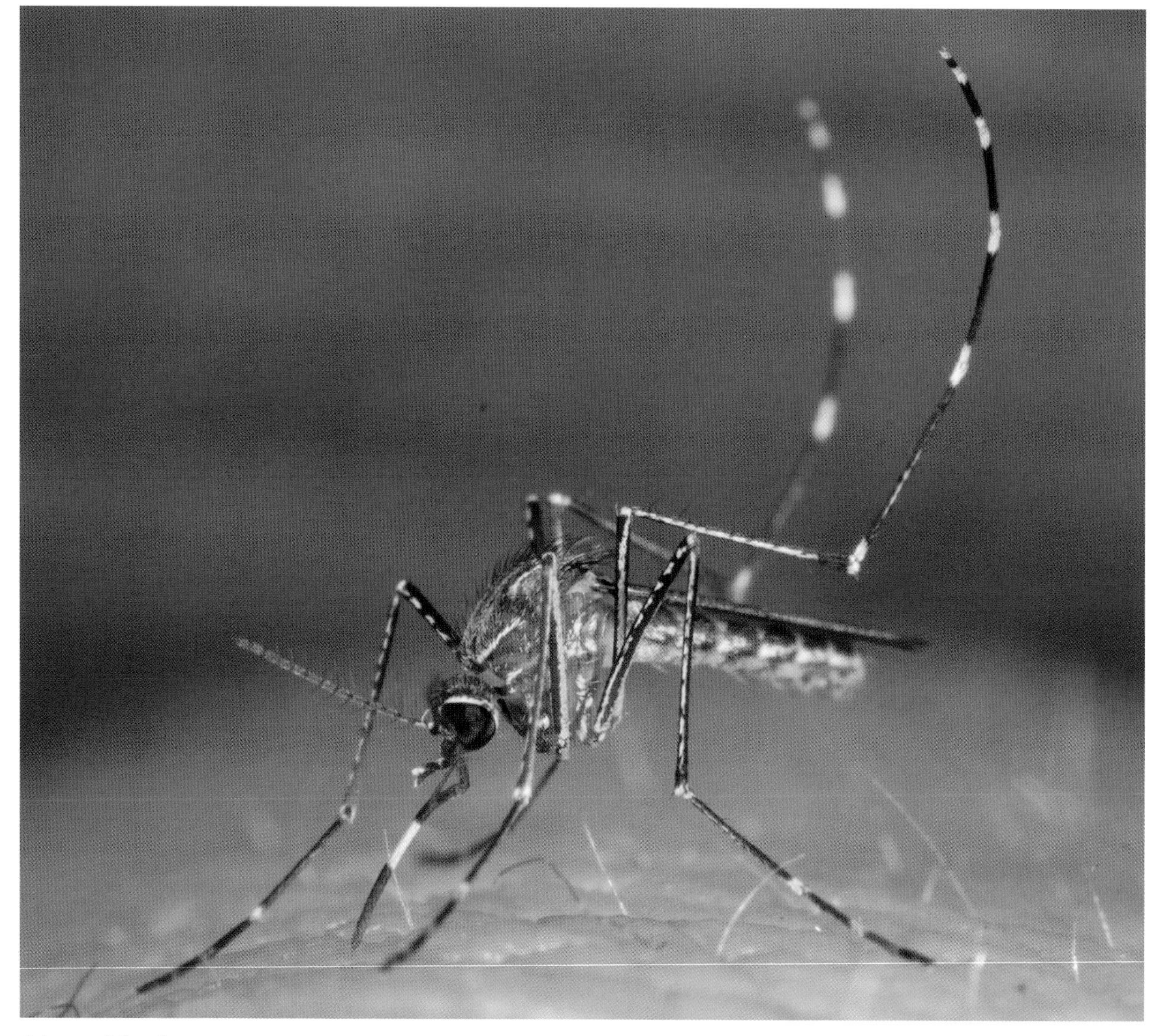

Culex tarsalis female

Culex tarsalis larvae

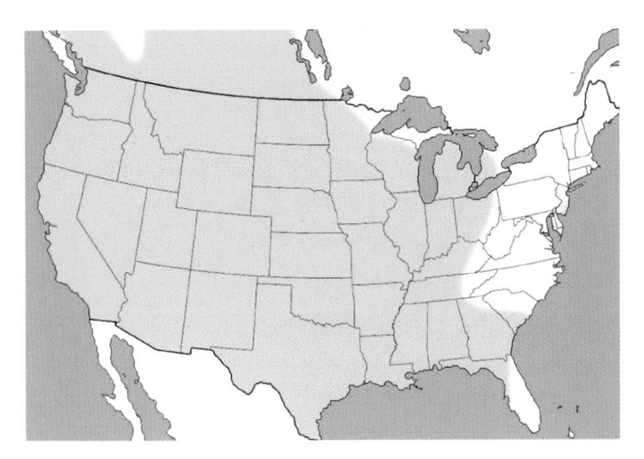

Distribution of *Culex tarsalis* in the U.S.

found in a variety of aquatic habitats, including polluted water and water of high organic content.

Distribution *Culex tarsalis* is found throughout much of the United States and southern Canada but is much more common in the western portion of its range than it is in the east.

Culex territans Walker, 1856

Subgenus *Neoculex*

Northern frog-biting *Culex*

Adults Adults of *Culex territans* are small, with a light brown thorax and dark brown abdomen. The white scales on the abdomen are located on the apical part of each segment, either as a narrow band or as lateral patches, a pattern that is unique among the *Culex* species in our area. The underside of the abdomen is entirely white. The tarsi, palps, and proboscis are dark and the wings have only dark scales. Females feed almost exclusively on frogs but will occasionally take blood from other animals, including lizards, birds, and mammals. *Culex territans* is not known to be important in the transmission of any human pathogens in the United States.

Larvae Larvae of *Culex territans* are small. Middle head setae (5C and 6C) are unbranched or have 2–3 branches. The siphon is quite long and thin, with 4–5 pairs of branched setae in a line. Larvae can be found throughout the year but are most abundant in late winter and early spring in the southeastern United States. They occur in a wide variety of natural and man-made habitats, including ponds, marshes, swamps, vernal pools, tree holes, ditches, and discarded automobile tires.

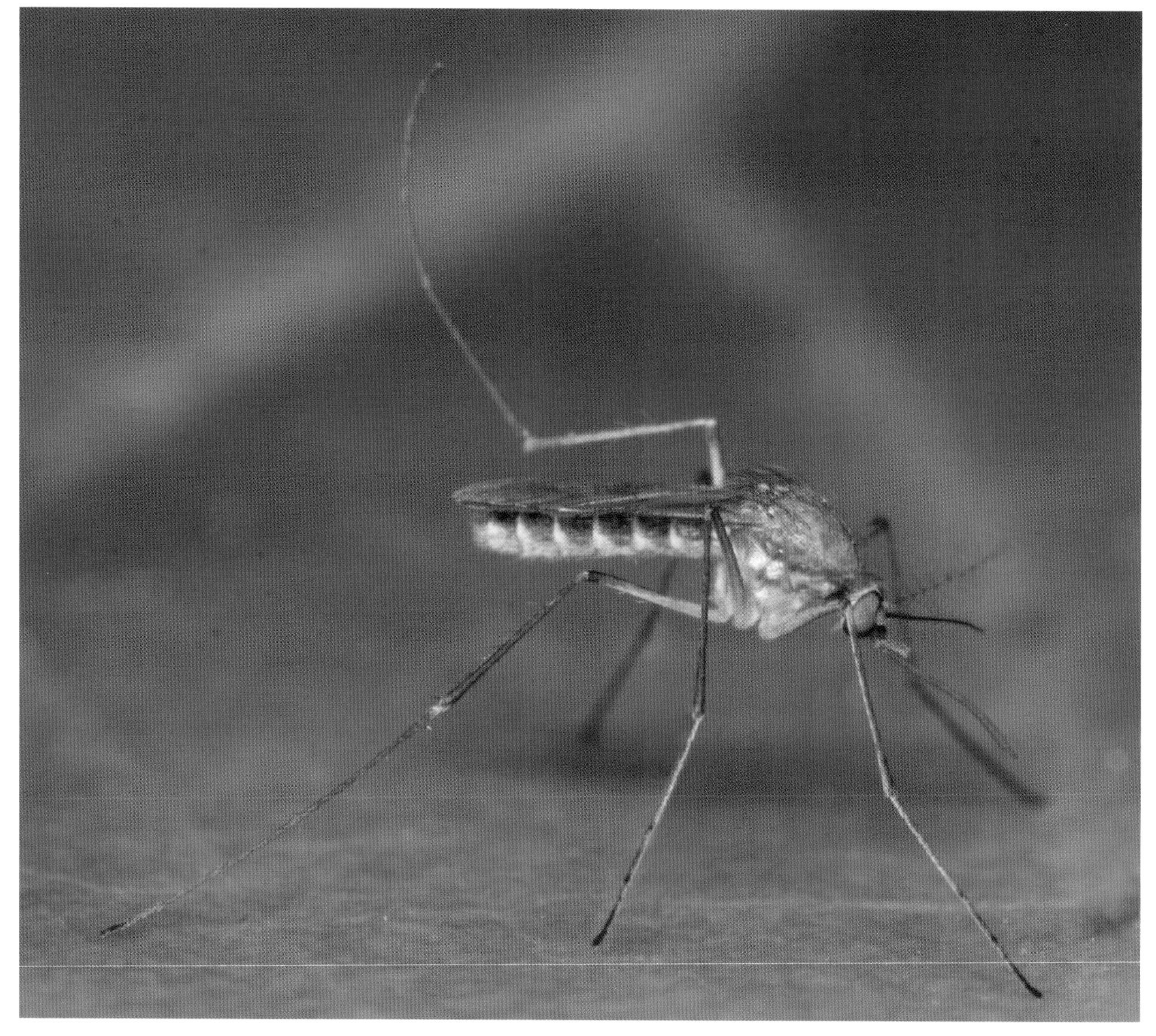

Culex territans female

Culex territans larva

Distribution *Culex territans* is found throughout the United States, excluding the Desert Southwest. It is also found throughout much of southern Canada.

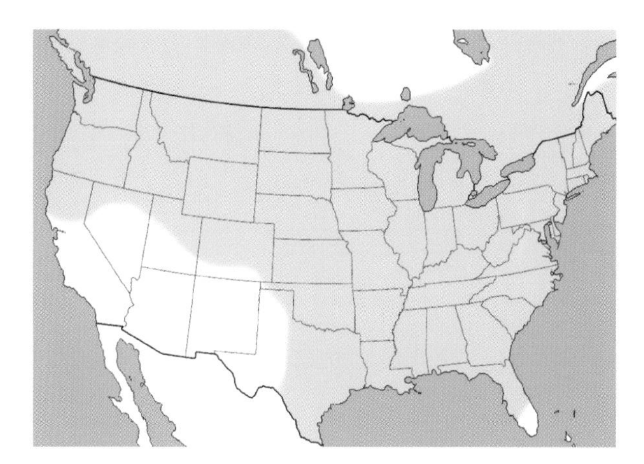

Distribution of *Culex territans* in the U.S.

Culiseta

Unlike most other groups of mosquitoes, *Culiseta* reaches its greatest diversity in temperate regions. Most of the nearly 40 known species occur in northern North America, northern Asia, or Australia. A few species are known from the tropics of Africa, Asia, and Central America, but none occur in South America. Two species occur in the southeastern United States. Adults of most *Culiseta* species are fairly large and superficially resemble adults of the genus *Culex*. The underside of the wing has a patch of setae at its base. Females of most species feed on birds and/or mammals, and a few feed on reptiles. Several species will feed on humans and are directly or indirectly important in the transmission of encephalitis viruses in North America. The larvae of most *Culiseta* species inhabit marshes, vegetated ponds, or woodland pools. A few are found in rock pools, tree holes, or underground puddles. *Culiseta* larvae have a characteristic seta at the base of the siphon.

Culiseta inornata (Williston, 1893)

Subgenus *Culiseta*

Winter marsh mosquito

Adults *Culiseta inornata* adults are large, with a salt-and-pepper appearance. The body is clad in brown and cream-colored scales. The legs, proboscis, and palps are dark brown, speckled with pale scales. The front edges of the wings have dark and pale scales intermixed. Like other members of its genus, adults of *Culiseta inornata* have a patch of hairlike setae at the base of the wing (on the underside). Females feed almost exclusively on the blood of large mammals, such as deer, cows, and horses, and only occasionally bite humans. Adults are most active in the winter in the southern United States. *Culiseta inornata* is not known to transmit any human pathogens in the eastern United States.

Larvae Larvae of *Culiseta inornata* are large and have a stout siphon with a pair of large setae basally, and a

Culiseta inornata female. Courtesy of Treasure Tolliver and the author.

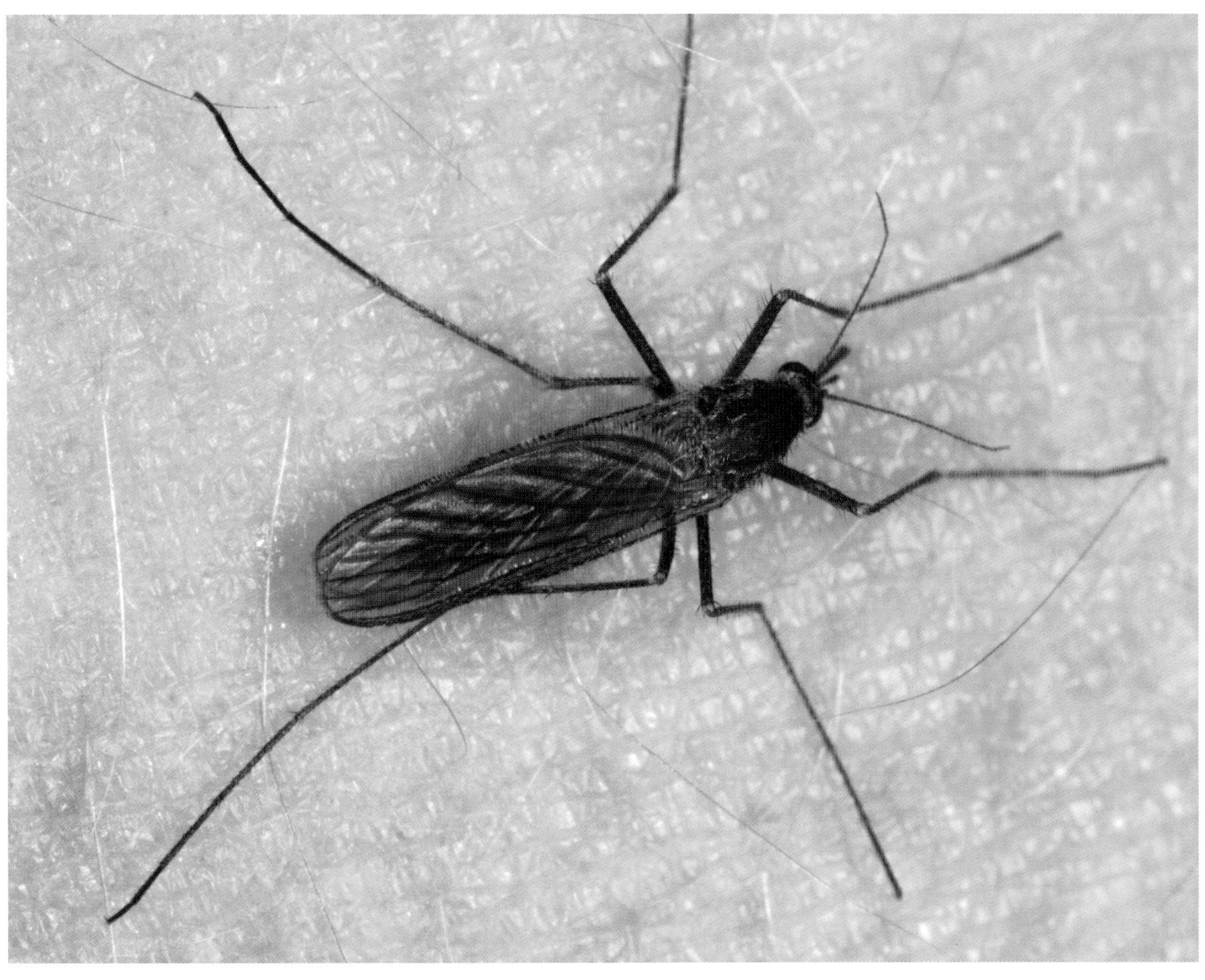

Culiseta inornata female. Courtesy of Treasure Tolliver and the author.

short row of pecten teeth followed by paired rows of hairlike setae. Antennae are short and thin. Larvae are most commonly encountered in rain-filled pools and ditches but are occasionally found in man-made containers and neglected swimming pools. They are usually encountered in fall and winter in our area.

Distribution *Culiseta inornata* is found throughout the United States and much of western Canada.

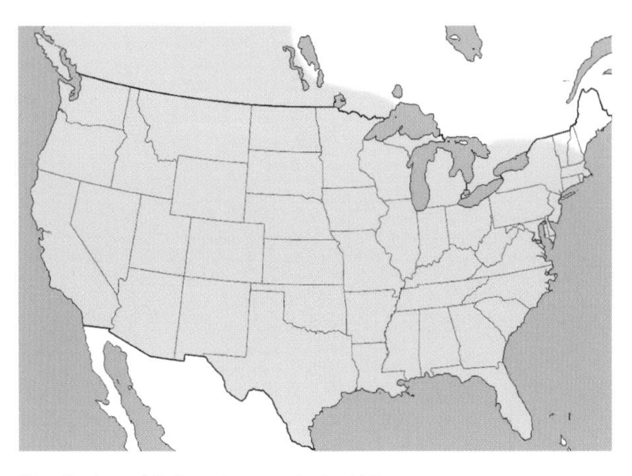

Distribution of *Culiseta inornata* in the U.S.

Culiseta melanura (Coquillett, 1902)

Subgenus *Climacura*

Dusky encephalitis mosquito

Adults Adults of *Culiseta melanura* are medium sized, dark brown, and have few distinctive markings. The basal portion of each abdominal segment is adorned with a small patch of pale scales on its lateral edge and occasionally has narrow pale basal bands. Pale scales may also be found on the underside of the abdomen. The remainder of the body is clad in dark brown (chocolate-colored) scales. The proboscis is particularly long and curves downward noticeably. Adults have a patch of hairlike setae at the base of the wing (on the underside). Females feed almost exclusively on the blood of songbirds, such as the northern cardinal, but occasionally feed upon reptiles and mammals. *Culiseta melanura* is important in the transmission of eastern equine encephalitis virus in the United States. Although this mosquito does not transmit the virus to humans directly, it efficiently transmits the virus between birds, enabling the virus to reach pre-epidemic levels.

Larvae Larvae of *Culiseta melanura* are medium sized and usually quite dark. The siphon is long and slender, with a pair of large setae basally and paired rows of very short, branched setae along the ventral midline. The larvae occur in lowland freshwater swamps and are most commonly encountered in water-filled holes in the ground and in depressions created by the upturned root masses of fallen trees.

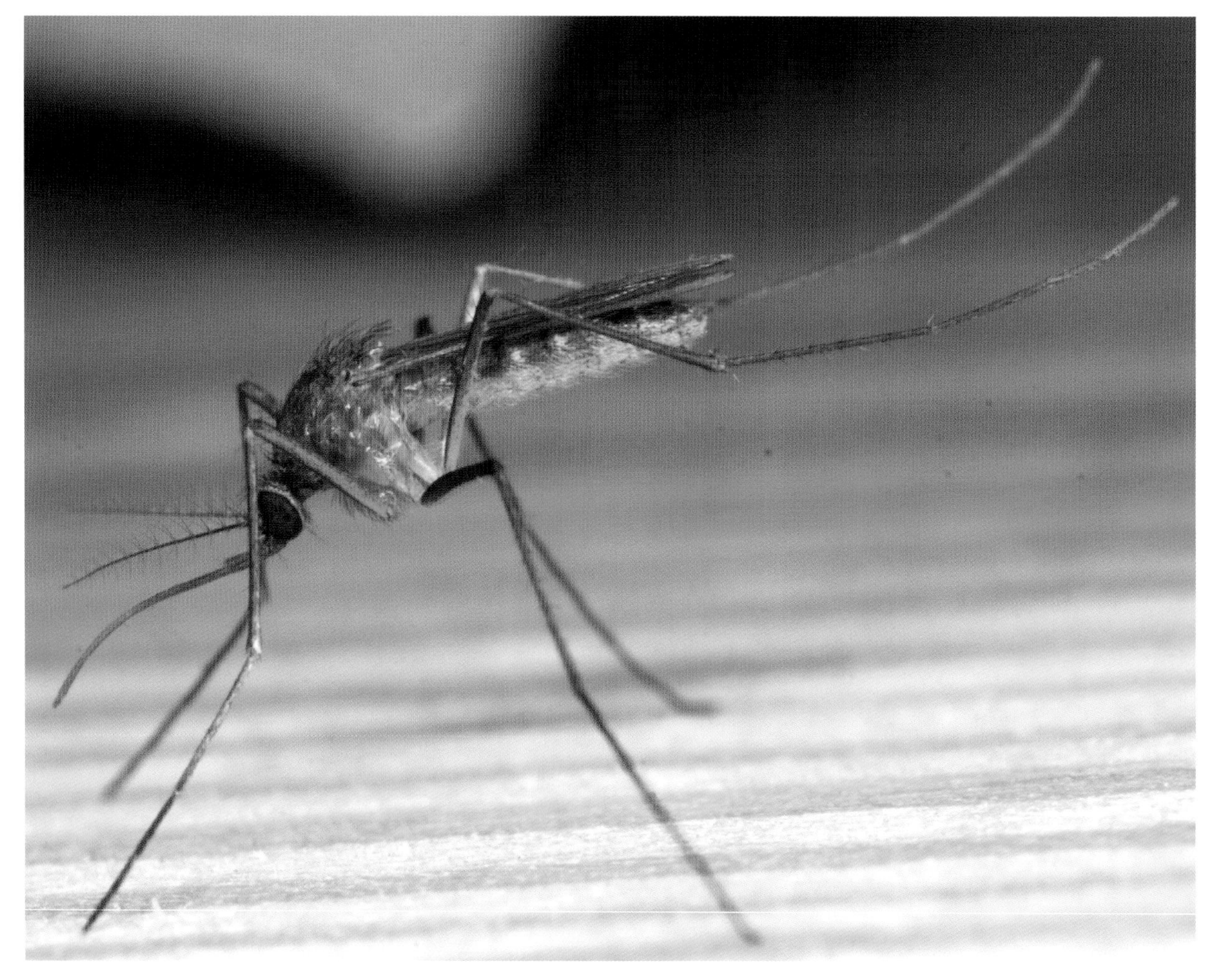

Culiseta melanura female

Culiseta melanura larva

Distribution *Culiseta melanura* is found primarily throughout low-elevation areas (coastal plains) of the eastern United States but is occasionally found at higher elevations.

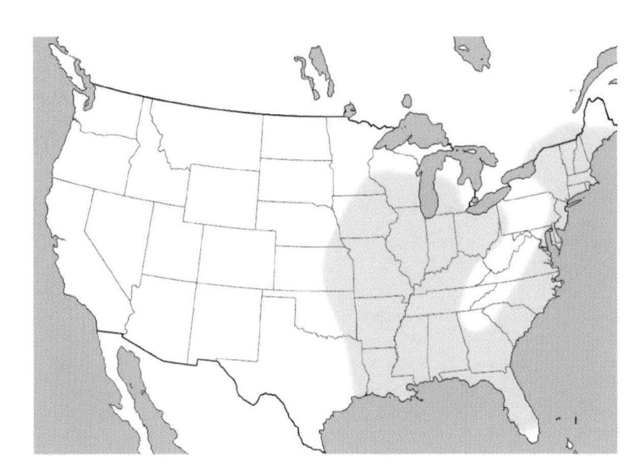

Distribution of *Culiseta melanura* in the U.S.

Mansonia

Mansonia species are found throughout most of the world but are most diverse in tropical regions. About 25 species are known, 2 of which occur in the southeastern United States. Adults are generally large and stout. Their wings have broad, asymmetrical, dark and pale scales in a speckled pattern. Some *Mansonia* species are important pests in tropical America and the southeastern United States. A few species are important in the transmission of human pathogens, including viruses (in the Americas) and filarial worms (in India and Southeast Asia). Females are persistent biters, usually attacking at night. *Mansonia* larvae occur in ponds, lakes, and marshes with aquatic plants. Their siphons are tapered to a point and used for piercing plant tissues.

Mansonia dyari Belkin, Heinemann, and Page 1970

Subgenus *Mansonia*

Dyar's mosquito

Adults Adults of *Mansonia dyari* are medium sized, stout bodied, and speckled. The body is clad in a mixture of dark and pale scales. The legs are dark brown with pale speckles and have white bands at the base of each tarsal segment. One other species of *Mansonia* occurs in our area (*Mansonia titillans*) and closely resembles *Mansonia dyari.* These 2 species may be distinguished by the presence or absence of a row of small spines at the apex of the seventh abdominal segment. *Mansonia titillans* has this row of small spines on abdominal segment 7, and *Mansonia dyari* does not. The shape of scales on the wings, the length of the palps, and the arrangement of pale scales on the proboscis are other characteristics often used to separate the 2 species. Females of *Mansonia dyari* are persistent and painful biters and can fly long distances from their breeding habitat in search of hosts. They feed mainly on the blood of mammals and birds but have also been reported to feed on alligators in Florida. Adults reach their greatest abundance in late summer

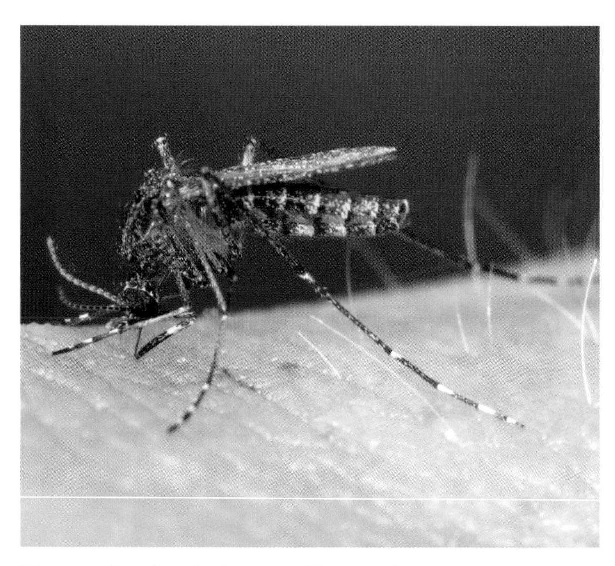

Mansonia dyari female. Courtesy of Sean McCann.

Water lettuce (*Pistia*)

and fall. *Mansonia dyari* is thought to be important in the transmission of Venezuelan equine encephalitis virus in the Caribbean but is not known to transmit any human pathogens in the United States.

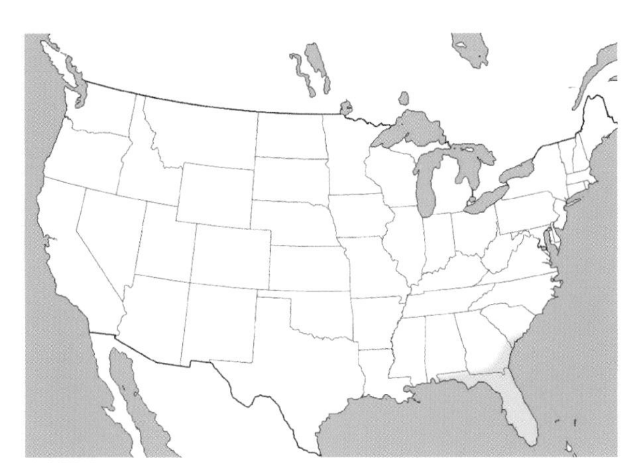

Distribution of *Mansonia dyari* in the U.S.

Larvae Larvae of *Mansonia dyari* are medium sized. The head is very broad and has numerous short, multibranched setae. The antennae are quite long and each has a large, profusely branched seta and a long, 2-branched seta. The siphon is short, pointed, and serrated on its dorsal tip. It lacks a pecten. The comb is a single row of 6–9 scales. Individual scales of the comb are broad, with several small spines of similar size. Larvae are found in permanent and semipermanent bodies of water with floating aquatic plants, which, in the United States, are most commonly water lettuce (*Pistia*) and water hyacinth (*Eichornia crassipes*). The pointed siphon is used to pierce the roots of these aquatic plants to reach the stores of air within.

Distribution In the United States, *Mansonia dyari* is found in Florida and parts of Georgia, South Carolina, and Alabama.

Mansonia titillans (Walker, 1848)

Subgenus *Mansonia*

Mansonia titillans larva

Adults Adults of *Mansonia titillans* are medium-sized, stout mosquitoes. Their bodies are dark and speckled with pale scales. The legs are dark, with pale speckles and narrow pale bands at the base of each tarsal segment. Another species of *Mansonia* in our area, *Mansonia dyari,* is very similar to *Mansonia titillans.* These 2 species may be distinguished by examining the tip of the seventh abdominal segment for the presence or absence of small spines; *Mansonia titillans* has these spines, whereas *Mansonia dyari* does not. Females of *Mansonia titillans* are persistent when pursuing a host and their bite is painful. They can fly long distances from their breeding site in search of hosts. They feed mainly on the blood of large mammals but also feed on birds. Adults reach their greatest abundance in late summer and fall. *Mansonia titillans* is known to trans-

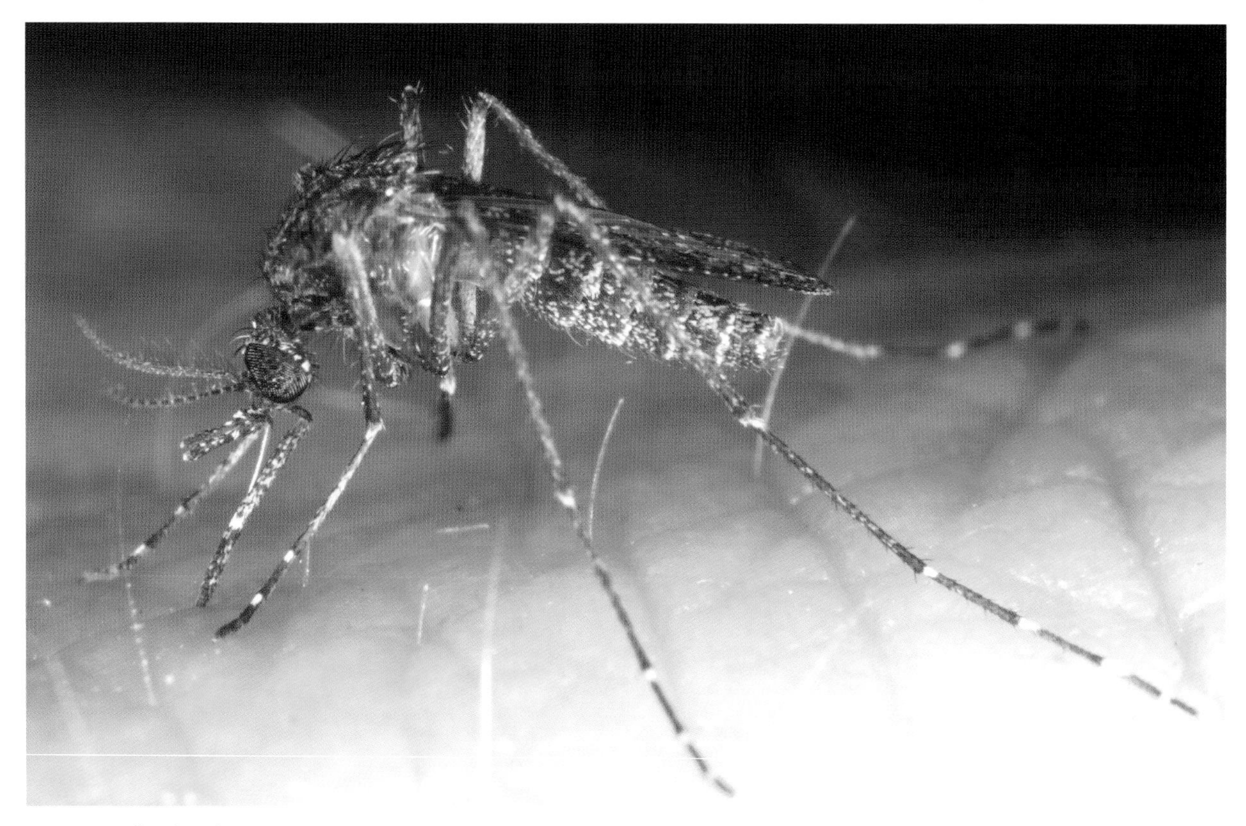
Mansonia titillans female

Water hyacinth (*Eichornia crassipes*)

mit eastern equine encephalitis virus in the United States, carrying the virus from birds to humans and horses.

Larvae Larvae of *Mansonia titillans* are medium sized. The head is very broad and has numerous short, multibranched setae. The antennae are quite long and each has a large, profusely branched seta and a long, 2-branched seta. The siphon is short, pointed, and serrated on its dorsal tip. It lacks a pecten. The comb is a single row of 6–8 scales. Individual scales of the comb are slender, with a single median spine. Larvae are found in permanent and semipermanent bodies of water with floating aquatic plants, such as water lettuce (*Pistia*) and water hyacinth (*Eichornia crassipes*). The pointed siphon is used to pierce the roots of these aquatic plants to reach the stores of air within.

Distribution *Mansonia titillans* is found in disjunct pockets in the southeastern United States, including southern Texas, Florida, southern Georgia, southeastern South Carolina, Mississippi, central Louisiana, and Alabama.

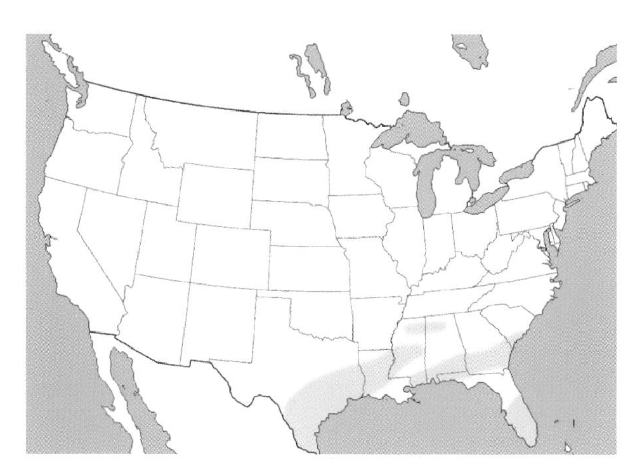

Distribution of *Mansonia titillans* in the U.S.

Orthopodomyia

Orthopodomyia includes 35 species found throughout tropical regions of the world, with 2 species in the southeastern United States. Adults are strikingly ornamented with bands and stripes of white, silver, or gold. They are readily distinguished from the adults of all other mosquitoes by having the fourth tarsomere of the foreleg shorter than the fifth tarsomere. Adults are encountered in forests and are active at night. Although little information is available on the feeding habits of females, most species are thought to feed on birds. *Orthopodomyia* larvae primarily inhabit tree holes, although some are found in bamboo or water-holding bromeliads. They may occasionally be found in man-made containers, as well. Larvae are usually reddish and lack a pecten. No species of *Orthopodomyia* are known to be important in the transmission of human pathogens.

Orthopodomyia alba Baker, 1936, and *Orthopodomyia signifera* (Coquillett, 1896)

Subgenus *Orthopodomyia*

Ornate treehole mosquitoes

Orthopodomyia alba and *Orthopodomyia signifera* are almost indistinguishable as adults and have overlapping ranges and quite similar life histories. They are therefore treated together in this book.

Adults *Orthopodomyia alba* and *Orthopodomyia signifera* are medium to large mosquitoes with quite striking appearances. Their slender black bodies have bold white markings on the thorax, abdomen, and legs. The markings of the scutum form a unique pat-

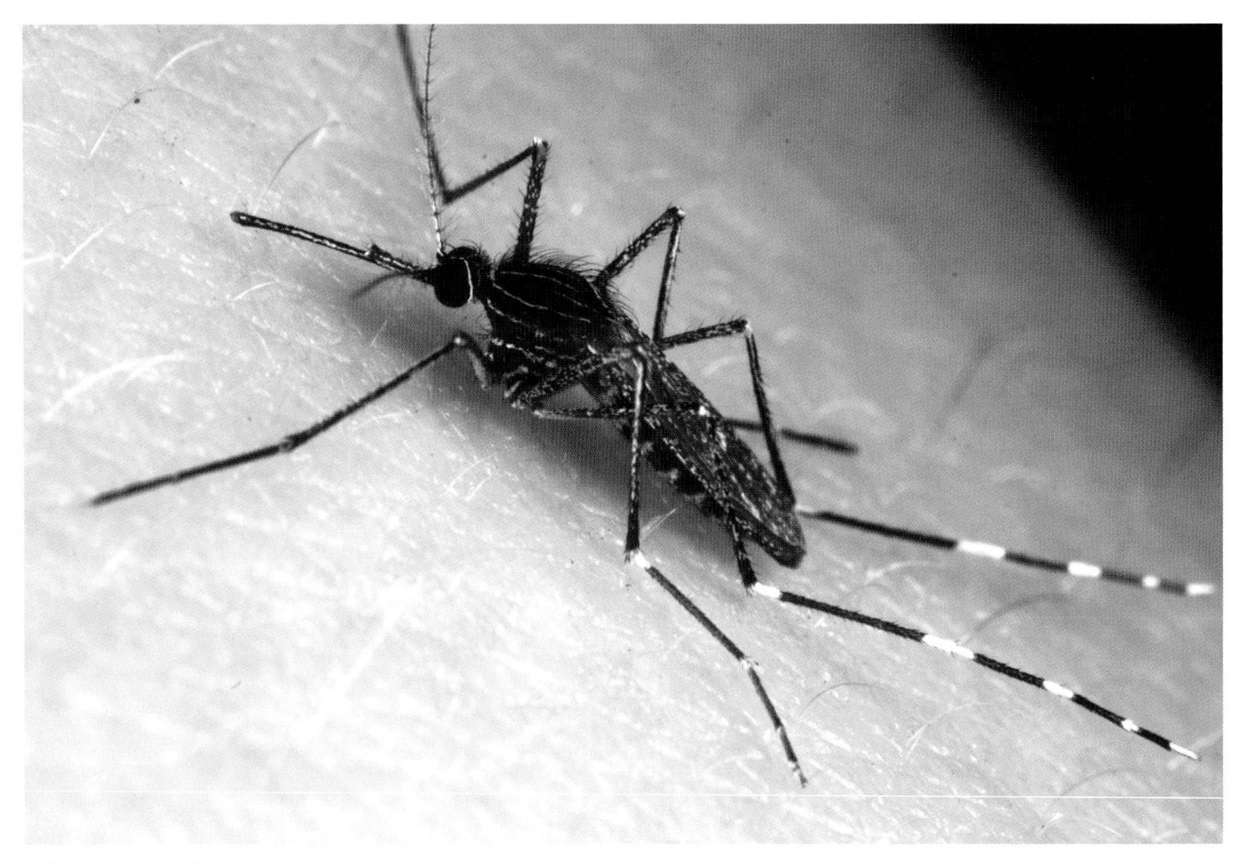

Orthopodomyia signifera female

Orthopodomyia signifera female

tern of thin white lines on a black background. Their hind legs have broad white bands crossing the junction of the tarsal segments. The tarsi are usually held very straight when at rest. The wings have broad scales of black and white, which form distinctive patterns. Females of these 2 species are thought to feed on the blood of birds, although few studies have confirmed this. Neither species is known to transmit any human pathogens.

Larvae Larvae of *Orthopodomyia alba* and *Orthopodomyia signifera* are medium sized. The body is dark reddish dorsally but pale on the ventral surface. The head capsule of both species is very dark and spherical and has numerous densely branched setae. The siphon is rather short and lacks a pecten. Larvae of *Orthopodomyia signifera* have a large sclerotized plate on abdominal segment 8 that covers the dorsal and lateral portions of that segment. Larvae of *Orthopodomyia alba*

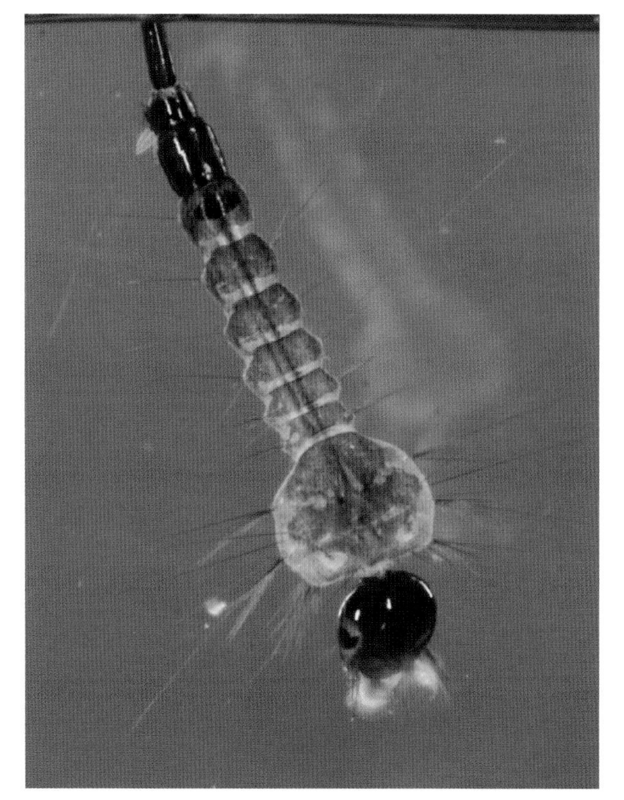

Orthopodomyia signifera larva

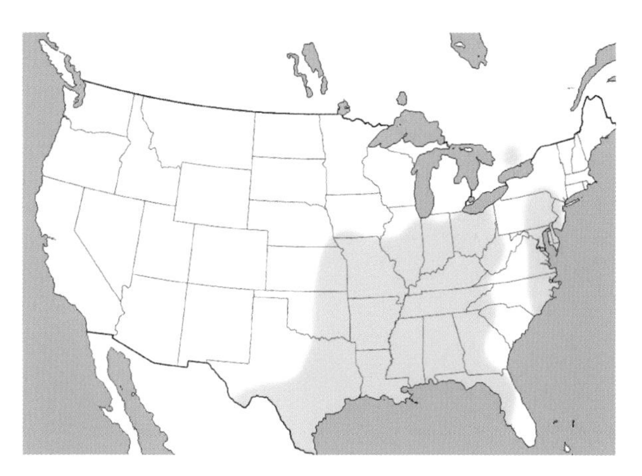

Distribution of *Orthopodomyia alba* in the U.S.

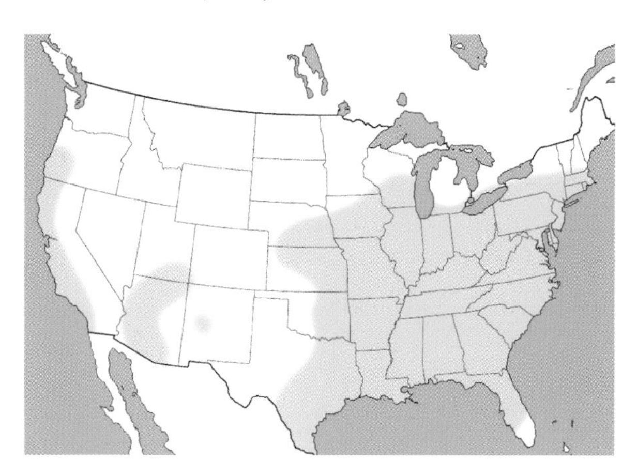

Distribution of *Orthopodomyia signifera* in the U.S.

do not have a plate on abdominal segment 8. Larvae of both species are most commonly encountered in water-filled rot holes of tree stumps and cavities of deciduous trees but may occasionally be found in water-filled man-made containers.

Distribution *Orthopodomyia alba* and *Orthopodomyia signifera* are found throughout the eastern United States and in scattered locations in the west.

Psorophora

Psorophora is a genus of small to large mosquitoes, many of which have striking coloration. *Psorophora* is found only in the Americas, and about 50 species are known, 8 of which occur in the Southeast. Most *Psorophora* females feed on mammals but will occasionally feed on other animals. Although their bite is often painful, few *Psorophora* species are important in the transmission of human pathogens. *Psorophora* larvae are found in temporary aquatic habitats, especially rainwater pools and overflow pools of streams and rivers. Most larvae have a large respiratory siphon. *Psorophora* females lay single eggs in depressions in ground that will one day be flooded with water. The eggs can remain dormant for months or even years.

Psorophora ciliata (Fabricius, 1794)

Subgenus *Psorophora*

Gallinipper

Adults *Psorophora ciliata* is the largest of the biting mosquitoes in our area. Only *Toxorhynchites rutilus,* which does not bite, is larger. The body and legs of *Psorophora ciliata* are quite stout and have a yellowish hue. Patches of black, white, and golden scales adorn the body. The legs have tufts of dark, erect scales on the femur and tibia, resulting in a shaggy appearance. The hind tarsi have broad pale bands at the base of each segment. Only 1 other mosquito in our area, *Psorophora howardii,* is likely to be mistaken for *Psorophora ciliata.* These 2 species can be distinguished by scale patterns on the scutum. *Psorophora ciliata* females have a golden stripe down the center of the scutum, whereas females of *Psorophora howardii* have a stripe of dark scales down the center of the scutum. In addition, the palps of *Psorophora ciliata* are short compared to those of *Psorophora howardii,* which are nearly half the length of the proboscis. Females of *Psorophora ciliata* feed on mammals, including deer, rabbits, armadillos, livestock, and humans. Although their bite can be quite painful, they are not known to transmit any disease agents to humans. They can be quite pestiferous and persistent biters in places where they are abundant, particularly in coastal areas.

Larvae Larvae of *Psorophora ciliata* are quite large. The thorax and abdomen are darker dorsally than ventrally. The head is square, with very small antennae. The siphon is long and bears a long pecten of numer-

Psorophora ciliata female

Psorophora ciliata female

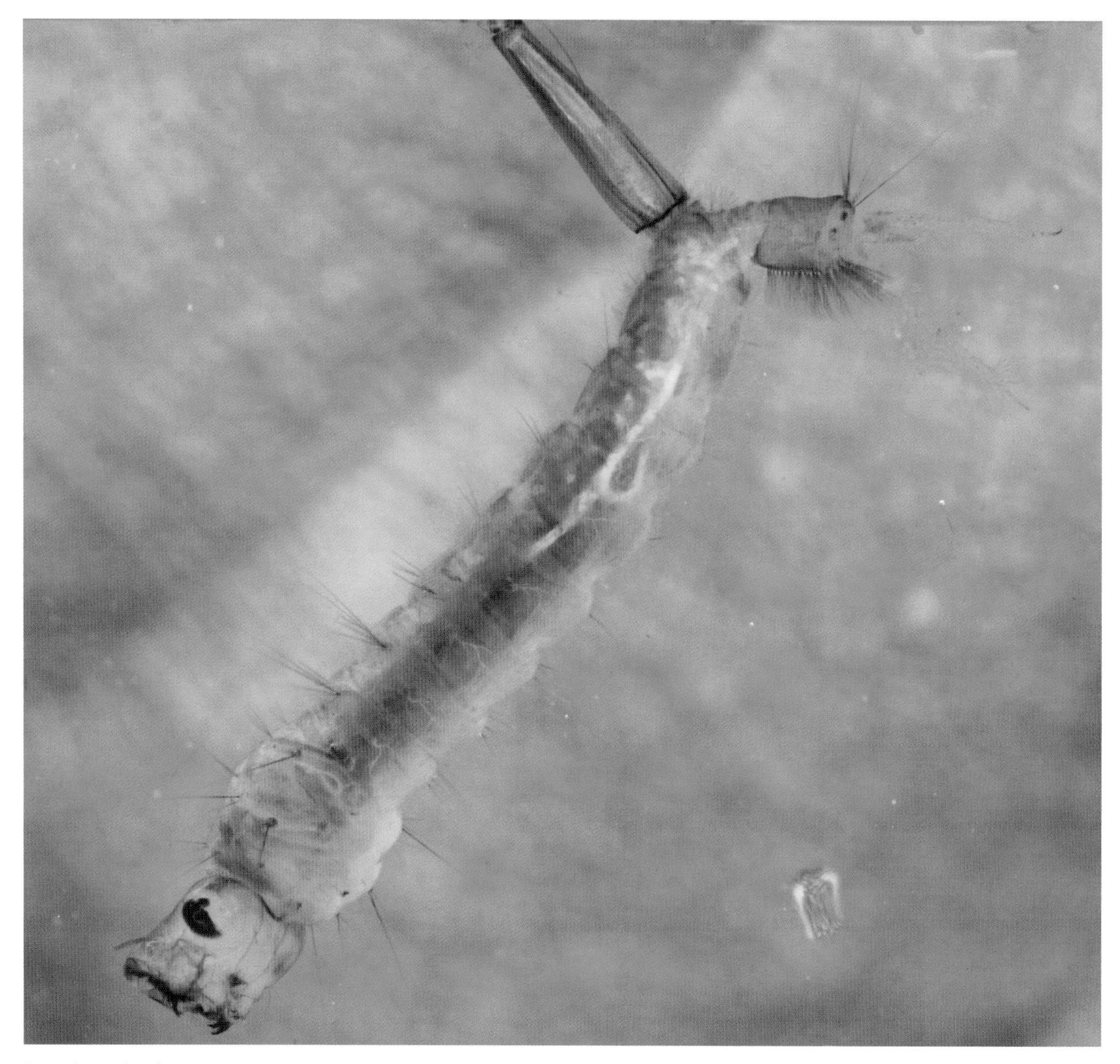

Psorophora ciliata larva

ous thin, almost hairlike teeth. The seta of the saddle is branched at its base. Larvae are most commonly encountered in temporary rainwater pools exposed to full sun. They are predaceous and develop rapidly, completing development in as little as 4–5 days.

Distribution *Psorophora ciliata* is found across the eastern half of the United States and portions of southern Canada.

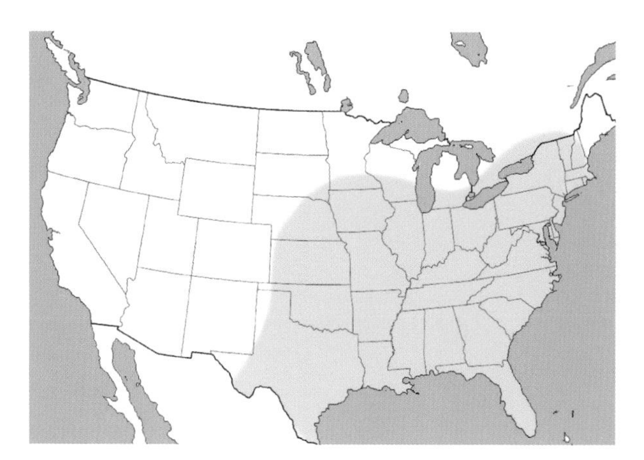

Distribution of *Psorophora ciliata* in the U.S.

Psorophora columbiae (Dyar and Knab, 1906)

Subgenus *Grabhamia*

Dark rice-field mosquito *or* Florida glades mosquito

Adults Adults of *Psorophora columbiae* are medium sized and rather dark and have a salt-and-pepper appearance. The dark body is speckled throughout with pale scales. The legs have white bands at the base of each tarsal segment, in the middle of the first tarsal segment, and near the apex of the femur. A broad band of pale scales is also present on the proboscis. The wings have dark and pale scales scattered throughout, in no discernible pattern. Adults of 1 other *Psorophora* species in our area, *Psorophora discolor,* resemble those of *Psorophora columbiae. Psorophora columbiae* is generally darker than *Psorophora discolor.* Also, the dark and light scales on the wings of *Psorophora columbiae* females are intermixed, whereas the dark and light scales on the wings of *Psorophora discolor* are grouped together, forming a discernible pattern. Females of *Psorophora columbiae* feed mainly on mammals, such as deer, livestock, rabbits, armadillos, and humans. In places where they are abundant, females of *Psorophora columbiae* are very aggressive biters and can occur in numbers so great that they can cause livestock to die from exsanguination and calves to die of suffocation. Although females feed readily on humans, they are not known to be important in the transmission of any human pathogens.

Wing of *Psorophora columbiae*

Psorophora columbiae female

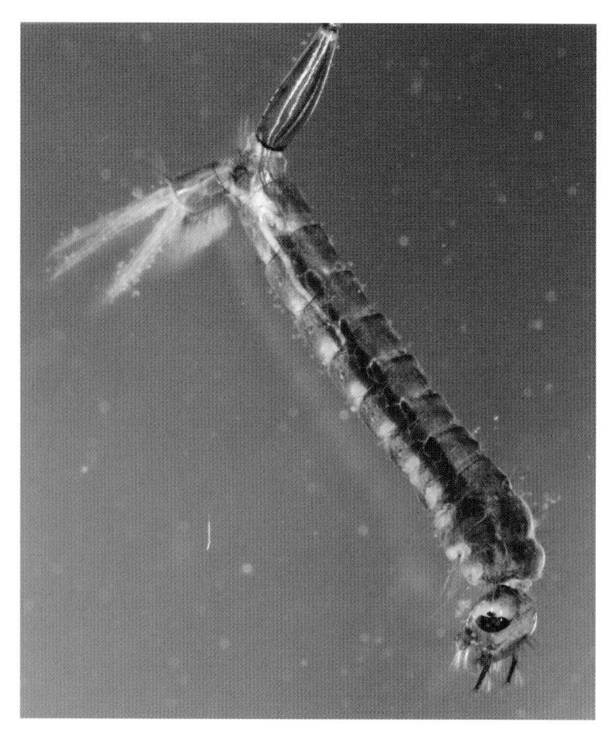

Psorophora columbiae larva

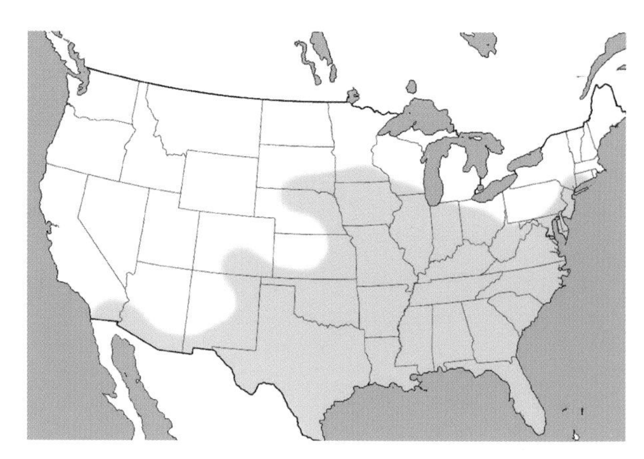

Distribution of *Psorophora columbiae* in the U.S.

Larvae Larvae of *Psorophora columbiae* are medium sized to large, with robust bodies. Setae of the head and antennae are multibranched. The antennae are noticeably shorter than the head. The siphon is large and slightly swollen at its base. The pecten has 3–6 teeth that are distantly spaced from one another. The comb, composed of a single row of 5–6 scales, is often borne on a weakly sclerotized plate. The anal papillae are long and pointed. Larvae are most commonly encountered in sunlit, water-filled puddles. They proliferate in areas of rice production and also occur in tire ruts, ditches, and water-filled depressions in pastures and other sunny places.

Distribution *Psorophora columbiae* is found throughout the eastern half of the United States, as well as parts of the central plains and the Desert Southwest.

Psorophora cyanescens (Coquillett, 1902)

Subgenus *Janthinosoma*

Purple-legged mosquito

Adults Adults of *Psorophora cyanescens* are medium-sized, colorful mosquitoes. Much of the body is clad in iridescent purple or golden scales. The scutum is black, with scattered golden or greenish-gold scales. The dorsum of the abdomen has apical median patches of pale scales on segments 2–6. The tarsi are entirely purple. Three other *Psorophora* species in our area (*Psorophora ferox, Psorophora mathesoni,* and *Psorophora horrida*) also have purple tarsi, but these species also have at least 1 pale-scaled tarsal segment on the hind leg. Of the *Psorophora* species with "purple legs," only *Psorophora cyanescens* has tarsi that are entirely purple. Females of *Psorophora cyanescens* feed mainly on the blood of large mammals, such as livestock, deer, and humans. They may bite during the day or night, but their main period of feeding is in the evening. *Psorophora cyanescens* is not known to be important in the transmission of any human pathogens.

Larvae Larvae of *Psorophora cyanescens* are medium sized to large. Setae of the head and antennae are small

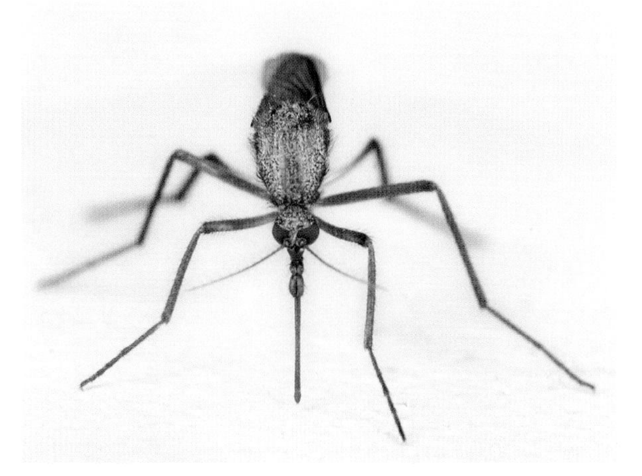

Psorophora cyanescens female. Courtesy of Gayle and Jeanell Strickland.

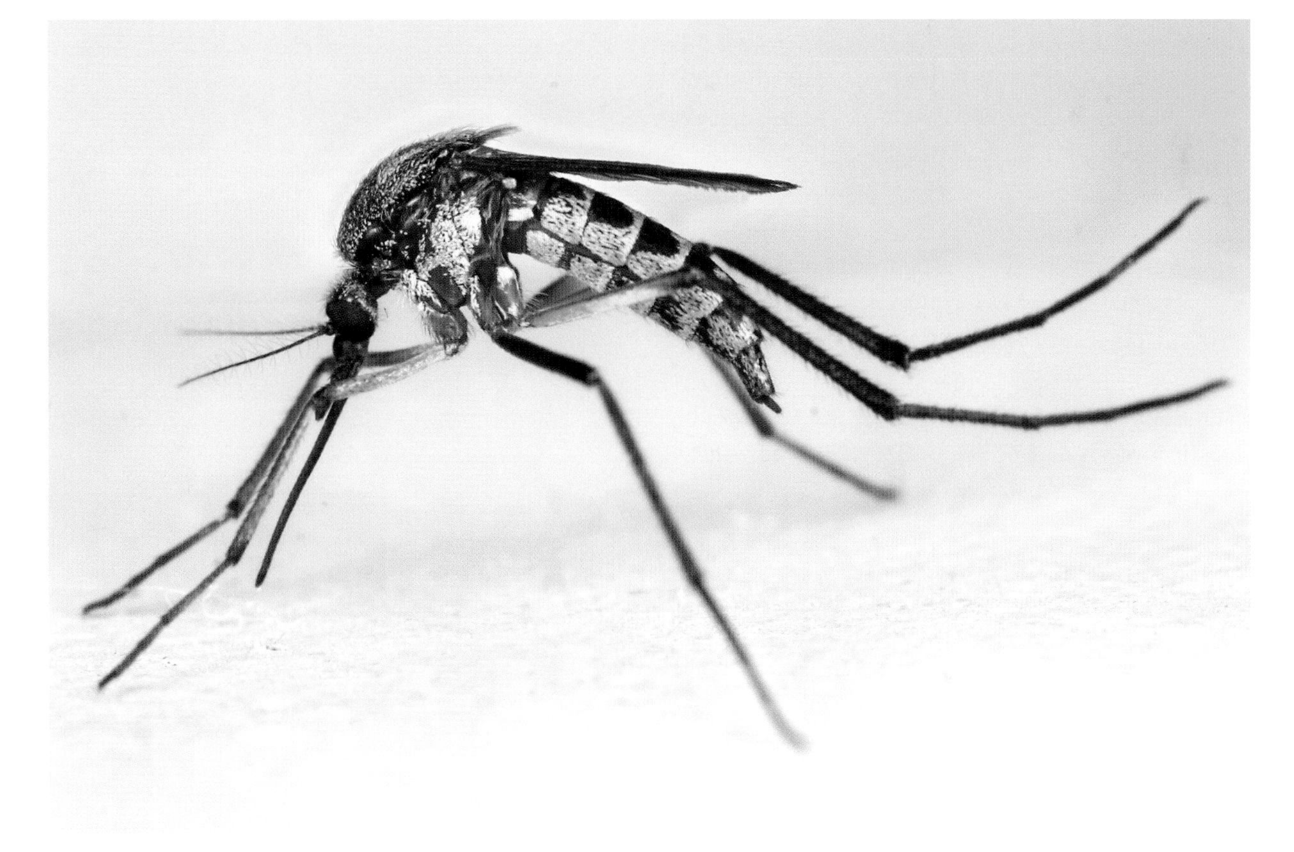

Psorophora cyanescens female. Courtesy of Gayle and Jeanell Strickland.

Psorophora cyanescens male

and unbranched or have 2–3 branches. The antennae are shorter than the head. The siphon is short, very stout, and swollen near its base and has a long seta at its tip. The pecten, which is borne on the basal third of the siphon, has 3–5 teeth. The comb is a single row of 4–5 scales, on the edge of a weakly sclerotized area. The anal papillae are long and pointed. Larvae are most commonly collected in flooded, sunny grassland depressions and ditches.

Distribution *Psorophora cyanescens* is found throughout much of the eastern United States, south of the Great Lakes.

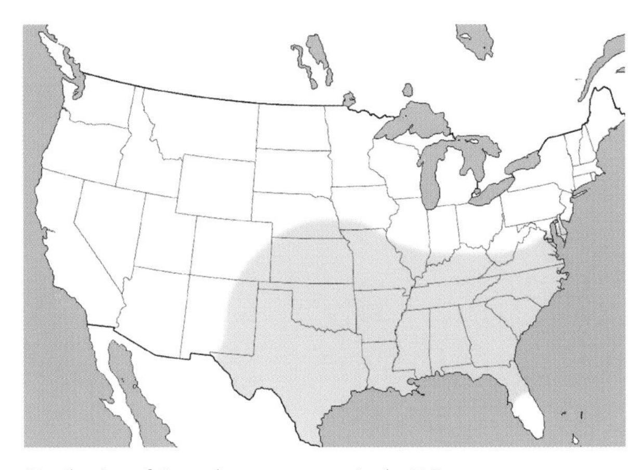

Distribution of *Psorophora cyanescens* in the U.S.

Psorophora discolor (Coquillett, 1903)

Subgenus *Grabhamia*

Pale rice-field mosquito

Adults Adults of *Psorophora discolor* are medium sized, with a pale salt-and-pepper appearance. The legs have very broad white bands at the base of each tarsal segment, and the first tarsal segment is mostly pale scaled. The proboscis has a very broad pale band and the palps are pale at their tips. The dorsal portion of the abdomen is nearly entirely pale scaled. Adults of only 1 other *Psorophora* species in our area, *Psorophora columbiae,* resemble those of *Psorophora discolor.* Adult females of *Psorophora discolor* are generally lighter in color than those of *Psorophora columbiae.* The brown and white scales on the wings of *Psorophora discolor* females form definite patterns, whereas the black and white scales on the wings of *Psorophora columbiae* are intermixed and do not form a discernible pattern. Females of *Psorophora discolor* feed on large mammals, including livestock and humans. *Psorophora discolor* is not known to be important in the transmission of any human pathogens.

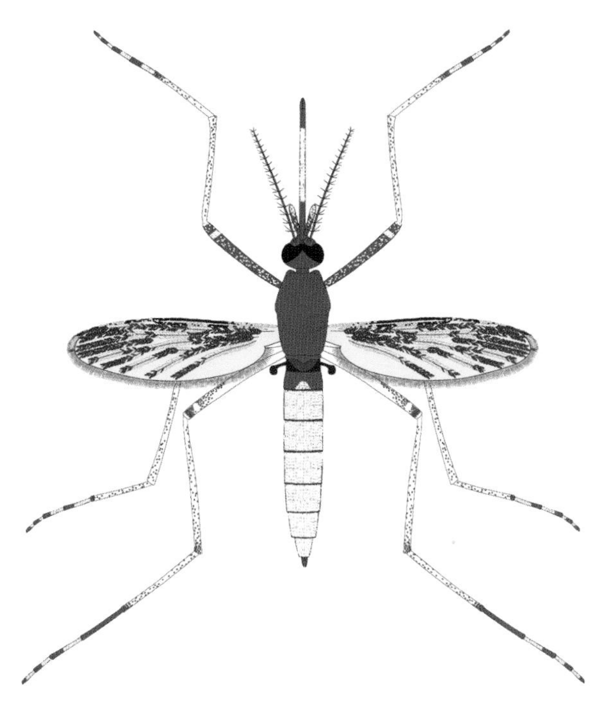

Psorophora discolor female

Larvae Larvae of *Psorophora discolor* are medium sized. Setae of the head are unbranched or have just 2 branches. The antennae are quite long and curved and appear inflated. Each antenna bears a large, multibranched seta. The siphon is quite small compared to that of other *Psorophora* species. The pecten, which extends to the midpoint of the siphon, is composed of 6–8 teeth. The tuft (branched seta) on the siphon is long and reaches well past the end of the siphon. The comb is a single row of about 6 scales, borne on the edge of a weakly sclerotized plate. The anal papillae are extremely long and pointed. Larvae are found in sunlit, rain-filled puddles, ditches, overflow pools along streams, and rice fields.

Distribution *Psorophora discolor* is a relatively rare mosquito but has been encountered throughout the eastern half of the United States, south of the Great Lakes. This mosquito reaches its greatest abundance in parts of Texas, Oklahoma, and Arkansas.

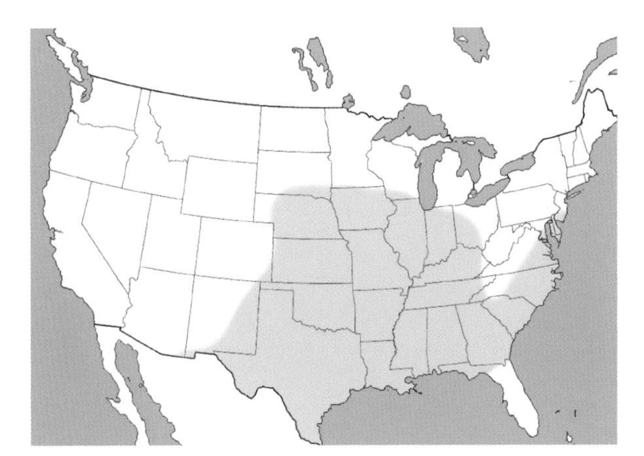

Distribution of *Psorophora discolor* in the U.S.

Psorophora ferox (Humboldt, 1820)

Subgenus *Janthinosoma*

Common white-footed mosquito

Psorophora ferox larva

Adults *Psorophora ferox* adults are medium-sized, colorful mosquitoes. Their legs are iridescent purple and the last 2 segments of the hind legs are bright white, giving the appearance of white socks. One other species in our area, *Psorophora horrida,* also has purple legs with "white socks." These 2 species may be distinguished by the scale patterns on the scutum. *Psorophora ferox* has iridescent gold scales scattered evenly about the scutum, while the scutum of *Psorophora horrida* has a dark median stripe bordered by patches of bright white scales. The abdomen of *Psorophora ferox* adults is ornamented with iridescent purple scales dorsally, and apical patches of bright white scales laterally. Their eyes appear bright green in life. Females feed on mammals, especially deer, rabbits, livestock, and humans.

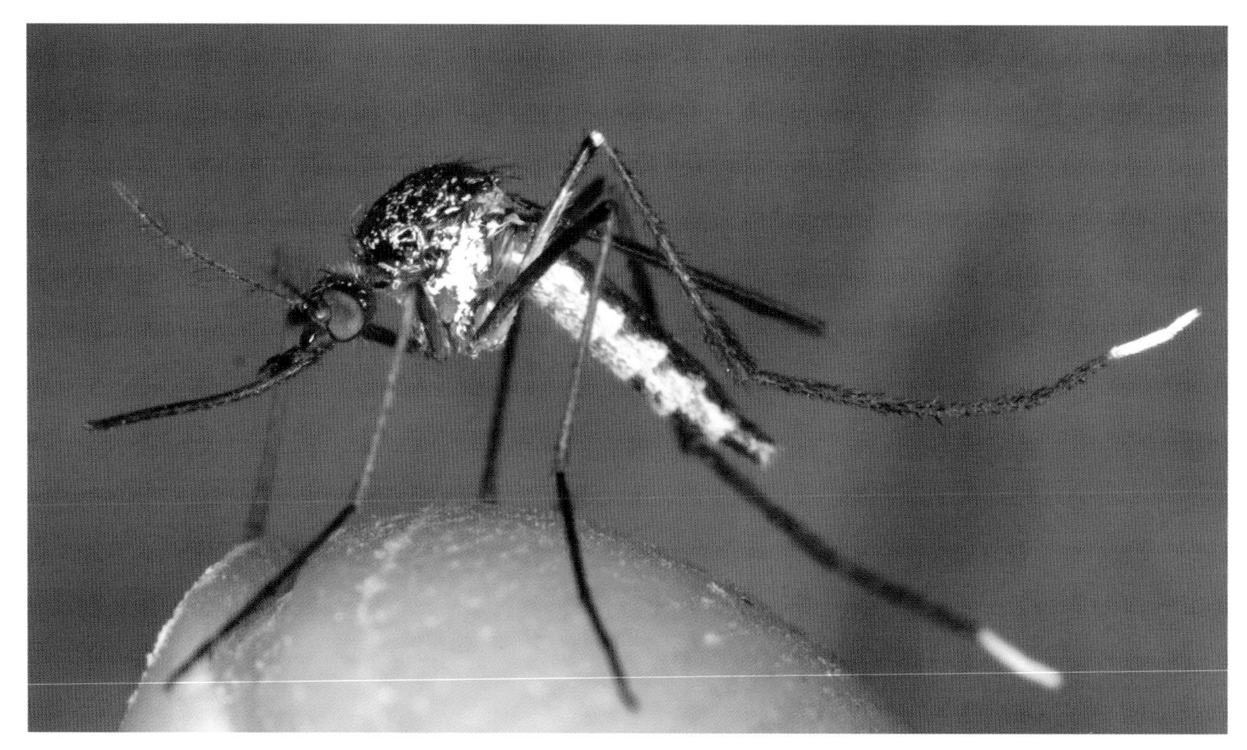
Psorophora ferox female

They can be quite bothersome, especially in late afternoon in shaded areas and in dense woods. *Psorophora ferox* is not known to be important in the transmission of any human pathogens in the United States.

Larvae Larvae of *Psorophora ferox* are medium sized and usually brownish. Setae of the head are branched. The antennae are quite long, usually much longer than the head, and often black. The siphon is large and appears slightly swollen just beyond its base. The pecten is borne on the basal third of the siphon and has 3–5 widely spaced teeth. The comb is a single row of 6–8 scales, borne on the edge of a weakly sclerotized plate. Larvae occur in woodland pools and overflow puddles along streams following rains.

Distribution *Psorophora ferox* is common throughout the eastern half of the United States, south of the Great Lakes. This mosquito is also found throughout Central and South America, as far south as Argentina.

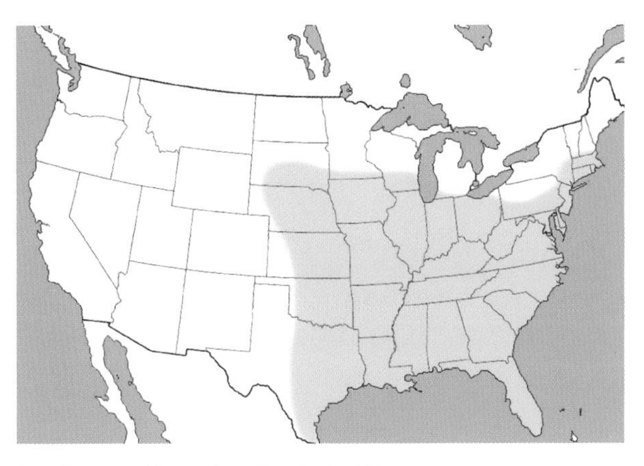

Distribution of *Psorophora ferox* in the U.S.

Psorophora horrida (Dyar and Knab, 1908)

Subgenus *Janthinosoma*

Adults Adults of *Psorophora horrida* are medium-sized, colorful mosquitoes. Their legs are iridescent purple and the last 2 segments of the hind legs are bright white, giving the appearance of "white socks." Another (more common) species in our area, *Psorophora ferox,* shares the same leg ornamentation as *Psorophora horrida.* These 2 species may be distinguished by the patterns of light and dark scales on the scutum. *Psorophora horrida* has a stripe of dark scales down the center of the scutum that is bordered on each side by a patch of silvery-white scales. *Psorophora ferox,* on the other hand, has iridescent gold scales scattered evenly about the scutum. The abdomen of *Psorophora horrida* is ornamented with iridescent purple scales dorsally, and apical patches of pale scales laterally. Females feed on a variety of mammals, including humans, but they rarely occur in numbers great enough to be a serious annoyance. *Psorophora horrida* is not important in the transmission of any human pathogens.

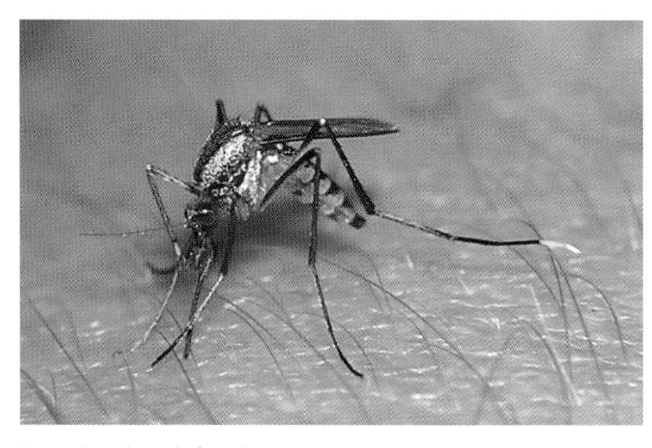

Psorophora horrida female

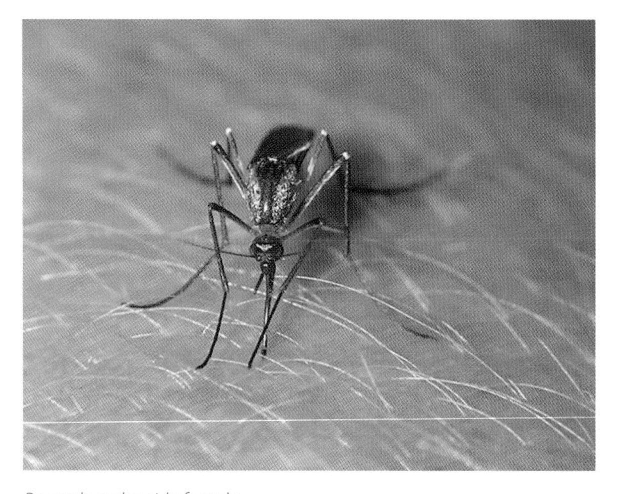

Psorophora horrida female

Larvae Larvae of *Psorophora horrida* are medium sized. Setae of the head are branched. The antennae are about as long as the head and each bears a small, multibranched seta. The siphon is large and swollen just beyond its base. The pecten is borne on the basal half of the siphon and has 3–5 widely spaced teeth. The comb is a single curved row of 6–9 scales. Larvae occur in riverine flood pools and woodland pools formed by heavy and prolonged summer rains.

Distribution *Psorophora horrida* is widely distributed across the eastern half of the United States, south of the Great Lakes, excluding much of Florida and higher elevations in the Appalachians.

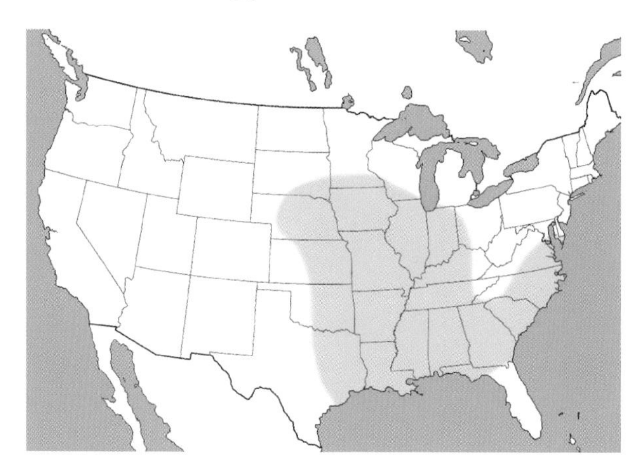

Distribution of *Psorophora horrida* in the U.S.

Psorophora howardii Coquillett, 1901

Subgenus *Psorophora*

Howard's gallinipper

Adults *Psorophora howardii* is the second largest of the biting mosquitoes in our area, being superseded in size by *Psorophora ciliata.* Both species have tufts of dark scales on the femur and tibia of their hind legs, giving the legs a shaggy appearance. The hind tarsi of both species have pale bands at the base of each segment, although the bands are usually much narrower in *Psorophora howardii.* While both species have long palps, the palps of *Psorophora howardii* are longer and reach nearly to the center of the proboscis. *Psorophora howardii* females have a stripe of dark brown or black scales down the center of the scutum, while females of *Psorophora ciliata* have a golden stripe down the center of the scutum. Females of *Psorophora howardii* feed mainly on mammals, including deer, rabbits, livestock, and humans. *Psorophora howardii* is not known to be important in the transmission of human pathogens.

Larvae Larvae of *Psorophora howardii* are quite large. The thorax and abdomen appear darker dorsally than ventrally. The head is square, with quite small anten-

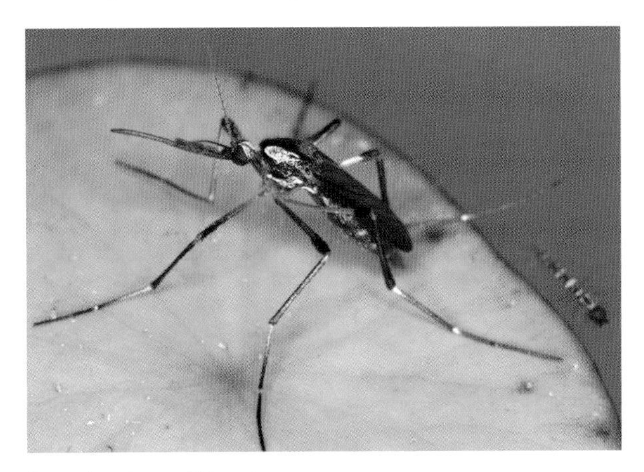

Psorophora howardii female

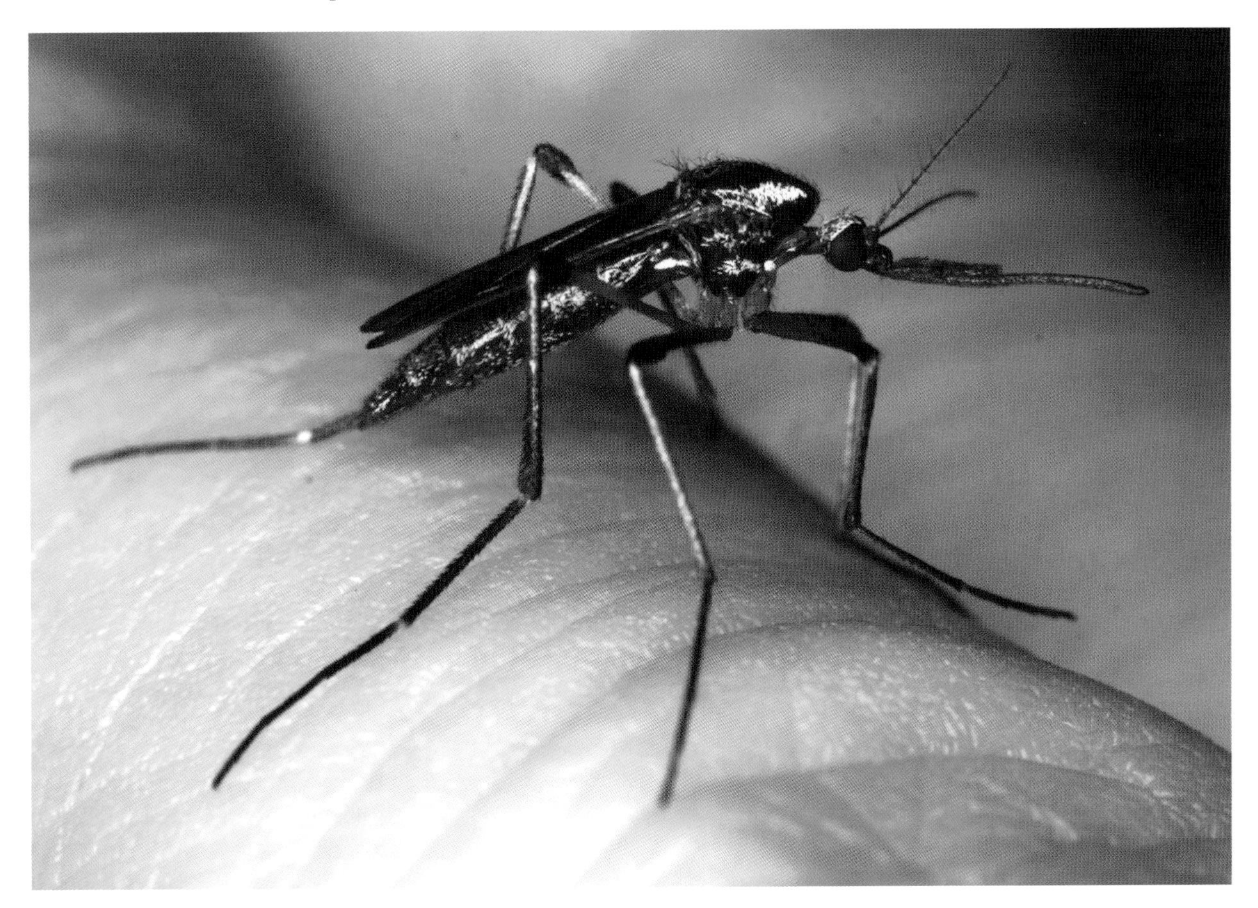

Psorophora howardii female

Psorophora howardii larva

nae. The siphon is long and bears a long pecten of numerous thin, hairlike teeth. The seta of the saddle is unbranched or branches beyond its midpoint. Larvae are found in sunlit temporary pools, where they prey upon aquatic arthropods (especially other mosquito larvae). They develop rapidly and can complete development in less than 1 week.

Distribution *Psorophora howardii* is found throughout the southeastern and east-central United States, excluding the Appalachian region. It is also found in the Caribbean and Central America.

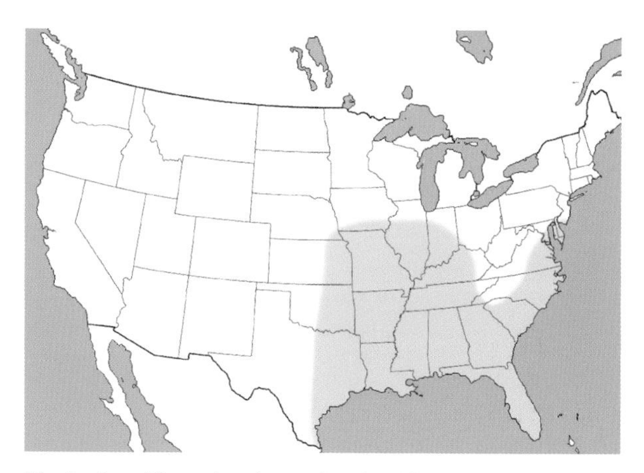

Distribution of *Psorophora howardii* in the U.S.

Psorophora mathesoni Belkin and Heinemann, 1975

Subgenus *Janthinosoma*

Matheson's mosquito

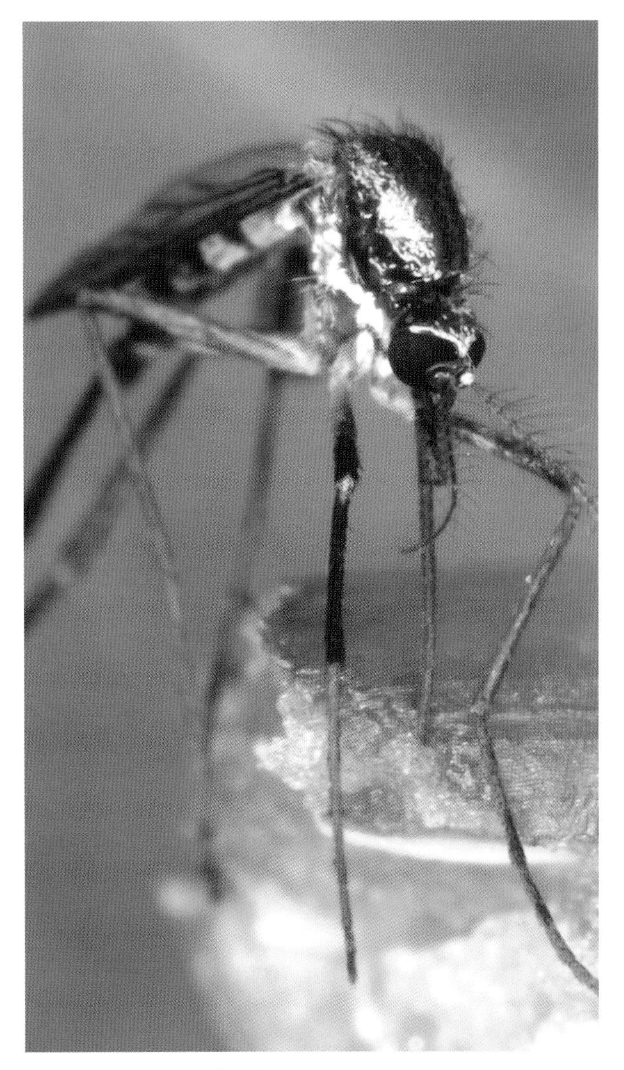

Psorophora mathesoni female

Adults Adults of *Psorophora mathesoni* are small to medium-sized, colorful mosquitoes. Their legs are mostly purple, but the hind legs have a small band of pale scales on the fourth tarsal segment. Three other *Psorophora* species in our area have purple legs, but these species either have entirely purple hind tarsi, or the last 2 segments (4–5) are pale scaled. The scutum of *Psorophora mathesoni* has a dark median stripe bordered by pale scales. Females will readily bite humans and can occasionally be abundant enough to be a real nuisance, especially in late afternoon in shaded areas along streams and in dense woods. *Psorophora mathesoni* is not known to be important in the transmission of any human pathogens.

Larvae Larvae of *Psorophora mathesoni* are medium sized. Setae of the head are branched. The antennae are

Psorophora mathesoni female

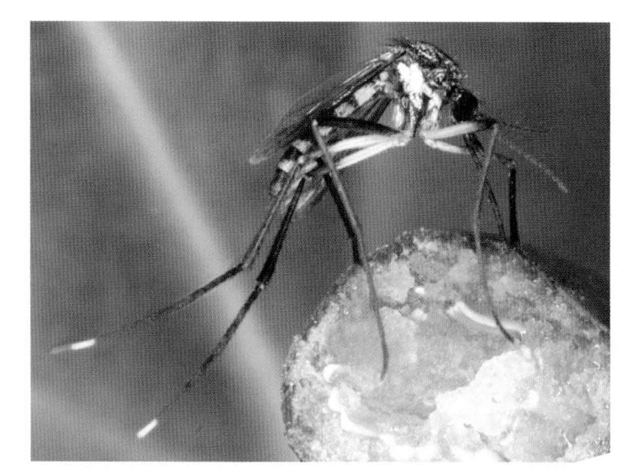
Psorophora mathesoni female

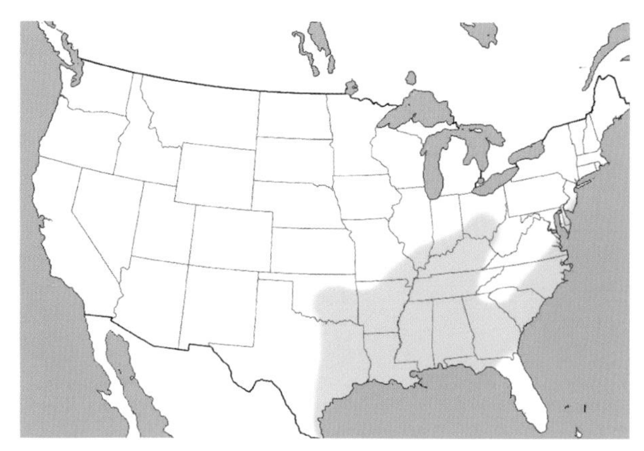
Distribution of *Psorophora mathesoni* in the U.S.

about as long as the head and each bears a small, multibranched seta. The siphon is large and swollen near its midpoint. The pecten is borne on the basal third of the siphon and has 3–4 widely spaced teeth. The comb is a single V-shaped row of 6–7 scales. Larvae are most commonly encountered in temporary rain-filled pools in dense woodlands.

Distribution *Psorophora mathesoni,* formerly known as *Psorophora varipes,* is widely distributed throughout the southeastern United States. This somewhat rare mosquito is occasionally abundant along streams after rains.

Toxorhynchites

Nearly 100 species of *Toxorhynchites* are recognized from around the world. Most are found in tropical regions, but a few species occur in temperate Asia and North America. One species is found in the southeastern United States. *Toxorhynchites* adults are very large (the largest of all known mosquitoes) and are easily recognized by their strongly bent and sharply pointed proboscis. Brightly colored, iridescent scales cover the body. The wing has a distinct notch on its posterior margin. Females do not feed on blood but, like males, feed on flower nectar. They are active during the daytime. *Toxorhynchites* larvae are very large and usually some shade of red. The larvae are predaceous, feeding on aquatic invertebrates, especially other mosquito larvae. Comb scales and pecten are absent. *Toxorhynchites* larvae are found in small aquatic habitats, such as tree holes, bamboo, water-holding bromeliads, pitcher plants, and rock pools. They are often found in man-made containers as well.

Toxorhynchites rutilus (Coquillett, 1896)

Subgenus *Toxorhynchites*

Cannibal mosquito

Adults Adults of *Toxorhynchites rutilus* are the largest of any mosquito species in North America. They are quite colorful, and their bodies are ornamented throughout with iridescent scales in patches of white, purple, green, and blue. The proboscis is long, downward curved, and pointed at its tip. The 2 subspecies are distinguished by differences in the adult males. Males of *Toxorhynchites rutilus rutilus* have a pale band on the tarsus of the foreleg. Males of *Toxorhynchites rutilus septentrionalis* have tarsi of the forelegs entirely dark scaled. Adults of *Toxorhynchites rutilus,* both males and females, feed exclusively on nectar and do not take blood meals, and thus they

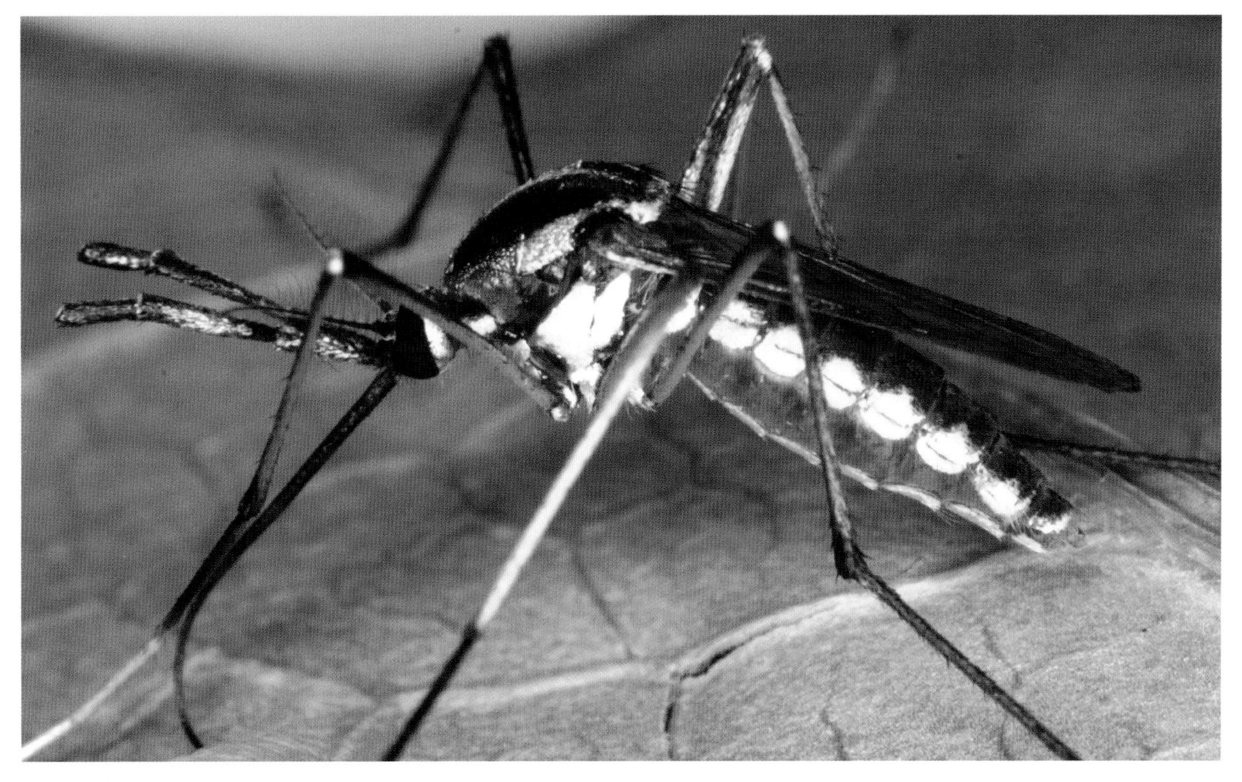

Toxorhynchites rutilus female

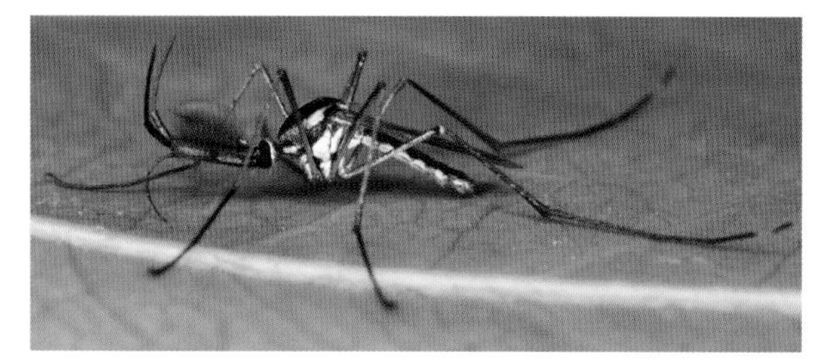

Toxorhynchites rutilus male

Toxorhynchites rutilus larva

do not bite. Because they do not feed on blood, *Toxorhynchites rutilus* females do not have the potential to transmit any human pathogens.

Larvae Larvae of *Toxorhynchites rutilus* are large and robust, their bodies colored a deep, dark red dorsally and pale grayish white ventrally. The head is square in front, with very short antennae. Setae of the head are small, but varied with respect to the number of branches. The thorax has numerous stout spines. The siphon is short and stout and lacks a pecten but bears a large branched seta that arises near the siphon's base. Comb scales are absent. The anal papillae are quite small. Larvae of *Toxorhynchites rutilus* are predators and prey upon the larvae of many species of mosquitoes (including those of their own species) and other aquatic animals found in their larval habitat. Their mandibles are stout, comblike structures used for grasping prey. Larvae are most commonly encountered in water-filled cavities in deciduous trees but are also found in water-filled man-made containers such as discarded automobile tires, birdbaths, paint cans, and troughs.

Distribution *Toxorhynchites rutilus* is widely distributed across the eastern half of the United States and southern Canada. Two subspecies occur in our area, *Toxorhynchites rutilus rutilus* and *Toxorhynchites rutilus septentrionalis,* the latter being more common and widespread.

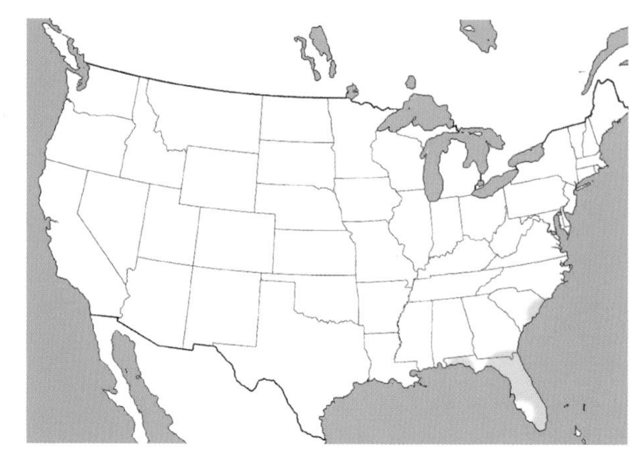

Distribution of *Toxorhynchites rutilus rutilus* in the U.S.

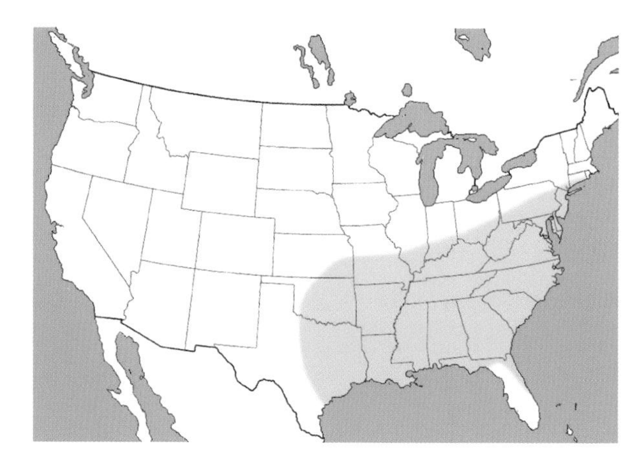

Distribution of *Toxorhynchites rutilus septentrionalis* in the U.S.

Uranotaenia

More than 250 species of *Uranotaenia* are found throughout the world, with 2 species occurring in our area. They reach their greatest diversity in tropical America, Africa, and Asia. Most species of *Uranotaenia* are very small, although a few are medium sized. The adults of many species have patches, stripes, or bands of iridescent blue scales on the body, legs, and/or wings. Females of many *Uranotaenia* species feed on the blood of reptiles and amphibians, while others may feed on birds or mammals. Very few are known to feed on humans and none are proven to transmit pathogens to humans. *Uranotaenia* larvae are found in a wide range of aquatic habitats. Many species are found in large bodies of water, including swamps and marshes, while others are found in small aquatic habitats, such as rock holes, tree holes, crab holes, and pitcher plants. Larvae of most *Uranotaenia* species have a comb plate on abdominal segment 8 and a single row of comb scales.

Uranotaenia lowii Theobald, 1901

Subgenus *Uranotaenia*

Pale-footed *Uranotaenia*

Adults *Uranotaenia lowii* is a very small mosquito with stripes and patches of iridescent blue scales on the head, thorax, abdomen, and wings. The thorax is light brown, with patches of blue scales along the sides. The abdomen is very dark and also has patches of blue scales along the sides. The palps are quite short and the proboscis is dark and swollen at its tip. Adults of *Uranotaenia lowii* are likely to be confused with only 1 other species in our area, *Uranotaenia sapphirina.* While both species are ornamented with iridescent blue scales, *Uranotaenia lowii* has blue scales on the abdomen, whereas *Uranotaenia sapphirina* does not. Also, the hind tarsi of *Uranotaenia lowii* have pale scales on the distal segments, while those of *Uranotaenia sapphirina* are all dark scaled. Females of *Uranotaenia lowii* feed on the blood of frogs and toads. They are not known to bite humans or transmit any human pathogens.

Uranotaenia lowii female

Uranotaenia lowii female feeding on a toad

Uranotaenia lowii larva

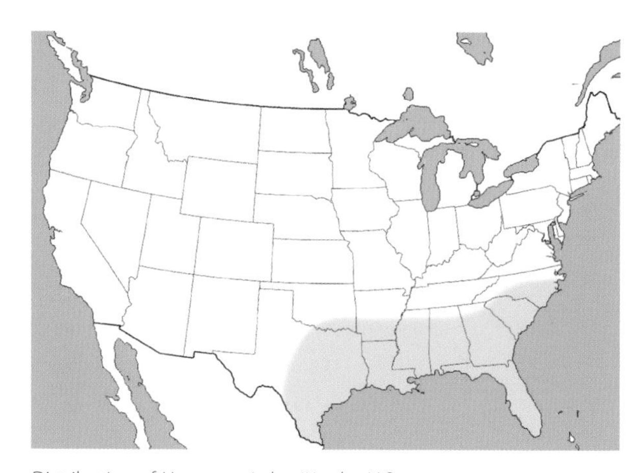

Distribution of *Uranotaenia lowii* in the U.S.

Larvae Larvae of *Uranotaenia lowii* are quite small. The head is black, longer than it is wide, and has 4 spikelike setae. The siphon is dark and has a pecten and a pair of multibranched setae near its middle. A large lateral plate on segment 8 bears the comb scales. Larvae are found in permanent and semipermanent ponds that support grassy vegetation and are exposed to direct sunlight.

Distribution *Uranotaenia lowii* is found in the southeastern Gulf and Atlantic states but is most common in the southernmost part of its range, particularly in Florida and southern Louisiana.

Uranotaenia sapphirina (Osten Sacken, 1868)

Subgenus *Uranotaenia*

Sapphire striped mosquito

Scutum of *Uranotaenia sapphirina* female

Adults *Uranotaenia sapphirina* is one of the smallest species of mosquitoes in our area. Despite their small size, adults of both sexes are easy to recognize due to the presence of iridescent blue scales about the head, thorax, and wings. These brilliant blue scales are arranged in stripes that adorn the dorsal and lateral faces of the thorax. Apart from the iridescent blue scales, the body is brown. In both sexes the palps are uncommonly short and the proboscis is swollen at its tip. Adults of *Uranotaenia sapphirina* are likely to be confused with only 1 other species in our area, *Urano-*

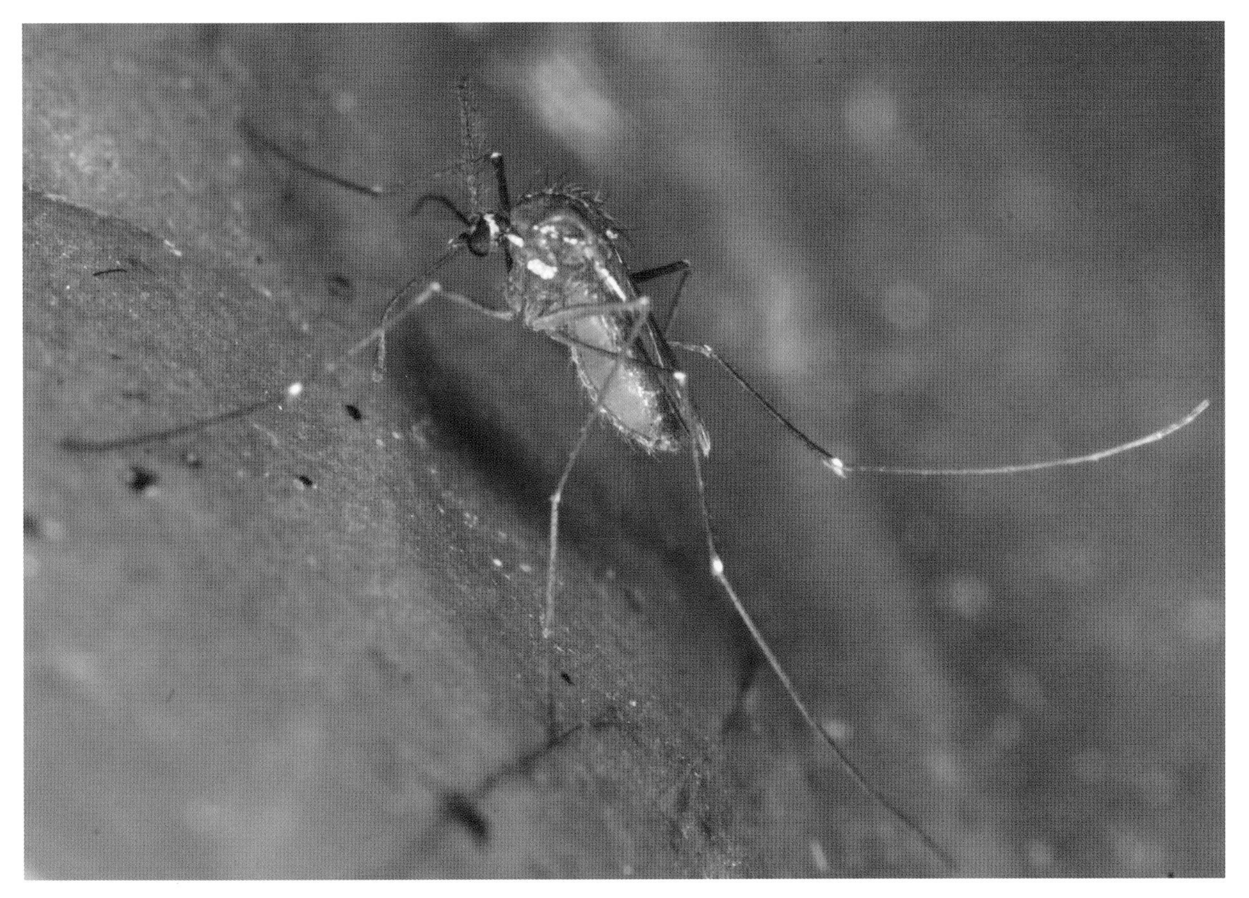

Uranotaenia sapphirina female

Uranotaenia sapphirina larva

Uranotaenia sapphirina larva

Uranotaenia sapphirina egg raft

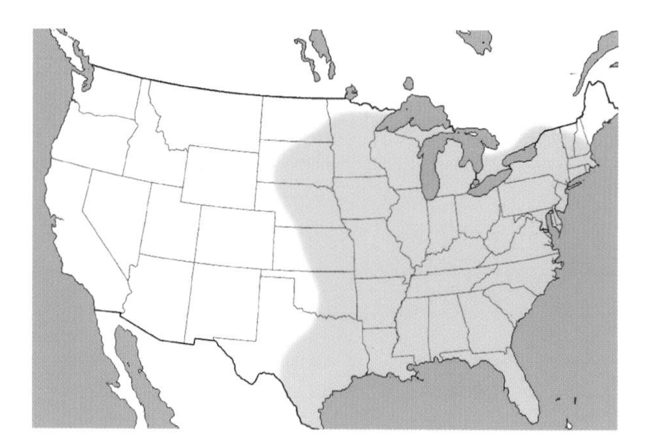

Distribution of *Uranotaenia sapphirina* in the U.S.

taenia lowii. While both species are clad in iridescent blue scales, *Uranotaenia sapphirina* has a bold stripe of iridescent scales down the center of the scutum but lacks blue scales on the abdomen. *Uranotaenia lowii* lacks the stripe of blue scales on the scutum and has blue scales on the abdomen. Also, the hind tarsi of *Uranotaenia sapphirina* are all dark scaled, while the terminal tarsal segments of *Uranotaenia lowii* are pale scaled. Females of *Uranotaenia sapphirina* feed on the blood of reptiles and amphibians, especially aquatic snakes and bullfrogs. They are not known to bite humans or to transmit any pathogens directly to humans.

Larvae Larvae of *Uranotaenia sapphirina* are quite small. The body color can vary tremendously; some individuals are very pale and others are almost black. The head is long and narrow and bears 4 spinelike setae. The siphon is stout and cylindrical, has a pecten, and bears a pair of large, multibranched setae near its midpoint. A large plate on segment 8 bears the comb scales. Larvae are found almost exclusively in permanent ponds, lakes, and marshes that support rooted and/or floating vegetation and are exposed to sunlight.

Distribution *Uranotaenia sapphirina* occurs throughout the eastern half of the United States.

Wyeomyia

The genus *Wyeomyia* is found only in the Americas and reaches its greatest diversity in the tropics. Nearly 150 species are known, only 1 of which occurs in our area (2 other species occur in peninsular Florida). Adults are active during the day and are rarely encountered far from their larval habitat. Although most species feed on blood, *Wyeomyia* females are generally not considered to be important in the transmission of any human pathogens. *Wyeomyia* larvae are adapted to inhabit small aquatic habitats, especially those associated with living plants. These include water in bromeliads, bamboo, and pitcher plants. Larvae of some species are found in tree holes and others are occasionally found in water-holding man-made containers.

Wyeomyia smithii (Coquillett, 1901)

Subgenus *Wyeomyia*

Pitcher-plant mosquito

Adults *Wyeomyia smithii* is a very small mosquito species. Adults are very dark dorsally and pale ventrally. The color of the dorsum is a deep dark brown that is almost black. Under certain lighting conditions, the dark portions of the abdomen appear to have a greenish-blue iridescence. The lateral parts of the thorax and the underside of the abdomen are silvery white. The dark and light portions of the abdomen meet along a distinct straight line. Adults have a very characteristic pose when resting. The proboscis points straight downward, and the hind legs arch forward and over the body. Females are not known to regularly bite humans. In fact, the protein-rich diet of the larvae enables adult females to lay their first batch of eggs without taking

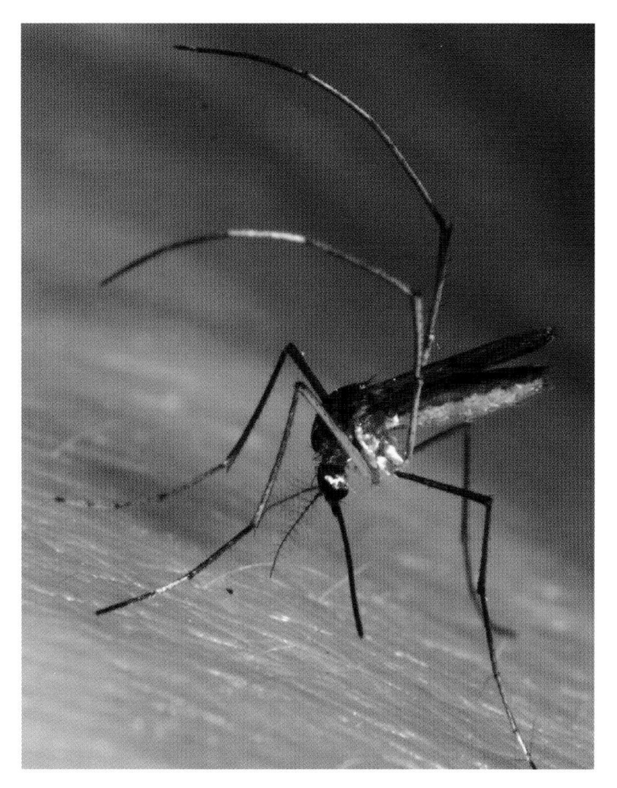

Wyeomyia smithii female

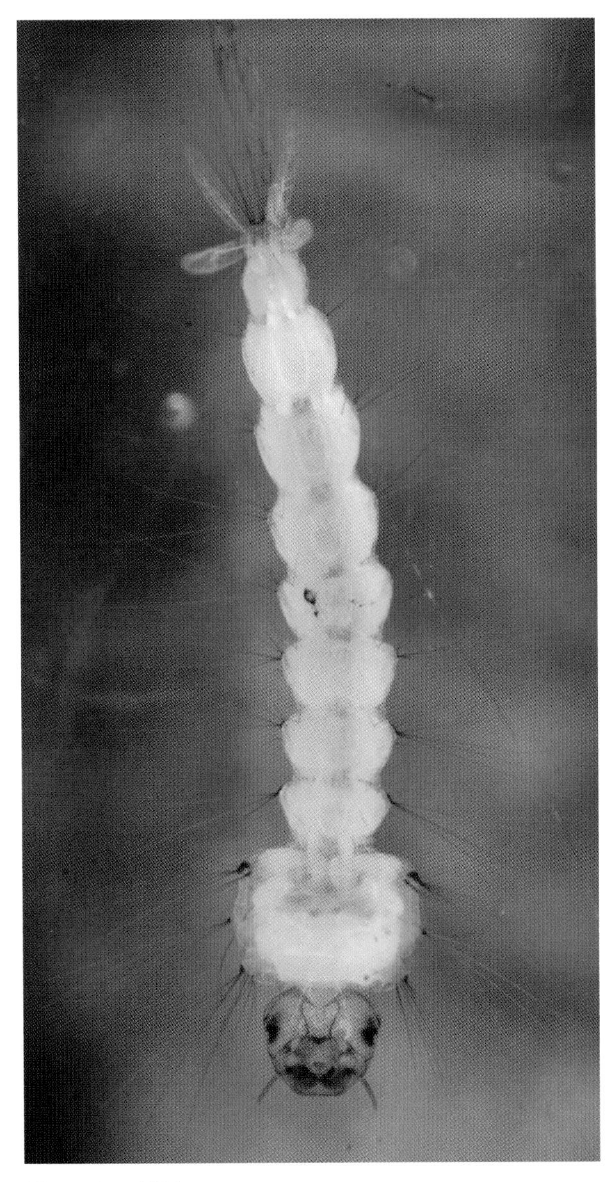

Wyeomyia smithii larva

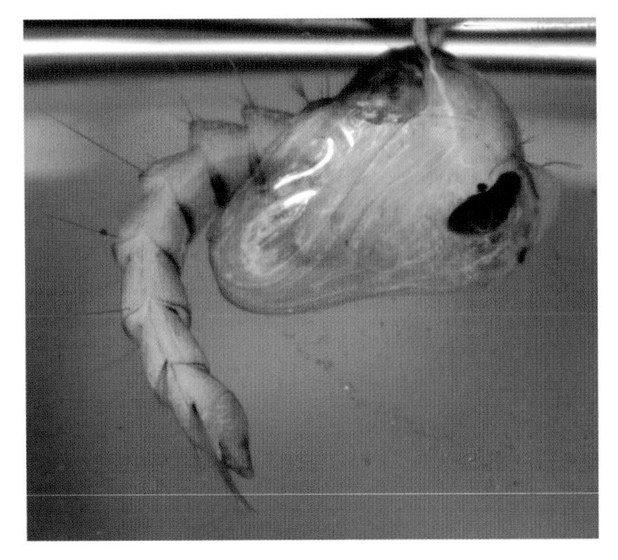

Wyeomyia smithii pupa

The purple pitcher plant, *Sarracenia purpurea*

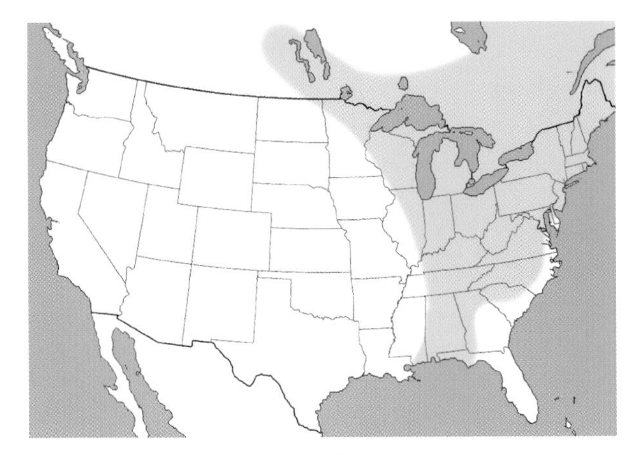

Distribution of *Wyeomyia smithii* in the U.S.

a blood meal. They may occasionally bite after laying their first batch of eggs. *Wyeomyia smithii* is not known to transmit any human pathogens.

Larvae Larvae of *Wyeomyia smithii* are small. The body is usually quite pale, with a straw-colored siphon and head capsule. The setae of the head are mostly small and unbranched. The thorax and abdomen have numerous long setae arising from the lateral margins. The siphon lacks a pecten and has many long, unbranched setae throughout. The comb is a single linear row of about 7–12 scales. Larvae are found exclusively in the water-holding leaves of the purple pitcher plant, *Sarracenia purpurea.* They feed on invertebrates that are trapped by the plant. Larvae are able to withstand freezing and may be found frozen along with the water inside the leaves of the pitcher plant.

Distribution *Wyeomyia smithii* occurs where the purple pitcher plant, *Sarracenia purpurea,* occurs (eastern North America). The larvae are uniquely adapted to survive and develop in the water-holding, pitcher-shaped leaves of the purple pitcher plant.

Glossary

abdomen The most posterior of the three main body regions of insects.

acus A small plate at the base of (and often attached to) the siphon of mosquito larvae.

adult In insects, the winged stage of dispersal and reproduction.

alveolus (pl., alveoli) A socket-like base of a seta or scale.

anal papilla (pl., papillae) Soft, elongate appendage(s) (either two or four) arising from the apex of the of the anal segment.

anal segment Also called segment 10. The tenth and terminal body segment of a mosquito larva, bearing the anus, saddle, and anal papillae.

antenna (pl., antennae) One of the paired segmented sensory appendages of the insect head, arising between the eyes in mosquitoes.

anterior Referring to a structure that is toward the front (head end).

apex The end or tip of a structure.

apical Referring to the apex of a structure or appendage.

appressed Lying flat against the body.

aquatic Referring to an organism that lives in water.

band A marking (usually pale) that is perpendicular to the long axis of the body.

basal Referring to the base of a structure or appendage; closest to the body.

capitellum The knob-like tip of the halter.

cercus (pl., cerci) One of the small paired appendages at the apex of the abdomen of insects.

comb A row or patch of scales on each side of abdominal segment 8.

comb scales The individual scales of the comb.

coxa (pl., coxae) The first (most basal) segment of the insect leg.

dorsal Referring to the upper surface of any structure.

femur (pl., femora) The third segment of the insect leg, usually the largest.

flagellomere An individual unit of the flagellum of the insect antenna.

flagellum The whiplike third segment of the insect antenna, divided into smaller units, called flagellomeres.

fore Toward the front or anterior portion of the body.

frontal setae Setae of the anterior margin of the head of a mosquito larva.

genitalia The structures used in mating.

genus (pl., genera) A taxonomic classification including one or a group of species having common characteristics.

halter (pl., halteres) The modified hind wing of flies, appearing as a small paired club-shaped structure.

head The most anterior of the three major body regions of insects.

hind Toward the back or posterior portion of the body.

holometabolous Having complete metamorphosis (larva, pupa, and adult).

integument The outer covering (exoskeleton) of insects.

labellum The tip of the proboscis.

larva The juvenile form in holometabolous insects.

lateral Referring to the side of the body.

median Referring to the middle .

mesepimeron The posterior plate of the lateral thorax of adult mosquitoes, below the wing base.

mesokatepisternum The large lateral plate of the thorax, above the base of the middle leg.

mesopostnotum A dorsal plate of the thorax, just posterior to the scutellum.

occiput The posterior portion of the head just behind the eyes.

palmate Palm-shaped.

palp or palpus (pl., palpi) One of the paired segmented appendages of the insect head, arising between the proboscis and the antenna in mosquitoes.

palpomere An individual segment of the palp.

pecten The comblike spines of the siphon.

pedicel The bulbous second segment of the insect antenna.

pleuron The lateral portion of the insect thorax.

plumose Feather-like.

posterior Referring to the rear of the body.

postspiracular setae Setae arising from the rounded, slightly raised area posterior to the anterior spiracle of the thorax.

prespiracular setae Setae of the small area (prespiracular area) just anterior to the anterior thoracic spiracle and posterior to a strong exoskeletal ridge.

proboscis The long, piercing mouthparts of the mosquito.

pupa The stage of transformation from larva to adult in holometabolous insects.

respiratory siphon The elongate, tubelike breathing apparatus of mosquito larvae, arising from abdominal segment 8.

saddle The exoskeletal plate of the anal segment of mosquito larvae.

scale In insects, a small, flat outgrowth of the exoskeleton.

sclerite Thickened (hardened) plates of the arthropod body.

sclerotized Hardened portion of the insect exoskeleton.

scutellum The posterior lobe of the dorsal thorax in insects.

scutum A sclerite of the dorsal thorax in insects; the most prominent sclerite of the dorsal thorax in mosquitoes.

seta (pl., setae) A hairlike growth of the insect exoskeleton.

siphon The respiratory siphon.

species The basic level in the classification of organisms; a group of individuals similar in structure and appearance and able to reproduce and bear fertile offspring.

spiculate Covered in many tiny spines, resulting in a rough appearance.

spiracle An opening in the exoskeleton, through which insects breathe.

sternite The ventral plate of a body segment.

sternum The ventral surface of the body.

stripe A longitudinal streak of color differing from the background color.

subcostal Arising from or appearing on the subcostal vein.

subcostal vein The second vein of the wing.

tarsomere An individual subdivision of the tarsus of the insect leg.

tarsus (pl., tarsi) The most distal segment of the insect leg, divided into five units, called tarsomeres.

tergite The dorsal plate of a body segment.

tergum (pl., terga) The dorsal (upper) surface of the body.

thorax The major body region of insects between the head and the abdomen, from which the legs and wings arise.

tibia The fourth segment of the insect leg.

transverse Perpendicular to the long axis of the body.

trochanter The second segment of the insect leg.

ventral Referring to the underside of the body (opposite of dorsal).

vertex The top of the head, posterior to the eyes.

References

Carpenter, Stanley J., and Walter J. LaCasse
1955 The Mosquitoes of North America (North of Mexico). University of California Press, Berkeley.

Carpenter, Stanley J., Woodrow W. Middlekauff, and Roy W. Chamberlain
1946 The Mosquitoes of the Southern United States East of Oklahoma and Texas. University of Notre Dame Press, Indiana.

Darsie, Richard F., Jr., and Charlie D. Morris
2000 Keys to the Adult Females and Fourth Instar Larvae of the Mosquitoes of Florida (Diptera, Culicidae). E. O. Painter, De Leon Springs, Florida.

Darsie, Richard F., Jr., and Ronald A. Ward
2005 Identification and Geographical Distribution of the Mosquitoes of North America, North of Mexico. University of Florida Press, Gainesville.

Gaffigan, Thomas V., Richard C. Wilkerson, James E. Pecor, Judith A. Stoffer, and Thomas Anderson
2011 Systematic Catalog of Culicidae. Walter Reed Biosystematics Unit (WRBU), Division of Entomology, Walter Reed Army Institute of Research (WRAIR), Silver Spring, Maryland, USA. http://www.mosquitocatalog.org.

Harbach, Ralph E.
2011 Mosquito Taxonomic Inventory. The Natural History Museum, London. http://mosquito-taxonomic-inventory.info/.

Harbach, Ralph E., and Kenneth L. Knight
1980 Taxonomists' Glossary of Mosquito Anatomy. Plexus, Marlton, New Jersey.

King, Willard V., George H. Bradley, Carroll N. Smith, and William C. McDuffie
1960 A Handbook of the Mosquitoes of the Southeastern United States. Agriculture Handbook No. 173. Agricultural Research Service, U.S. Department of Agriculture, Washington, D. C.

Means, Robert G.
1979 Mosquitoes of New York: Part I. The Genus *Aedes* Meigen with Identification Keys to Genera of Culicidae. University of the State of New York, Albany.

Silver, Jonathan B.
2008 Mosquito Ecology: Field Sampling Methods. 3rd ed. Springer, Netherlands.

Index